Spatial Analytical Perspectives on GIS

Also in the GISDATA Series

I ***GIS and generalization: methodology and practice***
edited by J.C. Müller, J.-P. Lagrange and R. Weibel

II ***Geographic Objects with Indeterminate Boundaries***
edited by P.A. Burrough and A.U. Frank

III ***GIS diffusion: the adoption and use of geographical information systems in local government in Europe***
edited by I. Masser, H. Campbell and M. Craglia

Series Editors

I. Masser and F. Salgé

Spatial Analytical Perspectives on GIS

EDITORS

MANFRED FISCHER, HENK J. SCHOLTEN and DAVID UNWIN

GISDATA IV

SERIES EDITORS

I. MASSER and F. SALGÉ

CRC Press
Taylor & Francis Group
Boca Raton London New York

CRC Press is an imprint of the
Taylor & Francis Group, an **informa** business
A TAYLOR & FRANCIS BOOK

CRC Press
Taylor & Francis Group
6000 Broken Sound Parkway NW, Suite 300
Boca Raton, FL 33487-2742

First issued in hardback 2018

ISBN 13: 978-1-138-43061-7 (hbk)
ISBN 13: 978-0-7484-0340-0 (pbk)

Visit the Taylor & Francis Web site at
http://www.taylorandfrancis.com

and the CRC Press Web site at
http://www.crcpress.com

British Library Cataloguing in Publication Data

A catalogue record for this book is available from the British Library.

Library of Congress Cataloging Publication Data are available

Cover design by Hybert Design and Type, Waltham St Lawrence, Berkshire.

Typeset in Times 10/12pt by MHL Typesetting Ltd, Coventry.

Contents

The GIS Data Series

Editors' Preface

Over the last few years there have been many signs that a European GIS community is coming into existence. This is particularly evident in the launch of the first of the European GIS (EGIS) conferences in Amsterdam in April 1990, the publication of the first issue of a GIS journal devoted to European issues (*GIS Europe*) in February 1992, the creation of a multipurpose European ground-related information network (MEGRIN) in June 1993, and the establishment of a European organisation for geographic information (EUROGI) in October 1993. Set in the context of increasing pressures towards greater European integration, these developments can be seen as a clear indication of the need to exploit the potential of a technology that can transcend national boundaries to deal with a wide range of social and environmental problems that are also increasingly seen as transcending the national boundaries within Europe.

The GISDATA scientific programme is very much part of such developments. Its origins go back to January 1991, when the European Science Foundation funded a small workshop at Davos in Switzerland to explore the need for a European level GIS research programme. Given the tendencies noted above it is not surprising that participants of this workshop felt very strongly that a programme of this kind was urgently needed to overcome the fragmentation of existing research efforts within Europe. They also argued that such a programme should concentrate on fundamental research and it should have a strong technology transfer component to facilitate the exchange of ideas and experience at a crucial stage in the development of an important new research field. Following this meeting a small coordinating group was set up to prepare more detailed proposals for a GIS scientific programme during 1992. A central element of these proposals was a research agenda of priority issues groups together under the headings of geographic databases, geographic data integration, and social and environmental applications.

The GISDATA scientific programme was launched in January 1993. It is a four-year scientific programme of the Standing Committee of Social Sciences of the European Science Foundation. By the end of the programme more than 300 scientists from 20 European countries will have directly participated in GISDATA activities and many others will have utilised the networks built up as a result of them. Its objectives are:

- to enhance existing national research efforts and promote collaborative ventures overcoming European-wide limitations in geographic data integration, database design and social and environmental applications;

- to increase awareness of the political, cultural, organisational, technical and informational barriers to the increased utilisation and inter-operability of GIS in Europe;
- to promote the ethical use of integrated information systems, including GIS, which handle socio-economic data by respecting the legal restrictions on data privacy at the national and European levels;
- to facilitate the development of appropriate methodologies for GIS research at the European level;
- to produce output of high scientific value;
- to build up a European network of researchers with particular emphasis on young researchers in the GIS field.

A key feature of the GISDATA programme is the series of specialist meetings that has been organised to discuss each of the issues outlined in the research agenda. The organization of each of these meetings is in the hands of a small task force of leading European experts in the field. The aim of these meetings is to stimulate research networking at the European level on the issues involved and also to produce high quality output in the form of books, special issues of major journals and other materials.

With these considerations in mind, and in collaboration with Taylor & Francis, the GISDATA series has been established to provide a showcase for this work. It will present the products of selected specialist meetings in the form of edited volumes of specially commissioned studies. The basic objective of the GISDATA series is to make the findings of these meetings accessible to as wide an audience as possible to facilitate the development of the GIS field as a whole.

For these reasons the work described in the series is likely to be of considerable importance in the context of the growing European GIS community. However, given that GIS is essentially a global technology most of the issues discussed in these volumes have their counterparts in research in other parts of the world. In fact there is already a strong UK dimension to the GISDATA programme as a result of the collaborative links that have been established with the National Center for Geographic Information and Analysis through the United States National Science Foundation. As a result it is felt that the subject matter contained in these volumes will make a significant contribution to global debates on geographic information systems research.

Ian Masser
François Salgé

Editors' Preface

In our increasingly global society, the future success of GIS technology in fields such as environmental monitoring and urban and regional planning depends on its ability to support decision making at all levels. On the one hand, this requires the development of new methods of spatial analysis that are sufficiently powerful to cope with the complexity of analysis in data-rich GIS environments. On the other hand, it necessitates a much greater integration of GIS with techniques to explore and explain observed spatial patterns in large datasets and to model future events. Although in Europe research on GIS and spatial analysis is being carried out in many different countries and disciplines, the need for a more concerted effort prompted the selection of this topic for the first specialist meeting of the socio-economic and environmental applications cluster of the GISDATA programme.

Initially, a small task force, made up of Professors Henk J. Scholten (Free University of Amsterdam), Manfred M. Fischer (Wirtschaftsuniversität Wien), Stan Openshaw (University of Leeds) and Andrea Fabbri (ITC, Enschede), met in Amsterdam on 1 July, 1993 to organise a specialist meeting on GIS and spatial analysis. The task force invited twenty-five participants to the meeting, which was held in Amsterdam on 1–5 December, 1993, and which had the primary objective of stimulating fundamental research in this area by bringing together scholars from across Europe and the USA. This book presents a major outcome of that gathering. Another has been the establishment of a specialist interest group which held a follow-up workshop organised by Professor Antony Unwin (University of Augsburg) on the application of exploratory spatial data analysis tools. Finally, all these developments were reported to the general community of geographers in special sessions organised in co-operation with the IGU Commission on Mathematical Models held during the 28th International Geographical Congress in The Hague in August 1996.

The chapters included in this volume overview the state-of-the-art in research in spatial analysis using GIS but, in addition, they provide numerous directions for future work. As more and more disciplines have realised the power of a spatial perspective in their work, so the last decade has seen an enormous increase in interest in spatial analysis. The years to come will see even greater interest as more and more spatially-located data become available for analysis. Above all, the contributors to this book make it abundantly clear that GIS technology has to be enriched with more sophisticated spatial analysis tools to enable it to meet the analytical needs created by this explosion of interest.

Finally, we wish to express our gratitude for support from our publishers, our home institutions, the Austrian Science Foundation (FWF), the Dutch Science Foundation (NWO) and, in particular, the European Science Foundation (ESF).

Manfred M. Fischer
Henk J. Scholten
David Unwin

Vienna–Amsterdam–London, August 1996

Notes on Editors

Manfred M. Fischer is Professor and Chair of the Institute of Economic and Social Geography at the University of Economics and Business Administration, Vienna, and Director of the Institute for Urban and Regional Research at the Austrian Academy of Sciences. He has a wide scope of research interests, and has published in fields including geoinformation processing and artificial intelligence, spatial econometrics and spatial modelling, pattern recognition, decision theory and micro-behavioural modelling. He is the author of 13 monographs, has 63 publications in refereed journals, 17 contributions in conference proceedings and 43 chapters in books, and has edited 14 books in geography, regional science and related fields. An acknowledged authority on geoinformation processing, spatial modelling and spatial decision behaviour, Dr Fischer is on the editorial boards of many prestigious journals in geography and regional science, and the GISDATA steering committee of the ESF.

Henk J. Scholten is Professor in Spatial Informatics at the Faculty of Economics at the Free University in Amsterdam, and Director of the Research Institute for GIS in Amsterdam, GEODAN. As Project Leader for Geographical Information Systems at the National Institute of Public Health and Environmental Protection (RIVM), he was responsible for the constitution of the Dutch environmental geographical information system, and its application for environmental protection and public health. As well as being a member of the steering board for GIS activities of the European Science Foundation, he is the project leader and consultant on several large international GIS projects.

David Unwin has held the Chair in Geography at Birkbeck College, University of London since 1993, and prior to that held university appointments at Aberystwyth and Leicester. In 1994, he was Erskine Fellow in Computer Science and Geography at the University of Canterbury, New Zealand. His research interests are in spatial statistical analysis and geocomputation (including GIS). He has written over 100 research papers in these areas as well as a number of textbooks, and is currently working on advanced tools for the interactive analysis and visualisation of spatial data for research and teaching. David was the first Director of the Computers in Teaching Initiative Centre for Geography, Geology and Meteorology at Leicester, and was also a member of the GISDATA Strategic Review Panel.

Contributors

Luc Anselin
Regional Research Institute, West Virginia University, PO Box 6825, 511 North High Street, Morgantown, West Virginia 26506-6825, USA

T. A. Arentze
Department of Architecture and Urban Planning, Eindhoven University of Technology, PO Box 513, 5600 MB Eindhoven, The Netherlands

R. S. Bivand
Institute of Geography, Norwegian School of Economics and Business Administration, University of Bergen, Breiviken 2, N-5035 Bergen-Sandviken, Norway

A. W. J. Borgers
Department of Architecture and Urban Planning, Eindhoven University of Technology, PO Box 513, 5600 MB Eindhoven, The Netherlands

P. A. Burrough
Department of Physical Geography, Institute of Geographical Science, University of Utrecht, PO Box 80.115, 3508 TC Utrecht, The Netherlands

António S. Câmara
Environmental Systems Analysis Group, New University of Lisbon, 2825 Monte de Caparica, Portugal

Paulo Castro
Via Department of Civil and Environmental Engineering, Virginia Tech, Blacksburg, Virginia 2061, USA

Chang-Jo F. Chung
Geological Survey of Canada, Ottawa, Canada K1A 0E8

Graham Clarke
School of Geography, Leeds University, Leeds LS2 9JT, UK

Andrea G. Fabbri
ITC, Hengelosestraat 99, 7514 AE Enschede, The Netherlands

Bianco Falcidieno
Consiglio Nazionale delle Ricerche, Via De Marini, 6, Torre di Francia, 16149 Genova, Italy

Enrico Feoli
International Centre for Science and High Technology (UNIDO), Trieste 34100, Italy

Francisco Ferreira
Environmental Systems Analysis Group, New University of Lisbon, 2825 Monte de Caparica, Portugal

Manfred M. Fischer
Department of Economic and Social Geography, Vienna University of Economics and Business Administration, A-1090 Vienna, Austria, and
Institute for Urban and Regional Research, Austrian Academy of Sciences, A-1010 Vienna, Austria

J. L. Gunnink
Department of Physical Geography, Institute of Geographical Science, University of Utrecht, PO Box 80.115, 3508 TC Utrecht, The Netherlands

Robert Haining
Department of Geography, University of Sheffield, Sheffield S10 2TN, UK

M. A. Oliver
Institute of Public and Environmental Health, School of Chemistry, University of Birmingham, Edgbaston, Birmingham B15 2TT, UK

Stan Openshaw
School of Geography, Leeds University, Leeds LS2 9JT, UK

Caterina Pienovi
Consiglio Nazionale delle Ricerche, Via De Marini, 6, Torre di Francia, 16149 Genova, Italy

Geoffrey G. Roy
School of Physical Sciences, Engineering and Technology, Murdoch University, Perth, Australia 6150

Lena Sanders
Equipe Paris, CNRS, Université Paris 1, 13 rue de Four, 75006 Paris, France

Henk J. Scholten
Department of Regional Economics, Free University Amsterdam, NL-1081 HV Amsterdam, The Netherlands

Folke Snickars
Department of Infrastructure and Planning, Royal Institute of Technology, Stockholm, Sweden

Michela Spagnuolo
Consiglio Nazionale delle Ricerche, Via De Marini, 6, Torre di Francia, 16149 Genova, Italy

U. Streit
Institute of Geography, Department of Environmental Informatics, University of Münster, D 48149 Münster, Germany

H. J. P. Timmerman
Department of Marketing and Economic Analysis, University of Alberta, Edmonton, Canada and *Department of Architecture and Urban Planning, Eindhoven University of Technology, 5600 MB Eindhoven, The Netherlands*

Antony Unwin
Lehrstuhl für Rechnerorientierte Statistik und Datenanalyse, Institut für Mathematik der Universität Augsburg, D-86135 Augsburg, Germany

David Unwin
Department of Geography, Birkbeck College, London W2P 2LL, UK

K. Wiesmann
Institute for Geography, Department of Environmental Informatics, University of Münster, D-48149 Münster, Germany

Vincenzo Zuccarello
International Centre of Theoretical and Applied Ecology, CETA, Gorizia 34170, Italy

The **European Science Foundation** is an association of its 55 member research councils, academies, and institutions devoted to basic scientific research in 20 countries. The ESF assists its Member Organisations in two main ways: by bringing scientists together in its Scientific Programmes, Networks and European Research Conferences, to work on topics of common concern: and through the joint study of issues of strategic importance in European science policy.

The scientific work sponsored by ESF includes basic research in the natural and technical sciences, the medical and biosciences, and the humanities and social sciences.

The ESF maintains close relations with other scientific institutions within and outside Europe. By its activities, ESF adds value by cooperation and coordination across national frontiers and endeavours, offers expert scientific advice on strategic issues, and provides the European forum for fundamental science.

This volume is the first of a new series arising from the work of the ESF Scientific Programme on Geographic Information Systems: Data Integration and Database Design (GISDATA). This 4-year programme was launched in January 1993 and through its activities has stimulated a number of successful collaborations among GIS researchers across Europe.

Further information on the ESF activities in general can be obtained from:
European Science Foundation
1 quai Lezay Marnesia
67080 Strasbourg Cedex
tel: + 33 88 76 71 00
fax: +33 88 37 05 32

EUROPEAN SCIENCE FOUNDATION

This series arises from the work of the ESF Scientific Programme on Geographic Information Systems: Data Integration and Database Design (GISDATA). The Scientific Steering Committee of GISDATA includes:

Dr Geir-Harald Strand
Norwegian Institute of Land Inventory
Box 115, N-1430 Ås
Norway

Dr Antonio Susanna
ENEA DISP-ARA
Via Vitaliano Brancati 48
00144 Roma
Italy

PART ONE

Spatial Analysis Procedures

CHAPTER ONE

Geographic information systems, spatial data analysis and spatial modelling: an introduction

MANFRED M. FISCHER, HENK J. SCHOLTEN and DAVID UNWIN

1.1 Introduction

Today, geographic information systems (GIS) incorporate many state-of-the-art principles such as relational database management, powerful graphics algorithms, interpolation, zoning and simplified network analysis, yet what is termed spatial analysis and modelling is often no more than map data manipulation such as polygon overlay and buffering. The lack of analytical and modelling functions is widely recognised as a major deficiency of current systems.

There is a wide agreement in both the GIS and the spatial analysis communities that the future success of the GIS technology will depend to a large extent on incorporating more powerful analytical and modelling capabilities. In recent years, this theme of GIS and spatial analysis has been addressed in several conferences, such as the European Congresses of the Regional Science Association (RSA) in Lisbon (1991) and Moscow (1993), and in workshops, such as that in Sheffield (1991) sponsored by the UK Economic and Social Research Council (ESRC) as part of its Regional Research Laboratory (RRL) initiative and an Expert Meeting in San Diego (1992) sponsored by the US National Center for Geographic Information and Analysis (NCGIA) and the International Geographical Union Commission on Mathematical Models. While the two workshops primarily focused on spatial data analysis techniques, the sessions at the RSA conferences paid attention to two major strands of development in spatial analysis concerned with spatial data analysis and spatial modelling.

Spatial data analytical techniques and spatial models can perform functions which, in the main, current GIS lack, but which are important for the sorts of question that decision makers in private and public organisations are interested in. It is into this context that the need for more basic research in the operation, development, and use of spatial data analytical techniques and spatial modelling was identified as the major focus of the GISDATA Specialist Meeting on GIS and Spatial Analysis, held in Amsterdam, December 1–5, 1993.

This introduction to the results of the Workshop is organised as follows. In the next section, the special characteristics of spatial data and some features of the field of spatial analysis are described. The history and current achievements in spatial analysis show that, despite its many and varied applications in both the environmental and the social and economic sciences, two main fields can be distinguished, spatial data analysis and spatial modelling. These two fields determined the basic structure of the GISDATA Expert Meeting and are described in Sections 1.3 and 1.4. In Section 1.5 we outline some implications for the future of the results of the meeting.

1.2 The nature of spatial data and spatial analysis

1.2.1 Why a spatial perspective?

Several arguments can be put forward to say why a spatial perspective should be adopted in data analysis (Goodchild *et al.*, 1992). First, space provides a simple, but useful, framework for handling large amounts of data. Secondly, the spatial perspective permits easy access to information on the relative location of objects and events. Thirdly, it allows objects or events of various types to be linked, in a process formalised in GIS as overlay. Finally, in both environmental and social applications, the distance between objects and events is often an important factor in determining the interaction between them.

1.2.2 What is special about spatial data?

The ability of GIS to handle and analyse spatial data is usually seen as the characteristic that most distinguishes them from other information, computer-aided design, and map production, systems. Spatial data sets provide two types of information:

- data describing the specific locations of objects in space (and their topological relationships), the so-called positional and topological data, and
- data describing non-spatial attributes of the objects recorded, the attribute or thematic data.

For the positional data, spatial (or geo-) referencing can take several forms. It may locate a single point at an exact location, such as the position of a firm. It can also be a set of references locating a more complex entity in space, such as a line object like a road or an area object such as a forest. Often positional references used for compiling spatial data include administrative units such as census tracts, land parcels, municipalities or counties. These primitive elements of spatial data are usually classified geometrically as point, line and area objects.

Major problems encountered in dealing with spatial data include the issue of errors in spatial databases, data integration, confidentiality, and appropriate spatial analysis. Errors can arise in measuring both locations and attributes and are related to the computerised process responsible for storing, retrieving and manipulating spatial data. This issue has received increasing attention (Goodchild and Gopal, 1989).

A frequent problem for GIS users with different types of data is that of data integration, by which we mean the process of making different data sets compatible with each other. Data incompatibilities may be caused by the use of different spatial referencing systems, different statistical or temporal coverage, different degrees of generalisation, and locational errors. This problem is fundamental to socio-economic GIS applications and has also received increasing attention (Flowerdew and Green, 1991).

As the size of a geo-reference unit decreases, so the number of objects located within its boundaries also tends to decrease. At some small unit size, the information needed would violate confidentiality laws under which many data providers operate. If a sufficient number of individual facts are aggregated together, such that one cannot ascertain information about individuals, confidentiality is maintained. In some socio-economic contexts, such as, for example, industrial geography and housing research, the events are sufficiently rare that even when data are aggregated in this way confidentiality issues might still arise.

Finally, because objects or events distributed over space tend not to be independent, problems are encountered in statistical analysis which often assumes such independence. Complications are caused by two special features of spatial data, related to spatial dependence and spatial heterogeneity. Spatial dependence refers to the relationship between geo-referenced data due to the nature of the variable under consideration, and the size, shape and configuration of the spatial units used as geo-referencing framework. The smaller the size of spatial units, the greater is the probability that nearby spatial units will be spatially dependent. Spatial heterogeneity arises when spatial uniformity of the effects of spatial dependence and/or of the relationships between the variables at hand are lacking.

These spatial effects may invalidate many of the standard statistical procedures and lead to errors in statistical inference, misleading indications of model validity, and so on. Consequently, spatial data analysis must go beyond standard statistical analysis (Anselin and Getis, 1993). Unfortunately, as yet there seems to be no readily available technology which is suited to deal with these problems in the typically very large data sets now being used.

1.2.3 Spatial analysis is more than statistical spatial data analysis

The origins of spatial analysis lie in the development in the early 1960s of quantitative geography and regional science. The use of quantitative (mainly statistical) procedures and techniques to analyse patterns of points, lines, areas and surfaces depicted on maps or defined by co-ordinates in two- or three-dimensional space characterised the initial stage. Later on, more emphasis was placed on the indigenous features of geographical space, on spatial choice processes and their implications for the spatio-temporal evolution of complex spatial systems.

As it has evolved over the past three decades, spatial analysis is more than spatial statistics. A closer look at the development and current achievements of spatial analysis shows that, despite the very large number of rather diverse contributions, two main fields of study can be identified (Figure 1.1):

- Statistical spatial data analysis providing more adequate and specialised frameworks and methodologies to deal with a wide range of spatial effects and spatial process models.
- Spatial modelling including a wide range of different models such as, for example, deterministic and stochastic process models as well as policy models in the environmental sciences, and location-allocation models, spatial interaction models, spatial choice models and regional economic models in the social sciences.

These both have the potential to enrich GIS technology and will be considered in some more detail in the following two sections.

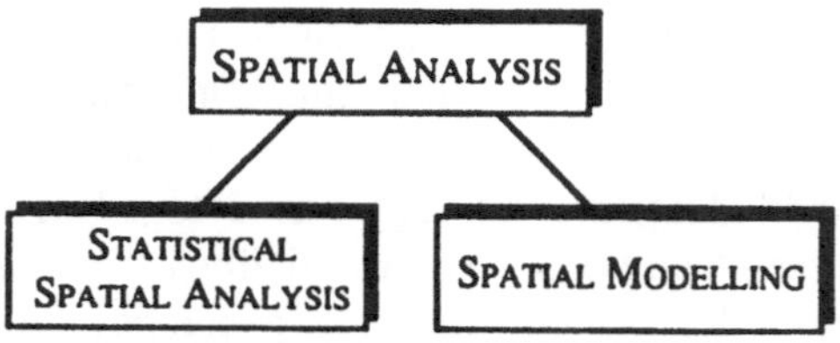

Figure 1.1 Two major fields of research in spatial analysis.

1.3 Spatial data analysis and GIS

1.3.1 Basic functions of a GIS

It is widely recognised that GIS have four basic functions related to spatial data (see, for example, Goodchild, 1987; Anselin and Getis, 1993; Fischer and Nijkamp, 1992):

1. Data input (data model, data measurement).
2. Data storage, retrieval and database management.
3. Data analysis (data manipulation, exploration and confirmation).
4. Output (display and product generation).

The way in which geographic space in the 'real' geographical system is measured and structured in a spatial database depends on the data model. There are two fundamental approaches to represent the spatial component of geographic information, the 'vector' and 'raster' models. After adequate generalisation of the form of the objects, the first sees initially empty geographic space populated by objects that are represented as points, lines or areas. This model seems to be more important for spatial analysis in the social sciences where discrete entities are considered as interacting over space. The second approach subdivides the space into a set of fields which may be modelled in different ways such as, for example, a raster of cells in a remote-sensing scene, a grid of regularly spaced points in a digital elevation model, a set of non-overlapping space-exhausting polygons in a soils map, or a map of digitised isolines in a contour map. This approach seems to be more prevalent in the environmental sciences. It is important to note that the choice of the data model also determines the set of analyses that can be carried out (Goodchild *et al.*, 1992).

Based on Anselin and Getis (1993), the interactions between the four basic functions of a GIS may be displayed in the simplified form shown in Figure 1.2. At the one end there is the 'perceived reality' of geographic space and at the other the GIS user concerned with the research or policy problems to be solved. In between there are the four functions input, storage, analysis and output. Although all of these are available in most existing GIS, analytical functions are, however, generally limited to data manipulations such as partitioning, aggregation, overlay and interpolation which are what GIS vendors generally understand by the term 'spatial analysis'.

Two issues arising from this will be briefly addressed in the next section. First, there is the question as to which statistical spatial data analysis techniques should be considered for inclusion in a GIS to increase its analytic capabilities. Secondly, there is the technical issue of how this integration might best be achieved.

1.3.2 Which statistical spatial data analysis techniques?

Despite the recognition that spatial analysis is central to the purpose of many GIS, the

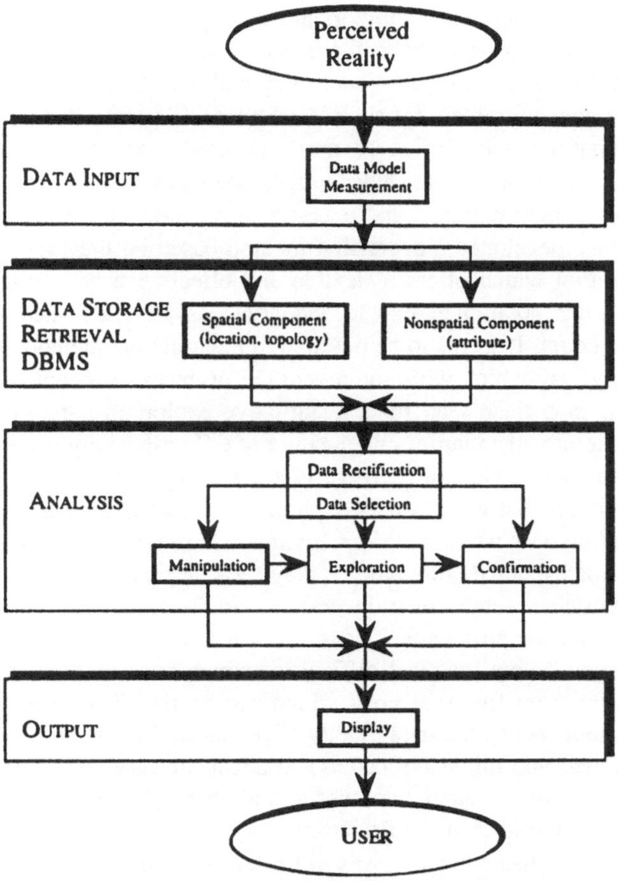

Figure 1.2 Basic functions of a GIS from a simplified and idealised perspective (adapted from Anselin and Getis, 1992).

lack of integration of the technology and spatial data analysis, and the relative simplicity of many GIS, are seen as a major impediment to their full utilisation. The need to develop GIS-relevant statistical spatial analysis tools is currently the subject of research projects at both the NCGIA in the USA and the Regional Research Laboratories in the UK (see Openshaw, 1991a,b; Anselin and Getis, 1993; Goodchild *et al.*, 1992; Fotheringham and Rogerson, 1993). There are four major areas where statistical spatial data analysis techniques can strengthen current GIS practice.

1. Sampling objects from the database and the choice of an adequate spatial scale of analysis.
2. Data rectification to compare variables which are defined for the same study area, but for a different and incompatible set of zones.
3. Exploratory spatial data analysis (ESDA) aiming to explore and exploit the GIS database to arrive at new insights, including the search for data characteristics such as trends, spatial outliers, spatial patterns and associations.
4. Confirmatory or explanatory spatial data analysis (CSDA) concerned with systematic analysis of data and hypothesis testing based upon specified assumptions.

First, methods of data sampling, such as the Dependent Areal Units Sequential Technique, to identify areas to be included in a sample so as to maximise its information content (Arbia, 1989) may have an important role to play for extracting information from databases. Similarly, geostatistical methods can be used to assist in deciding where and how densely to sample for a given model of spatial variation.

Secondly, a frequent problem for socio-economic GIS users is the need to transfer data between incompatible sets of areas, more generally between incompatible spatial referencing systems. Some aspects have already been analysed, but more attention needs to be paid to developing practical and operational procedures. Recently, Flowerdew and Green (1991) have developed a promising method based on the expectations maximum likelihood algorithm which offers a flexible and effective way to improve the various methods of areal interpolation at least for the large class of cases where the variable to be interpolated is a count. It seems to be possible to integrate the method with GIS software in a user-friendly way which will circumvent the problem of incompatible areal units.

Thirdly, there is a clear need for a quantitative exploratory style of spatial analysis which can complement the map-oriented nature of GIS, especially in data rich and theory poor GIS environments. Generally, there is a broad agreement in the GIS and spatial analysis communities that general data exploratory procedures developed in mainstream exploratory data analysis would be of great value within GIS. The techniques considered as primary candidates include box plots, Tukey stars, nearest neighbour, *k*-function plots, scatter plot matrices, autocorrelation tests, classification and spatial classification procedures for data simplification, and outlier and spatial outlier detection methods. There are, however, many practical difficulties which have to be faced in this context. Spatial data suffer from forms of error which non-spatial data do not, and these often propagate to further reduce the data quality (Openshaw, 1991a and b). Some techniques appear to be limited and might not be easy to adapt for very large GIS data sets.

Fourthly, a lack of *a priori* hypotheses and theory seems to make ESDA more important in a GIS world than confirmatory analysis. Nevertheless, although confirmatory data analysis is also relevant, very little has been achieved in this field. Spatial autocorrelation problems have been well analysed, but most applications of explanatory data analysis are non-spatial applications of regression and fail to exploit the information on the topology of the observations. This is primarily a result of the inadequacy of the software used (Haining, 1990). Recently, a number of macros and specialised software to carry out spatial data analysis within standard statistical/econometric packages have been developed (Anselin, 1989; Griffith, 1989; Bivand, 1991) but the linking of this to GIS is very limited (Anselin and Getis, 1993).

Finally, there seems also to be a need to develop new innovative spatial data-analysis tools which are especially suited to the data-rich and theory-poor GIS environment. Neurocomputing offers a basis for such approaches and may establish a new *modus operandi* in spatial data analysis in general, and ESDA and CSDA in particular. Neural approaches seem to be particularly well suited to cope with problems such as very large volumes of spatial data, missing and noisy data, and fuzzy information, for which conventional statistical techniques may be inappropriate or too cumbersome to use (Openshaw, 1993; Fischer, 1993 and 1994; Fischer and Gopal, 1994).

1.3.3 Linking spatial data analysis and GIS

The problem of linking new spatial analytical functions to standard GIS is beginning to emerge as an important research area (Anselin and Getis, 1993; Goodchild *et al.*, 1992). Various logical ways of coupling spatial data analysis and GIS can be identified and there is work under way to explore them. In principle, Goodchild (1992) distinguishes three major approaches:

1. Full integration of spatial analytic procedures within the GIS.
2. Close coupling between statistical spatial data analysis software and GIS.
3. Loose coupling where an independent spatial data-analysis module relies on a GIS for its input data, and for functions such as graphic display, via the import and export of data in a common format.

The first approach, aiming to integrate spatial analysis techniques fully within the GIS, is essentially not yet realised. It is perhaps significant that the richest GIS in this area, IDRISI, was developed, and finds most use, in a university environment. Possibly, Openshaw's idea of developing a spatial analysis tool kit comes close to this approach (Openshaw, 1990). This approach has several important advantages. Software would be well documented and supported by the vendors, making the techniques available to all the software's users. However, it will be difficult to persuade GIS vendors to adopt such an approach without corresponding pressure from the market-place (Goodchild *et al.*, 1992).

The second approach seems to be much more realistic. The unresolved problem here relates to the nature of the user interface involved in the linkage. Possibilities range from source code through subroutine libraries, high-level languages, sequences of commands, and menus, to navigation through a virtual reality. These offer the developer access to the standard user-interface facilities of the GIS package in question and to (locational and attribute) data structures held by the GIS. Even though this approach links a statistical package to a GIS, to date it has been limited to simple descriptive measures (Kehris, 1990). An effective form of close coupling in which data can be passed between a GIS and a spatial analysis module without loss of higher structures, such as topology, object identity, metadata, or various other relationships, is still missing (Goodchild, 1992). However, use of modern CASE tools and interface builders such as the Tcl/TK scripting language now make this a far easier proposition than hitherto. A challenge for the future is to develop a standard spatial query language for spatial data which permits users to access them without knowing the particular data structures used (Goodchild *et al.*, 1992).

The third approach seems to be the strategy most often adopted in practice. It uses a GIS package for the things it is best suited to and, when spatial analysis facilities are required, takes the output from the GIS in some standard data format as input into a statistical software package. The major drawback of this lies in its failure to take the distinctive characteristics of the GIS database into account for use in spatial data analysis.

1.4 Spatial modelling, GIS and decision support

1.4.1 Spatial models

Spatial analysis goes beyond sampling, manipulation, exploratory and confirmatory analysis of data into spatial modelling which encompasses a large and diverse set of models. In the environmental sciences these range from models of physical dispersion (for example, for suspended particulates), chemical reactions (for example, photochemical smog), and biological systems (for example, for ecosystems in water) through deterministic process and stochastic process models (for example, for erosion and groundwater movement) to comprehensive models of regional environmental quality management and integrated models for environment – energy – economic assessment.

For many of these, geographical location is not considered to be of overriding importance and spatial variation in model results is obtained using data from different spatial entities (for example, polygons) as inputs and relating the model output back to the areas in question. Process models include deterministic versions, which attempt description using known physical laws, and stochastic versions, which describe process such as erosion, groundwater movement and absorption of pollutants using stochastic theory. For such processes there is an interaction between the spatial process and a substrate which provides a one-, two- or three-dimensional framework within which the process model can operate. The substrate itself may be unaffected by the process, as in the case of simple run-off models, or may be modified by it, as in the case of pollutant absorption, erosion or transport. The basic unit of the substrate (for example, a catchment or a slope facet) is modelled as a finite element and the space in which the process is modelled is disaggregated into a set of such elements which are usually assumed to be internally homogenous. Two-dimensional models generally use small pixels as their finite elements. In contrast, the three-dimensional models used in groundwater modelling use a voxel (volume element) approach and quickly generate considerable data acquisition and storage problems. For example, even a 60 × 60 × 60 voxel array needs 216000 memory elements (Burrough, 1991).

Models of regional environmental quality are comprehensive tools used in assessing alternative options for managing environmental quality according to some economic criteria. Typically, they include five major components, an economic model of production and consumption, a model of waste production and emissions, an environmental model which translates the space–time pattern of environmental discharges into spatio-temporal states of natural environment, a model of effects of such concentrations or receptors in terms of damage functions, and, finally, an analytical component on management strategies which compares the benefits and costs associated with alternative environmental management strategies (Lakshmanan and Bolton, 1986).

Integrated models for environment–energy–economic assessment link models of the energy system, national–regional economy and environmental emissions into integrated impact models which assess interrelated energy supply, environmental quality, and economic policies. They usually share some common characteristics. First, they contain some version of a macroeconomic model; secondly, they represent production and consumption; thirdly, there are clear linkages between the energy sector, the rest of the economy and the environment; fourthly, they have a reference energy supply system which depicts energy technologies and demands, and, finally, they represent space in a discrete multiregional framework (Lakshmanan and Bolton, 1986). These models seem to draw their components not only from the environmental and ecosystems modelling traditions, but also from different analytical traditions in the social sciences.

In the social sciences in general, and in quantitative geography and regional economics in particular, a wide range of spatial models has been developed in the past decades for the purposes of describing, analysing, forecasting, and policy appraisal of economic developments within a set of localities or regions. Such models deal not only with internal structures of regions and relationships within one region, but also with interregional interrelationships. The tradition of models in quantitative geography and regional science has shown a strong orientation towards location-allocation problems in space (Baumann *et al.*, 1983, 1988; Haynes and Fotheringham, 1988) dealing with flows of people, commodities and resources between regions. Most of these models were static in nature and regarded space as a set of discrete points (areas and grids) rather than as a continuum. Problems relating to ecological fallacies and spatial autocorrelation in such models, caused essentially by the sometimes arbitrary spatial demarcation, have led to various mis-specifications (Baxter, 1987). Early studies in spatial choice processes were dominated by spatial interaction models justified by probability and entropy maximising formulations. The lack of a behavioural context began to be criticised in the 1970s and gave rise to the study of individual choice behaviour in various contexts (journey-to-work, journey-to-shop, migration, and so on.) This, in conjunction with the parallel development of discrete choice models, made it possible to propose new disaggregate choice based alternatives. Unlike spatial interaction models, these could explicitly link individual decisions at the micro level with population flows and other observables at the macro level (Fischer *et al.*, 1990).

The development of regional and inter-regional economic models occurred almost at the same time as a growth of interest in national input–output modelling. The regional input–output model may be clearly seen as a model of the production side of a regional economic system in which demand, and changes in demand, create the signals for production to take place. One of the most promising developments in (inter)regional input–output modelling in recent years has been the extension of the modelling framework in two major ways. First, there have been extensions of the input-output model itself, and, secondly, it has been linked with input–output analysis to take account of the wider socio-economic context in which individual activity takes place (Hewings and Jensen, 1986). Various modifications have been suggested, for example to take into account labour markets and migration (Batey, 1985). Other models are concerned with the regional labour and housing markets, land use and transportation, and multiobjective decision analysis (see Nijkamp, 1986).

In general, spatial models can be used for three purposes: forecasting and scenario generation, policy impact analysis and policy generation and/or design. A major hurdle in the way of their practical application and interpretation is the specificity of each practical situation. The spatial arrangement, size, shape and organisation of the basic spatial units for which data are gathered affect the spatial process which can be identified in the parameters which are thus not independent of the spatial framework used. This problem, which has been extensively recognised in spatial statistics, has only relatively recently been approached in spatial modelling and there are now several simulation results to illustrate these effects (for spatial interaction models, see Openshaw, 1977; for multiregional labour market modelling, see Baumann *et al.*, 1983 and 1988). All these models have functions which proprietary GIS almost totally lack, but which are important for the sorts of question that policy decision makers are interested in.

1.4.2 Interfacing spatial modelling and GIS

GIS and spatial modelling can be considered complementary in two respects (Fischer and Nijkamp, 1993b):

- In studies of spatial pattern and flows, where spatial differences in many dimensions can be shown by statistical representations or by GIS-generated displays. The choice of one or other representation depends on the complexity of the pattern to be represented.
- In explanatory and predictive analysis, spatial models are usually much more powerful than GIS in carrying out precise numerical experiments, but the final results can again be used by a GIS as an input for a user-friendly computer presentation.

The main problem is that current GIS are not as suitable for exploratory or predictive modelling as conventional spatial modelling. Nevertheless, progress has been made, for example in spatial location-allocation modelling where traditional spatial interaction tools have been linked to a GIS representation of the resulting patterns and flows. To some extent one may claim that GIS seems to be a more proper tool for perception/visualisation methods for impact analysis rather than a direct analytical tool.

In dynamic modelling the same remarks hold. A GIS is able to produce dynamic maps displaying the results of a dynamic descriptive, explanatory, or predictive model. In this context GIS is a meaningful vehicle for dynamic scenario analysis. Good examples can be found in Grossmann and Eberhardt (1993) who suggest linking complex aggregated dynamic feedback models, simple generic dynamic models (in particular object-oriented models), and physical dispersion models (for example, for heat conduction, diffusion of noise or transport of gaseous pollutants) with GIS to produce a time series of maps.

In recent years, much attention has been given to the coherent and joint use of GIS and spatial models in spatial planning. This raises the question of how the interface between GIS and spatial modelling might be achieved and the solutions adopted are essentially the same as those discussed previously for the integration of spatial statistical analysis and GIS. At one extreme, the GIS can be incorporated in its entirety into the modelling framework (Figure 1.3(a)). This is, however, inefficient in that most GIS contain many more functions than are relevant to any single spatial model or set of spatial models. At the other extreme, GIS designers often embed spatial models entirely within the GIS and consider them simply as additional functions (Figure 1.3(b)). However, any system which is extensive enough to embrace a set of complex planning models is unlikely to be easily incorporated into an integrated package and must be a set of relatively independent modules which may be interlinked in diverse ways (Figure 1.3(c)) (Batty, 1993).

At the moment, GIS and spatial models have come closest in addressing problems of decision making in the form of spatial decision-support systems (SDSS). The term decision-support system was coined in the late 1970s to embody a framework for integrating database management systems with analytical and operational research models, graphical display, tabular reporting capabilities, and the expert knowledge of decision makers to address business problems.

It is this use which has been adopted by researchers working with operational research models in a spatial setting (Batty, 1993). Models based on the location of facilities,

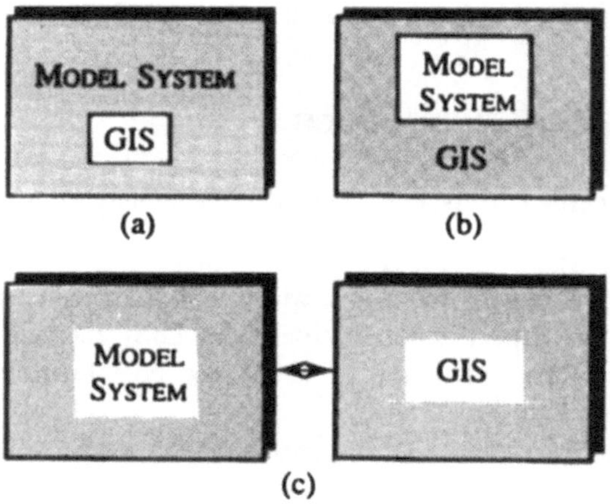

Figure 1.3 Linking spatial models and GIS (adapted from Batty, 1992).

ranging from those based on formal mathematical programming and optimisation to those based on more informal heuristics, have been quite widely developed as part of SDSS (Densham and Rushton, 1988; Densham, 1991).

Comparatively less comprehensive, but more GIS based, is Despotakis' (Despotakis, 1991; Despotakis *et al.*, 1993) hybrid GIS–DSS model for sustainable development planning. In a large-scale regional development study, Despotakis (1991) focused attention on principles and policies for coevolutionary planning and development of the Greek Sporades islands. A GIS to assist sustainable development for an area was developed and linked to a decision-support system in order to evaluate its spatio-temporal results in the context of selected criteria. The whole system was then applied to the test area of the Sporades islands, for which relevant data at various spatial levels and with different degrees of reliability were selected. The results showed that GIS may be successfully employed to assist, monitor and control ecologically sustainable regional economic development for a region. This is achieved by applying strategic policies which are based on optimum (from the sustainability point of view) development scenarios generated by the system. The underlying model system was based on a combination of meso-economic models for the area and micro land-use developments. The GIS can create the intermediate link between these layers of the model system, while the results from the scenario and simulation experiments can be visualised in an attractive way.

Several other examples, such as the ISP computer system for interactive spatial planning for generating and evaluating land-use plan alternatives (Roy and Snickars, 1993) and a prototype GIS for urban program impact appraisal in St. Helens, Northwest England (Hirschfield *et al.*, 1992) may be found in Fischer and Nijkamp (1992a).

1.5 Structure of the book

1.5.1 Part 1: Spatial analysis procedures

In Chapter 2 Stan Openshaw and Graham Clarke start with the discussion how to integrate Spatial Analysis and GIS. Chapter 3, written by Roger Bivand, argues the case for the construction of spatial analysis tools as clearly interfacing modules, under the control of suitable command languages. In the discussion the author has chosen to concentrate on Unix system tools and languages, partly because they represent the mainstream in GIS platform operating system environments and partly because the software tool approach has characterised UNIX since its beginning. The contribution considers examples taken from GIS, and other systems for geographical data manipulation and presentation. It concludes that a new class of 'script' coupled spatial analysis tools may be easier to create than either to elaborate a range of loose-coupled programs, or to convince GIS-software houses of the commercial wisdom of incorporating spatial analytical tools into their already over-complex products.

Chapter 4, written by Robert Haining, describes some aspects of a collaborative project with the Department of Public Health of the Sheffield Area Health Authority, and illustrates the use of statistical spatial data analysis techniques in a GIS world. The chapter considers a wide range of techniques, such as those for constructing an appropriate spatial framework, constructing incidence rates and facilitating the sensitivity analysis of these rates to the choice of time period, spatial framework and possible database errors. The chapter also includes a criticism of the current design criteria and describes how the system might be improved in the future to include interactive graphics. The research shows that, with effort, techniques useful to this application area can be 'hooked' on to a GIS and a facility developed that has considerable potential in an area that all too often does not exploit the spatial nature of its data or does so in sometimes misleading ways. The view of the author has been that, given the wide range of problems to be tackled, application specific developments are a useful way forward and in certain circumstances may be more useful than trying to construct general purpose systems.

Exploratory spatial data analysis (ESDA) is one way of dealing with the complexity of spatial variation. Geostatistics, based on the theory of regionalised variables, embraces a suite of spatial analytical tools. By some kind of explicit link between methods of ESDA and GIS an integrated system for spatial analysis could be created. This would enhance the analytical power of GIS considerably. The variogram and kriging are central in geostatistics. Two case studies in Chapter 5, by Margaret Oliver, illustrate their application. One describes how geostatistics has been adapted to search for spatial pattern in a rare disease and to estimate the risk of developing it. The other illustrates the potential of disjunctive kriging for aiding decision making in environmental management.

When preparing data, such as point observations of soil attributes for geostatistical interpolation and to determine areas where values may exceed required thresholds, it is useful first to examine the data for spatial homogeneity, stationarity and normality. Exploratory Data Analysis (EDA) is not only useful in the pre-interpolation phase, but also in validating the results. The quality of kriging predictions is largely determined by the choice of variogram model, which can be seriously affected by a few extreme attribute values if these are closely located in space. Dynamic interaction with the data, in which several views of the data such as histograms, scattergrams, maps and variogram clouds can be examined simultaneously, permits the relations between extreme data values or trend residuals and geographical location to be easily seen. The study described in Chapter 6 by Leo Gunnink and Peter Burrough was carried out using the REGARD software package developed at Trinity College, Dublin, with modifications made by the authors. The use of EDA for detecting spatial anomalies that can affect geostatistical interpolation is demonstrated using data on polluted flood-plain soil in the Netherlands.

ESDA is powerful, but tools appropriate for many areas of application have still to be developed. Spatio-temporal data is one such case discussed in Chapter 7 by Antony Unwin, one of the developers of REGARD. Three distinct cases of spatio-temporal data are considered: a few long series of many points each (for example, regional unemployment data); many short series of a few points each (for example, annual trade data); and many long series (for example, pollution monitoring). It is shown that the first two can be successfully analysed using existing interactive graphical software, though of different kinds. Initial proposals are made for tackling the challenge of the third case.

An important component of ESDA is to measure the spatial association between observations for one or several variables. Most measures of spatial autocorrelation can easily be incorporated in a framework that combines spatial analysis with a GIS. In Chapter 8 Luc Anselin suggests a simple tool to visualise and identify the degree of spatial instability in spatial association by means of Moran's spatial autocorrelation coefficient. It is based on the interpretation of this statistic as a regression of the spatially lagged variable on the original variable. This scatterplot may be used in isolation or may be integrated as an additional view of the data in a system of dynamic or interactive graphics, to allow for so-called scatterplot brushing. The technique is illustrated by an analysis of the spatial pattern of conflict between African countries.

1.5.2 Part 2: Spatial integration issues

Map overlay is a basic GIS function, used in many operations. In both raster and vector data models most systems implement it as an error-free Boolean operation. For some types of objects, for example known census tracts where category membership is certain and where the boundaries are known exactly, this may be reasonable. In almost all work involving the national environment it is not. Boundary uncertainty is common, subscale 'inclusions', such as those recognised by soil surveyors, and plain classification error in assignments in the multinomial fields characteristically used, mean that overlay must be seen probabilistically. Moreover, there is almost always additional sample evidence available that might be used to inform the overlay. To date, several alternative approaches have been adopted, including expert systems, multicriteria analysis, the application of Bayes theorem, and standard OLS regression. Chapter 9, written by David Unwin, develops a typology of GIS operations in which the concept of overlay is central. He describes the usual implementation of the method, summarises the case that almost always this should be regarded as an operation in statistical modelling, and gives an overview of some of the methods that have been adopted. A selection of these methods is illustrated by a study of landslide hazard prediction in a part of China. It is suggested that categorical data modelling provides a very general means by which to select which maps should be overlain and to estimate the necessary weights.

Chapter 10 by Bianca Falcidieno, Caterina Pienovi and Michela Spagnuolo emphasises the need for abstraction mechanisms which generate high-level descriptions based on prominent features. In particular, the authors consider descriptive and declarative modelling which broadens the idea of spatial data modelling from relatively passive inquiry to much more active intent. Descriptive modelling corresponds to a bottom-up approach to spatial data organisation, and provides tools to extract prominent features from the low-level geometric model of the GIS. Declarative modelling involves two aspects: exploration and generation. Exploration of a high-level description of spatial data is performed to allocate stated geographical configurations, while the generation of low-level descriptive models supports the simulation of geographical phenomena.

In Chapter 11, Andrea Fabbri and Chang-Jo Chung propose an analytical conceptualisation of GIS, in which processes are modelled through the integration of information. Modelling of spatial data is seen as a predictive process in which positional coincidences are associated with relationships between non-spatial attributes. The predictive and prescriptive value of GIS models is not restricted to the power of advanced modelling tools, it is also the result of converting information on processes and phenomena into a computational form in which conceptual and quantitative models are represented. This chapter analyses a spectrum of predictive models, including Bayesian probabilistic measures, the Dempster–Shafer belief function, and the membership function in fuzzy sets. Such models are of fundamental importance in the design of expert procedures for the construction and subsequent verification of decision processes and strategies.

The application of GIS techniques to modelling hydrological processes fosters an optimal acquisition and use of spatially varying basic data, a more efficient management of local or regional model parameters and visualisation of the model output. Therefore, the integration of GIS and hydrological models is an essential condition for the further development from lumped to spatially distributed models. Several conceptual frameworks for the coupling of spatial analytical models with GIS are possible. A severe problem of actual GIS applications for hydrological studies is the missing time domain of existing GIS systems. Urgently needed are GIS functions for the management of time series as basic input data type and statistical methods of time series analysis. Chapter 12, written by Streit and Wiesmann, discusses conceptual issues and practical difficulties in integrating GIS and hydrological models. Particular emphasis is laid on the coupling of the Modules Hydrological Modelling System (MHMS) with a multiple vendor GIS platform consisting of Arc/Info and GRASS, and the development of GIS tools for spatial hydrological modelling.

Chapter 13 by Enrico Feoli and Vincenzo Zuccarello illustrates the use of probabilistic similarity functions for landscape-ecological analysis based on choropleth maps in a GIS environment. The final chapter of Part 2, by Arentze, Borgers and Timmerman, discusses some ways to improve the modelling capabilities and functionalities of GIS in a socio-economic context.

1.5.3 Part 3: Spatial dynamic modelling

Differential equations, partial differential equations and, in some instances, empirical equations have been the underlying mathematical tools behind spatial simulation models. In Chapter 15, by António Câmara, Francisco Ferreira and Paolo Castro, approaches based on cellular automata are proposed to replace the conventional tools. Issues such as the definition of transition rules, computer implementation with raster GIS and model verification are discussed. Applications to predator–prey, water quality and fire propagation modelling are included for illustrative purposes. Future developments related to extension of the cellular automata paradigm to handle multidimensional cells and interacting bit planes are also presented.

Chapter 16 by Geoffrey Roy and Folke Snickars proposes CityLife as modelling environment of the cellular automata type to study urban dynamics. CityLife is a modelling framework based on a grid of cells, just like the Game of Life. Each cell is intended to represent a unit of space which can contain some particular urban activity, typically dwellings, workplaces, transport infrastructure, or green space. The spatial arrangement of cells is intended to reflect the spatial organisation of an urban system. Actors may cooperate or compete with each other about locations in space. From some initial state, and through the application of a set of simple behavioural rules and selection criteria, the system will grow and evolve spatially. The authors attempt to study these changes in land use patterns. They discuss how such a simple approach to modelling can contribute to the understanding of the emergence of complexity in urban systems.

In Chapter 17, Lena Sanders attempts to compare different applications of dynamic models of urban systems, to show their respective advantages and disadvantages, and to analyse how they can be coupled with GIS. Although there has been a lot of research in dynamic systems, most of it focuses on theoretical developments and most models have been transferred into geography from other disciplines, such as mathematics, physics, and biology. Improvements in the treatment of complex systems in these fields have led to useful applications in urban geography where sets of spatial units such as cities, at an inter-urban scale, or districts, at an intra-urban level, often have a systematic organisation. Geographers have successfully incorporated mathematical tools such as differential equations, cellular automata and distributed artificial intelligence into their models. Parallel improvements in GIS can give a new impetus to the development and application of dynamic models and increase their usefulness as simulation and prediction tools.

REFERENCES

ANSELIN, L. (1989). Spatial regression analysis on the PC: Spatial econometrics using GAUSS. Discussion Paper No. 2, Ann Arbor, Mich; Institute of Mathematical Geography.

ANSELIN, L. (1991a). Quantitative methods in regional science: Perspectives on research directions, in Boyce, D., Nijkamp, P. and Shefer, D. (Eds) *Regional Science: Retrospect and Prospect*, pp. 403–424. Berlin: Springer-Verlag.

ANSELIN, L.(1991b). SpaceStat: A Program for the Analysis of Spatial Data. Department of Geography, Santa Barbara, CA: University of California.

ANSELIN, L. and GETIS A. (1993). Spatial statistical analysis and geographic information systems, in Fischer, M.M. and Nijkamp, P. (Eds) *Geographic Information Systems, Spatial Modelling, and Policy Evaluation*, pp. 35–49. Berlin: Springer-Verlag.

ARBIA, G. (1989). Spatial Data Configuration in Statistical Analysis of Regional Economic and Related Problems. Dordrecht: Kluwer.

ARBIA, G. (1991). GIS-based sampling design procedures, in Harts, J., Ottens, H. and Scholten H. (Eds) *Proceedings, EGIS'91 Second European Conference on Geographic Information Systems*, **1**, pp. 27–35. Utrecht: EGIS Foundation.

BATEY, P.W.J. (1985). Input-output models for regional demographic-economic analysis: Some structural comparisons. *Environment and Planning*, **A**(17), 73–99.

BATTY, M. (1992). Geographic information systems in the social and policy sciences: The research program of the National Center for Geographic Information and Analysis. Paper presented at the ESF Workshop on Socio-Economic Applications of Geographical Information Systems, Sintra, Portugal, June 10–14.

BATTY, M. (1993). Using geographic information systems in urban planning and policy-making, in Fischer, M.M. and Nijkamp, P. (Eds) *Geographic Information Systems, Spatial Modelling, and Policy Evaluation*, pp. 51–69. Berlin: Springer-Verlag.

BAUMANN, J., FISCHER, M.M. and SCHUBERT, U. (1983). A multiregional labour supply model for Austria: The effects of different regionalisations in multiregional labour market modelling. *Papers of the Regional Science Association*, **52**, 53–83.

BAUMANN, J., FISCHER, M.M. and SCHUBERT, U. (1988). A choice-theoretical labour-market model: Empirical tests at the mesolevel, *Environment and Planning A*, **20**, 1085–1102.

BAXTER, M.J. (1987). Testing for misspecification in models of spatial flows, *Environment and Planning A*, **19**, 1153–60.

BIVAND, R. (1991). SYSTAT – Compatible software for modelling spatial dependence among observations. Paper presented at the 7th European Colloquium on Theoretical and Quantitative Geography, Hasseludden, Sweden.

BURROUGH, P.A. (1990). Methods of spatial analysis in GIS, *International Journal of Geographical Information Systems*, **4**(3), 221–3.

BURROUGH, P.A. (1991). Intelligent geographic information systems: A case for formalizing what we already know. Proceedings, ESF-Workshop on European Research in Geographical Information Systems (GIS), 24–26 January, Davos.

CLARKE, M. (1990). Geographical information systems and model based analysis: Towards effective decision support systems, in Scholten, H.J. and Stillwell, J.C.H. (Eds) *Geographical Information Systems for Urban and Regional Planning*, pp. 165–75. Dordrecht: Kluwer.

DENSHAM, P.J. (1991). Spatial decision support systems, in Maguire, D.J., Goodchild, M.F. and

Rhind, D.W. (Eds) *Geographical Information Systems, Volume 1: Principles*, pp. 403–12. Essex: Longman.

DENSHAM, P.J. and RUSHTON, G. (1988). Decision support systems for locational planning, in Golledge, R. and Timmermans, H. (Eds) *Behavioural Modelling in Geography and Planning*. pp. 56–90. London: Croom-Helm.

DESPOTAKIS, V. (1991). Sustainable development planning using geographic information systems. Unpublished Ph.D. Thesis, Department of Economics and Econometrics, Amsterdam: Free University.

DESPOTAKIS, V.K., GIAOUTZI, M. and NIJKAMP, P. (1993). Dynamic GIS models for regional sustainable development, in Fischer, M.M. and Nijkamp, P. (Eds) *Geographic Information Systems, Spatial Modelling, and Policy Evaluation*, pp. 235–61. Berlin: Springer-Verlag.

FEDRA, K., WEIGKRICHT, E. and WINKELBAUER, L. (1991). Decision support and information systems for regional development planning. Paper presented at the International Colloquium on 'Regional Development: Problems of Countries in Transition to a Market Economy', 21–24 April, Czech Republic.

FISCHER, M.M. (1993). Expert systems and artificial neural networks for spatial analysis and modelling: Essential components for knowledge-based geographical information systems, *Geographical Systems*, **1**, 221–35.

FISCHER, M.M. (1994). From conventional to knowledge based geographical information systems. *Computers, Environments and Urban Systems*, **18**(4), 233–42.

FISCHER, M.M. and GOPAL, S. (1994). Artificial neural networks. A new approach to modelling interregional telecommunication flows. *Journal of Regional Science*, **34**(4), 503–27.

FISCHER, M.M. and NIJKAMP, P. (1992). Geographic information systems and spatial analysis, *The Annals of Regional Science*, **26**(1), pp. 3–17.

FISCHER, M.M. and NIJKAMP, P. (Eds) (1993a). *Geographic Information Systems, Spatial Modelling, and Policy Evaluation*. Berlin: Springer-Verlag.

FISCHER, M.M. and NIJKAMP, P. (1993b). Design and use of geographic information systems and spatial models, in Fischer, M.M. and Nijkamp, P. (Eds) *Geographic Information Systems, Spatial Modelling, and Policy Evaluation*, pp. 1–13, Berlin: Springer-Verlag.

FISCHER, M.M., NIJKAMP, P. and PAPAGEORGIOU, Y.Y. (Eds) (1990). *Spatial Choices and Processes*, Amsterdam: North-Holland.

FLOWERDEW, R. and GREEN, M. (1991). Data integration: Statistical methods for transferring data between zonal systems, in Masser, I. and Blakemore, M. (Eds) *Handling Geographical Information: Methodology and Potential Applications*. pp. 38–54. Essex: Longman.

FOTHERINGHAM, A.S. and ROGERSON, P.A. (1993). GIS and spatial analytical problems, *International Journal of Geographical Information Systems*, **7**(1), 3–19.

GOODCHILD, M. (1987). A spatial analytical perspective on geographical information systems, *International Journal of Geographical Information Systems*, **1**(4), 327–34.

GOODCHILD, M. (1992). Integrating GIS and spatial data analysis: Problems and possibilities, *International Journal of Geographical Information Systems*, **6**(5), 407–23.

GOODCHILD, M. and GOPAL, S. (1989). *Accuracy of Spatial Databases*. London: Taylor & Francis.

GOODCHILD, M., HAINING, R., and WISE, S. (1992). Integrating GIS and spatial data analysis: Problems and possibilities, *International Journal of Geographical Information Systems*, **6**(5), 407–23.

GRIFFITH, D.A. (1989). Spatial regression analysis on the PC: Spatial statistics using Minitab. Discussion Paper No 1, Ann Arbor, Mich: Institute of Mathematical Geography.

GROSSMANN, W.D. and EBERHARDT, S. (1993). Geographical information systems and dynamic modelling – Potentials of a new approach, in Fischer, M.M. and Nijkamp, P. (Eds) *Geographic Information Systems, Spatial Modelling, and Policy Evaluation*, pp. 167–80. Berlin: Springer-Verlag.

HAINING, R. (1990). *Spatial Data Analysis in the Social and Environmental Sciences*, Cambridge University Press, Cambridge.

HAYNES, K.E. and FOTHERINGHAM, A.S. (1988). *Gravity and Spatial Interaction Models*, Beverley Hills: Sage Publications.

HEWINGS, G.J.P. and JENSEN, R.C. (1986). Regional, interregional and multiregional input–output analysis, in Nijkamp, P. (Ed.) *Handbook of Regional and Urban Economics. Volume 1: Regional Economics*, pp. 295–355. Amsterdam: North-Holland.

HIRSCHFELD, A.F.G., BROWN, P.J. and MARSDEN, J. (1993). Information systems for policy evaluation: A prototype GIS for urban programme impact appraisal in St. Helens, North West England, in Fischer, M.M. and Nijkamp, P. (Eds) *Geographic Information Systems, Spatial Modelling, and Policy Evaluation*, pp. 213–34. Berlin: Springer-Verlag.

KEHRIS, E. (1990). Spatial autocorrelation statistics in ARC/INFO. North West Regional Research Laboratory, Research Reports, 16, Lancaster University, UK.

LAKSHMANAN, T.R. and BOLTON R. (1986). Regional energy and environmental analysis, in Nijkamp, P. (Ed.) *Handbook of Regional and Urban Economics. Volume 1: Regional Economics*, pp. 581–628. Amsterdam: North-Holland.

MASSER, I. and BLAKEMORE, M. (Eds) (1991). *Handling Geographical Information: Methodology and Potential Applications*. Harlow: Longman.

NIJKAMP, P. (Ed.) (1986). *Handbook of Regional and Urban Economics. Volume 1: Regional Economics*, Amsterdam: North-Holland.

OPENSHAW, S. (1977). Optimal zoning systems for spatial interaction models, *Environment and Planning A*, **9**, pp. 169–84.

OPENSHAW, S. (1990). Spatial analysis and geographical information systems: A review of progress and possibilities, in Scholten, H.J. and Stillwell, J.C.H. (Eds) *Geographical Information Systems for Urban and Regional Planning*, pp. 153–63. Dordrecht: Kluwer.

OPENSHAW, S. (1991a). Developing appropriate spatial analysis methods for GIS, in Maguire, D.J., Goodchild, M.F. and Rhind, D.W. (Eds) *Geographical Information Systems, Volume 1: Principles*, pp. 389–402. Harlow: Longman.

OPENSHAW, S. (1991b). A spatial analysis research agenda, in Masser, I. and Blakemore, M. (Eds) *Handling Geographical Information: Methodology and Potential Applications*, pp. 18–37. Harlow: Longman.

OPENSHAW, S. (1993). Some suggestions concerning the development of artificial intelligence tools for spatial modelling and analysis in GIS, in Fischer, M.M. and Nijkamp, P. (Eds) *Geographic Information Systems, Spatial Modelling, and Policy Evaluation*, pp. 17–33. Berlin: Springer-Verlag.

OPENSHAW, S., CHARLTON, M., WYMER, C. and CRAFT, A. (1987). A Mark1 geographical analysis machine for the automated analysis of point data sets, *International Journal of Geographical Information Systems*, **1**(4), 335–58.

RIPLEY, B.D. (1981). *Spatial Statistics*. New York: John Wiley.

ROY, G.G. and SNICKARS, F. (1993). Computer-aided regional planning – Applications for the Perth and Helsinki regions, in Fischer, M.M. and Nijkamp, P. (Eds) *Geographic Information Systems, Spatial Modelling, and Policy Evaluation*, pp. 181–97. Berlin: Springer-Verlag.

SCHOLTEN, H.J. and STILLWELL, J.C.H. (Eds) (1990). *Geographical Information Systems for Urban and Regional Planning*. Dordrecht: Kluwer.

TAYLOR, P.J. (1977). *Quantitative Methods in Geography*. Boston, MA: Houghton Mifflin.

WILSON, A.G. and BENNETT, R.J. (1985). *Mathematical Methods in Human Geography and Planning*. London: John Wiley & Sons.

WILSON, A.G., COELHO, J., MCGILL, S.M. and WILLIAMS, H.C.W.L. (1981). *Optimisation in Locational and Transport Analysis*. Chichester: John Wiley & Sons.

CHAPTER TWO

Developing spatial analysis functions relevant to GIS environments

STAN OPENSHAW and GRAHAM CLARKE

2.1 Introduction

The GIS revolution that started in many countries in the mid-1980s is creating a wealth of spatial information in a large number of application areas. Sooner or later, the analogue map data conversion process will be completed and nearly all the computer databases relevant to many aspects of environment and human society on a global scale will be available with locational referencing attached. In addition, the emphasis on GIS database creation and systems building will be replaced by a new concern with *analysis*. The proliferation of spatial information, a continuing downward trend in hardware costs, and the emergence of a new era of computation with computer speeds and memory capacities moving into 'terracomputing' domains will help create an environment where spatial analysis will not merely be of increasing interest and value but in some areas it will be an absolute necessity. More and more users will want to analyse their data for a range of good public, community, commercial and research reasons. An increasing number of these analyses will be real-time (or almost) and the majority will probably involve forms of spatial analysis that do not as yet exist except in nascent form. The drive behind these future analysis needs is threefold:

1. New requirements to make use of the information resources created by GIS;
2. Attempts to gain competitive or other benefits from IT, and
3. Hardware improvements that trivialise the computational requirements.

It may already be a scandal that so many of the key databases are not being properly analysed, for example those related to morbidity, mortality and crime. Commercial concerns and government agencies probably waste large sums of money by poor spatial analysis of needs, inefficient spatial planning of facilities, and poor targeting of resources (Openshaw, 1994a). If better, more relevant, spatial analysis tools were available then they would almost certainly be used. Some even claim that their non-availability threatens the viability of GIS itself (Openshaw and Brunsdon, 1991). The post-GIS world of the 1990s is quite different from the computing and data environments in which many of the existing spatial analysis and modelling technologies were developed. Are these old technologies still appropriate, or is a new period of basic research and development needed to create the spatial analysis tools likely to be required in the late 1990s and beyond? This chapter attempts to address some of these concerns by focusing on the general and generic aspects in an attempt to broadly define the types of new exploratory spatial data analysis that are likely to be most relevant for use within GIS whilst at the same time recognising the usefulness of some of the more traditional spatial analysis tools.

The mid-1990s seems a good time for a far-reaching and fundamental rethink on GIS functionality because spatial analysis is entering a new era in the development of quantitative geography and new analysis needs are being created and stimulated as a by-product of GIS. Birkin *et al.* (1996) argues that there are three broad responses to the poor spatial analysis routines currently available in proprietary packages. The first is 'simply' to add more functions as standard menu options. The second recognises that the needs for spatial analysis are increasingly becoming data driven; analysis is to be performed because data exist and a user wants some analysis as an adjunct to other GIS activities. This is quite different from more traditional quantitative geographic concerns performed as acts of scholarship in a research context. The third response is to recognise the usefulness of some existing methodologies and then attempt to couple this sort of analysis into current GIS packages. Birkin *et al.* (1996) describe this approach in some detail using spatial modelling methods. Their argument is that we now have a lot of experience of applied urban modelling which should not be lost in the new era of data driven spatial analysis, but it will be unless the tools become more widely available. Adopting the second and third strategies for GIS futures requires a major change in emphasis and is the subject for the rest of this chapter.

This change in emphasis and its implications have not been properly understood. The lack of recognition of these new needs can be seen in the continuing confusion as to what spatial analysis in GIS actually means. Whilst there is a general consensus that spatial analysis functionality is needed in GIS toolkits, there is no agreement about what kinds of spatial analysis methods are most relevant to GIS and which ones are not. The various 'Spatial analysis and GIS' initiatives have not only failed to clarify the situation but have probably, yet unwittingly, contributed to the confusion. It is useful to note at this point that, despite considerable publicity for the NCGIA I-14 Spatial Analysis and GIS initiative, it was later admitted to be primarily a public relations exercise. The total resources expended probably amounted to less than £35 000 (Fotheringham and Rogerson, 1993). Maybe the current state of spatial analysis confusion is, therefore, understandable. Despite many promisingly titled books on the subject of Spatial Analysis and GIS or vice versa (see for example Fotheringham and Rogerson (1994)), most of the problems have not been solved although there is now a much improved awareness of some of the issues.

A broader consensus about a set of GIS relevant spatial analysis methods is important for a number of strategic reasons. It defines in an explicit manner the nature of the analysis technology that needs to be provided, it sets a research agenda for developing relevant new methods, and it offers a focus of attention for international attempts to develop spatial analysis tools for GIS via collaborative exercises. The question 'What kind of spatial analysis do we want in GIS?' has to be tempered by feasibility (what kinds of spatial analysis can be implemented in GIS?) and sensibility (what kinds of spatial analysis is it sensible to provide for GIS environments?) constraints. Another set of general design constraints reflect other considerations such as who are the likely users, what is it they want, and what sort of analysis technology can they handle given likely fairly low levels of statistical knowledge and training in the spatial sciences. Table 2.1 summarises the principal design questions. It is possible that the answers to these sorts of

Table 2.1 Basic design questions

What kinds of spatial analysis:
are relevant to GIS data environments;
are sensible given the nature of GIS data;
reflect likely end-user needs;
are compatible with the GIS style;
are capable of being used by end-users;
add value to GIS investment;
can be an integral part of GIS;
offer tangible and significant benefits?

questions will shape the particular direction future GIS applications will take. It is noted that user-ability criteria are very important and the future viability of whatever technologies are proposed will ultimately depend on the extent to which the methods can be safely packaged for use by non-expert users. We often forget that in the future the users are unlikely to be a handful of highly skilled researchers performing basic or pure research but end-users interested in making or supporting decision making that impacts on real people in a broad range of application areas. If the criteria in Table 2.1 are applied to the set of available spatial analytical, statistical, and modelling tools, then it is quite clear many will never fit.

The debate as to whether these methods should be accessed from within or without a GIS package is really quite irrelevant. There is no reason to insist on only one form of integration or interfacing, except the obvious point that to be a GIS tool the spatial analysis operation has to be called from, and end within, a GIS environment but in an era of heterogeneous distributed computing there is no longer any need for all the systems to be on the same machine. Equally, the extent to which methods are perceived as having to be run within a GIS environment is often overplayed. What does GIS have to offer spatial analysis that is unique? In practice, it often amounts to little more than spatial data and consistently defined contiguity lists, in which case much system complexity can be avoided by the simple expedient of separating out the different components needed by the analysis process and developing a high-level system to call a GIS here, a model or analysis tool there, and a map drawer if one is needed. It is no longer the case that only those methods which can be physically put into a single GIS, or which have access to fully functional spatial data bases at a low level, can perform any kind of spatial analysis.

2.2 Statistical geographic hangovers

It is useful to start by trying to put many traditional spatial analysis methods in a proper historical context. Quite simply, many do not belong to the GIS era. Similar 'unkind' thoughts can also be applied to many largely aspatial statistical methods originally designed for use in survey sampling contexts using data collected by simple random sampling. These are seldom relevant because they treat spatial information as if it were equivalent to sampled survey data, and generally fail to handle many of the key features that make spatial data special (that is, spatial dependency, non-sample nature, modifiable areal unit effects, noise and multivariate complexity). The literature of historical statistical geographical books should, in general, be seen for what it is, a set of introductions to statistical methods that are of geographical interest mainly because they have been written for a geographical audience. The application of aspatial statistical methods in GIS might appear sensible, but it has to be recognised that the technology may often be inherently unsuitable and, whilst most methods may be harmless, they are not all safe. An example of the latter is ranking zones by a Z-score, a widely used procedure in targeting areas for resource allocation. This seemingly simple and safe procedure has a number of fundamental problems which include spatial autocorrelation which might inflate Z-score values; the zones being ranked are not comparable entities; the data for each variable are not measured with the same degree of precision (more extreme results are more easily obtained in small zones than in big ones), and the results are influenced to an unknown and varying degree by the nature and scale of the zones being used.

Many similar problems associated with the analysis of spatial data are not new and their existence has been known for at least 25 years. As a result, most of the statistical geographical literature is simply not relevant to GIS. We believe that key texts such as Ebdon (1977) Gregory (1963), Taylor (1977) Haining (1992), Griffith and Amrhein (1991) contain little or nothing relevant to GIS. They are simply not appropriate as a source of 'GISable' spatial analysis tools even if they are all excellent texts describing the use of statistical methods with spatial data. Anselin (1989) clearly recognises some of the statistical problems that exist when he argues:

> With the vast power of a user-friendly GIS increasingly in the hands of the non-specialist, the danger that the wrong kind of spatial statistics will become the accepted practice is great. Since the 'easy' problems have more or less been solved, a formidable challenge lies ahead. (pp.14–15)

Indeed, developing new and more relevant purely statistical methods suitable for use with spatial data looks like being too difficult for current technology particularly when it has to be performed in a rigorous fashion within a classical statistical framework. Is this really what GIS needs? Might it not be easier to contemplate more descriptive and less statistical theory dependent technologies? Maybe the statistical problems are just too hard to be resolved at present, so why not change the nature of the problem to make it easier to handle? It should be possible to do this because purely spatial statistical analysis is probably not what most GIS end-users want or need. Furthermore, in the traditional concept of the map, geographers have probably the best visualisation tool ever invented. Why not use it rather than feel embarrassed about it, as it seems many human geographers now tend to be? New map-based spatial analysis technologies are probably what is needed most.

The problem with much traditional statistical analysis in geography dates back to Cliff and Ord (1973). Based on the 1950s' spatial statisticians, the assumption is that it makes sense to summarise whole map patterns by a single numerical index or to test whole map statistics for departures from randomness. The classic Irish County ($N = 27$ observations) data was too small for the dependency of the results on a particular set of areal units and the fundamental limitations of interpreting whole map statistics to be evident. Scale the methods up by applying them to much larger sets of zones and all manner of geographical problems appear. The need now is to disaggregate the whole map statistics and to develop effective means of identifying localised spatial heterogeneities and localised patterning within, rather than over, maps.

The real challenge that GIS presents is to rethink what GIS relevant spatial analysis is now needed, free from a bias that insists on this or that method always being applied. It is becoming important to dump much of the obsolete baggage of the past and try to develop some new spatial analysis tools that are simultaneously GISable, relevant to the needs of GIS-based users, and appropriate for spatial data. In so doing it is particularly important to be aware of the very considerable sensitivities that exist in some areas about the quality of the spatial statistical analyses performed by geographers. Martin (1990) expresses them as follows:

> It seems unfortunate to me that non-statisticians who publish statistics, however flawed, are held in high esteem in their professions, whilst statisticians who publish statistics, however good, are seen as doing no more than trade. (p. 117)

The answer in many cases can be simply to replace the term statistical analysis by geographical analysis, but this really misses the point because geographical analysis involving numbers, by definition is statistical analysis. The problem of geographical common sense also needs to be addressed. Far too many statisticians seem to hold (or have once held) the view that the role of the geographer is to define the problem, pass them some data, and wait for the definitive statistically approved results to appear. However, it is not satisfactory to separate spatial analysis functions from an understanding of the geography of the problem. Neither statistical nor geographical knowledge are by themselves sufficient; both are necessary, and it is important that both sets of skills are brought together. This is particularly so because there is a growing recognition that there is no easy, theoretical solution to many of the problems of statistical spatial analysis.

2.3 Spatial modelling hangovers?

So far we have been particularly critical of statistical methods in GIS as we believe that adding a new regression or spatial autocorrelation procedure to this or that GIS is not the answer for many GIS users. In some respects, the same arguments apply to a number of mathematically-based modelling procedures. We note with some concern the current trend for each GIS to offer some kind of, usually pre-1970s, spatial interaction modelling capability. Whilst we are convinced that many such methods will remain applicable to GIS environments of the future (at least until we discover how to build better models), the whole question of how to access good mathematical models of spatial systems still needs to be resolved in a satisfactory fashion. Perhaps this is a classic example of a technology that does not need to be imprisoned in the straitjacket of a proprietary GIS where it may only utilise 1 per cent of the available GIS functionality. The major advantage of such models lies in their ability to address not only 'what-is?' questions, but also the crucial 'what-if?' questions that many organisations require for future planning. Spatial modelling procedures based on spatial interaction or location optimisation are generally very good at addressing these sorts of questions. However, those now appearing in proprietary GIS are both inadequate and often presented as black-box solutions which ignore 25 years of work on the problems of model specification, parameter estimation, and model evaluation. Benoit and Clarke (1995) provide an evaluation of the appropriateness and ease of use of spatial interaction models for retail analysis in proprietary GIS packages. Thus, whilst it is likely that spatial modelling procedures will continue to be coupled to GIS software, most of the specialist modelling will take place outside, but loosely linked to, the main package environment. The skills required to handle the complexities of applied spatial modelling makes 'generic' solutions difficult but it is important to stress that not all progress should be technology-led. If applications demand what-if? planning then geographers have a role to provide these solutions. These arguments are taken further in Birkin *et al.* (1996).

2.4 Ten basic rules for identifying future GISable spatial analysis technology

It is useful to try and be quite clear as to what is needed without being to concerned initially with how to achieve it. The aim here is to create a list of criteria able to discriminate between GISable and GIS-irrelevant technology (see also Openshaw, 1994b and Openshaw and Fischer, 1996).

2.4.1 A GISable spatial analysis method should be able to handle large and very large *N* values

In the GIS era, data sets typically have large *N* values. It is interesting that the *N* values in many databases will soon reach their theoretical maxima. For example, in the UK there are about 150 000 census enumeration districts, 1.6 million postcodes, and an ultimate limit of about 55 million persons whose homes will soon have a sub-1 m resolution grid-reference. This may seem a lot but computer hardware can now handle these numbers with ease. Hardly any GIS users will be interested in 20 or 50 zones; if they are, then they should realise that data of two or three orders of magnitude greater resolution exist and might be preferable even for local and subregional scale studies. Accordingly, any statistical method that involves the use and manipulation of several $N \times N$ matrices is probably not sensible to include in a GIS. Of course it is possible to perform operations on quite large matrices but perhaps not within a regular GIS environment. As computer power increases then the problem of matrix size in mathematical modelling also diminishes but it is likely to remain problematical for a long time yet.

2.4.2 Useful GISable analysis and modelling tools are study-region independent

Spatial pattern analysis results that depend on, or are determined by, the arbitrary definition of a study region are not useful. This observation is a simple extension of the modifiable areal unit problem. It is also geographical common sense. At one level it is a trivial and obvious problem. If you study data just for Scotland and then repeat the analysis for England and Scotland, or just the city of Leeds, then the conclusions obtained from the results will probably differ. The differences reflect variation in the phenomenon under study, the choice of study region, and interactions between the two and the effects of this are so obvious that they are often completely ignored. For example, if you wish to summarise the distribution of public houses in Leeds by a nearest neighbour statistic then the results depend on the definition of Leeds and on the patterning of pubs within Leeds. In the north, it might be systematically and spatially even, in the south it might be random, and in the city centre clustered. The result for the whole of Leeds averages out all these differences and will almost certainly be meaningless unless the spatial pattern is more or less the same everywhere.

Another example might help clarify this further. Imagine a cancer data set with one very strong cluster. If a small study region is selected that includes the cluster then virtually any spatial pattern statistic will detect its existence. If this is done deliberately then it is a form of gerrymandering the results by boundary tightening. If the study region is increased in size by expanding its geographic extent and adding randomly located cancer cases there comes a point when the cluster is lost in a sea of randomness. As the study region decisions may be arbitrary and are seldom part of the analysis process, so it follows that *whole* map statistics should not be used. They are not as helpful in GIS except where the whole map region is characterised by only one distinctive type of spatial pattern. The concern now is therefore to define and measure localised patterns rather than global ones. This distinction is of fundamental importance.

Scale effects provide an additional level of complexity. For example, if the answer to the question 'does a disease cluster at the district scale?' is no, this does not mean that some districts will not have localised clustering within them or that at a small area level there might be considerable clustering. What it does indicate is that, when viewed at an aggregate district scale, clustering is not a characteristic phenomenon of the disease under study. This is a useful, but limited, result with a potential greatly to mislead. From a geographical point of view it is the localised clustering of small scales that are of greatest interest. Indeed, it is hard to imagine why district level clustering (given their heterogeneous nature) should be of any great interest to anyone. The disease processes simply do not operate at that scale unless they are somehow related to causative agents that are specific to districts.

It is argued, therefore, that in most analysis applications spatial pattern detectors are needed which can look within the map to find patterns and relationships that are geographically localised in extent. This is important to avoid the modifiable study region effect and because geographic space is not characterised by uniform densities of anything. In a spatial context techniques that can detect or model only global patterns or relationships need to be used with great circumspection and, ideally, not at all.

2.4.3 GIS relevant methods need to be sensitive to the special nature of spatial information

To be relevant to GIS, spatial analysis tools should be able to handle spatial information in a reasonable manner, taking into account rather than ignoring any special features that matter. The corollary of this statement is that if an analysis tool has been developed for use with aspatial data then it will probably be inappropriate for use with spatial data without major modification.

Table 2.2 lists some of the features that characterise spatial data. Of particular importance is the spatially structured nature of data precision and errors. They are not spatially random. Database anomalies might be detected as patterns but could just as easily be data error and reflect other data inconsistencies. In addition, the data to be analysed are often known to be wrong but still have to be used. An example is the analysis of cancer data consisting of incidents for 1969–84 using denominator for either 1971 or 1981. The temporal resolution of the different components of the data differ. The data richness of the GIS world is often achieved at the expense of accuracy and the spatial analyst faces the prospect of what might be termed 'poor data analysis', of having to analyse data containing a mixture of different levels of error and uncertainty about which nothing can be done. Similarly, in relation to spatial data, more extreme results are often found in areas with the smallest denominators. Such areas tend to be geographically

Table 2.2 Features that characterise spatial data in the GIS era

Many cases/objects/points
Many variables
Large data volumes
Spatial autocorrelated values
Spatial data error of various types
Non-conformity with standard statistical distribution
Data precision can be spatially structured
Errors need not be random
Not samples in usual sense
Surrogate variables abound
Non-linearity is the norm
A high degree of complexity
Modifiable areal unit, scale and aggregation effects
Mixtures of scale and data resolution
Mixtures of measurement
Small number problems can be important

structured (for example, rural rather than urban). Likewise, the internal heterogeneity of all zoning systems varies and this must influence the results. If the data are converted into continuous form, as surfaces, then the problems do not disappear. They are simply hidden and appear in a different way.

Openshaw (1989) argues that a key principle in spatial analysis in GIS is to handle, rather than ignore, the problems. It is unacceptable to assume that the modifiable areal unit problem does not exist. In many instances, the problems can be addressed by Monte Carlo simulation, sensitivity analysis, boot-strapping, or seeking out high leverage data values. Some of the extra CPU power that is now available should be used to make the spatial analysis tools less 'naive' and more robust in the face of the known characteristics of spatial information.

2.4.4 The results should be mappable

GIS is a highly visual and graphics-oriented technology. The results of GIS spatial analysis operations should also be available in a graphic and mappable form. This is very important and in some ways represents a major design constraint on the forms of spatial analysis that might be considered as GIS relevant. The output cannot just be a set of statistics or model parameter values embedded in a text report, they have to be visual and mappable. This provides for a different spatial analysis paradigm in which the emphasis is firmly on visibility and visualisation (Batty, 1993). Figure 2.1 summarises what is intended and how the three basic components – human knowledge and intuition, analysis tool, and GIS – interlink.

The role of spatial analysis is essentially that of a filter designed to remove the rough from the smooth or to highlight areas of 'unusualness'. Far too many geographers seek complex, statistical solutions to problems that are best visualised, at least initially, so that when the time comes to apply more sophisticated methods there is at least some idea as to what to look for. The spatial analysis paradigm in Figure 2.1 ideally should be applied iteratively; each iteration has an input of human intuition and knowledge and with each successive iteration more insight is gained.

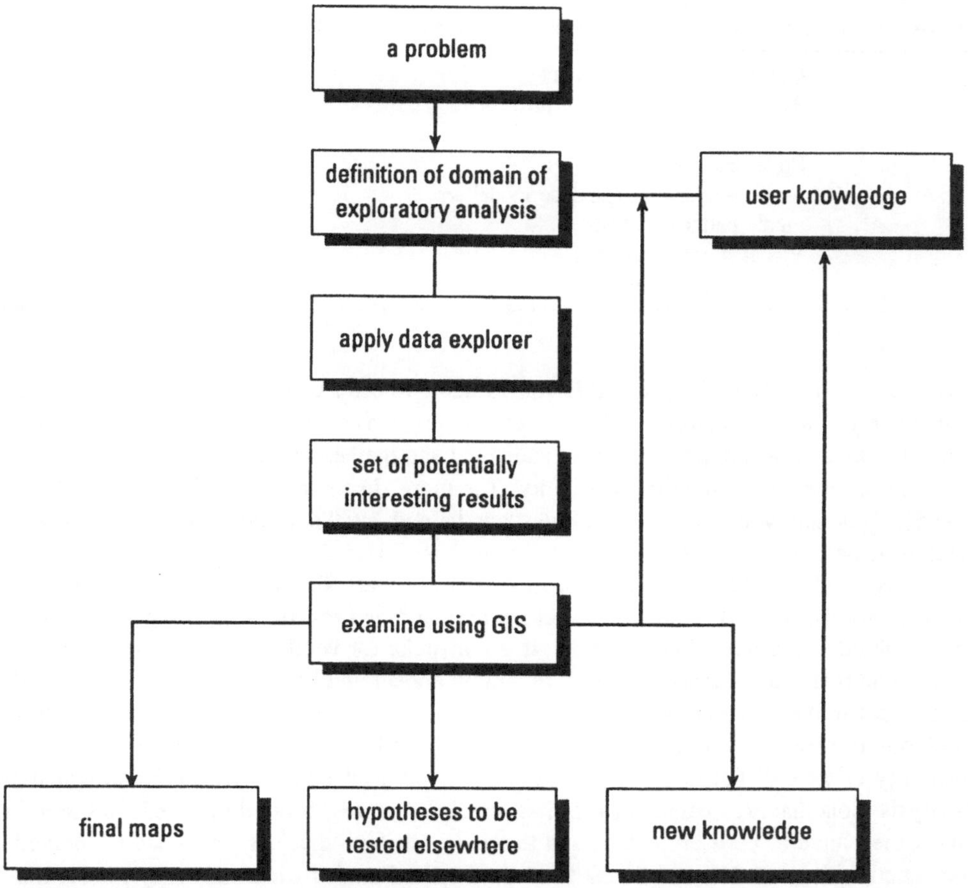

Figure 2.1 Exploratory spatial analysis paradigm.

2.4.5 GISable spatial analysis is generic

The GIS toolkit consists of a set of generic, application-independent tools for data capture, storage, manipulation, mapping and analysis. Very little, if anything, is application specific. The tools are generic. It follows that spatial analysis methods with the greatest claim for inclusion in proprietary GIS should also be application independent. It is claimed here that generic, general purpose, data invariant spatial analysis methods can in fact be defined. For example, methods that examine point data for clustering can be applied to disease data, crime data, earthquake, lightening, gas or water bursts, telephone faults, traffic accidents, and so on. The function is application independent. It was with this concept in mind that Openshaw (1991a, b) listed the set of general spatial analysis procedures that are reproduced in Table 2.3. At the present point in time it is probably more useful to agree on broad areas of analysis functionality than how to achieve it.

2.4.6 GISable spatial analysis methods should be useful and valuable

If users are to be lured into using spatial analysis there has to be a good and compelling reason. GIS relevant tools will not merely have to meet needs in academic research but

Table 2.3 Basic generic spatial analysis procedures relevant to GIS

Pattern spotters and testers
Relationship seekers and provers
Data simplifiers
Edge detectors
Automatic spatial response modellers
Fuzzy pattern analysis
Visualisation enhancers
Spatial video analysis

also have to have something tangible and valuable to offer end-users in applied contexts. The benefits have to outweigh the end-users' perceptions of the 'costs' of applying spatial analysis; an area where perceptions of statistical and mathematical complexity might give the impression of outstanding difficulty (for more discussion see Clarke and Clarke, 1995). This unfavourable impression is not helped by the usually complex and very technical nature of spatial analysis conferences and workshops.

It also follows that if GISable spatial analysis methods can be defined, then they should not be free. In the GIS world where all aspects of everything are owned, copyrighted, licensed and commercial, it is a mistake for vendors and system developers to assume that spatial analysis tools will simply appear and be 'free'. They cannot be left to the research sector to develop, because what they will concentrate on is the interesting methods in research that are probably least likely to be of general interest or meet end-usability criteria. What is needed is a new generation of explicitly user-friendly spatial analysis tools that are designed from first principles centred around the needs of users. In short they have to be user-centred and this is an attribute that is hard to add to methods designed without it. Not only must users find spatial analysis methods easy to use, they should also want to use them and then be able to communicate the results to others in a language relevant to decision makers and action (Openshaw and Perrée, 1996).

2.4.7 Interfacing issues are initially irrelevant and subsequently a problem for others to solve

In seeking to develop spatial analysis methods for GIS, the nature of the linkage should be irrelevant. At the present moment in time it is far more important to demonstrate the utility of GISable and GIS relevant spatial analysis tools. The philosophical rights and wrongs of programming in this or that language or system, of embedding or not embedding, of tight or loose coupling, of integrated or stand-alone, are largely an irrelevant distraction. The computing world is moving towards open, distributed, heterogeneous systems with multiple special or single-function hardware being invisibly interlinked. This is not the spatial analyst's prime concern. Their task is to create the most useful and relevant technologies for analysis. Likewise, computer intensiveness is no longer a critical issue. If it can be done at all at present and there is a good case for doing it, then it will one day become common practice on affordable hardware located somewhere on a global network. If the concern for interfacing reflects the need for access to low-level GIS databases, then that too is probably misplaced. Whatever is there that matters can be exported. The great benefit of GIS is that it provides useful data such as contiguity information for cleaned-up and topologically correct digital map databases.

Once this exists, the GIS has done its work, and we should use whatever system is most relevant. So many researchers seem to overlook the fact that, whilst the availability of generic spatial functions in a GIS is useful, they typically run three or more orders of magnitude slower than could be programmed in stand-alone software.

2.4.8 Ease of use and understandability are very important

GISability is not just about relevance to GIS environments and relevance to the needs of GIS users, it is also about the ability of the typical end-users to use it. For this reason, highly statistical technology and raw statistical methods will not usually be appropriate. They fail on ease of use and usability grounds. Statistical packages demonstrated long ago how even the most mathematically and complicated of methods can be presented in a fairly easy-to-use form. There are dangers in hiding complexity and allowing the statistically ignorant and unskilled to use highly advanced methods, but no more so here than in the other areas of GIS. The converse is also unproven. There is no guarantee that statistical experts will only perform sensible analysis with GIS data and GIS is based on the proposition that the complexity is hidden. For example, how many users understand the computational geometry that lies behind many of the methods in their GIS packages? There is no good reason why the complexity of any relevant spatial analysis technology cannot also be hidden behind suitable user interfaces but the real question is how to aid or foster a meaningful understanding of what the results mean. This understandability criteria is not a problem with many GIS because most basic GIS operations are almost instantly understandable independent of the technology used to perform them. The need now is for a similar type of almost instantly obvious and understandable spatial analysis toolkit.

It should be noted that there is no reason to assume that GISable spatial analysis tools must only involve statistically and mathematically complex methods. It is possible that the most relevant spatial analysis need not use any advanced forms of conventional statistical technology. It is also not clear that any conventional statistical analysis skills are really needed. Maybe the GIS vendors lack of interest in spatial analysis reflects the mistaken assumption that advanced statistical skills are a prerequisite before even the simplest method can be used. Indeed, it might be observed that quite often there is an inverse relationship between the degree of mathematical complexity present in a spatial analysis method and the associated level of applied geographical usefulness. On the other hand, the statement may be misleading because maybe those researchers who are most likely to indulge in statistical spatial analysis may not see themselves as doing anything particularly applied.

2.4.9 GISable analysis should be safe technology

The applied nature of GIS makes it important that naive methods (or those methods with a high innate propensity to generate spurious results) should be avoided. Spatial analysis results should be reliable, robust, resilient, error and noise resistant, non-parametric, and not based in any important way on standard inference. Monte Carlo significance tests, boot-strapping, jack-knifing and other computational statistics tools should be used wherever important analysis is being performed. GIS databases contain errors and uncertainties of various kinds and it is important that they do not mislead the innocent.

The basic null hypothesis in GIS is not of randomness but of database error. Only if this can be rejected is it worth applying anything sophisticated. The main problem with the analysis of GIS data is not that the data contain error and uncertainty, or are wrong in various ways, it is that the analysis technology does not know or expect or understand such problems. The errors and uncertainties inherent in spatial databases can be handled only if allowance is made for them (Openshaw, 1989). Methods should be developed to be self-checking numerically and the basic rule of the user having to validate any really interesting results via independent data needs to become an established practice. Conventional significance testing is a very minimalistic barrier to cross and this important message needs to be universally understood.

2.4.10 GISable methods should be useful in an applied sense

It has to be faced that GIS is primarily and predominantly an applied technology. Those spatial analysis methods most relevant to it will be applied and seek to meet generic, important and valuable goals. They should also focus on spatial analysis tasks that are relevant to GIS environments. Academic research needs to fall into a different category. However, we should not throw away the baby with the bath-water, there must be room for both new generic spatial analysis routines which solve applied problems and conventional spatial analysis and modelling which has a proven track record of applied problem solving (Birkin *et al.*, 1996). The problem is that of emphasis and balance.

2.5 Developing a spatial analysis paradigm that is appropriate for GIS

2.5.1 A basic typology

It would seem to follow from the previous discussion that future GISable spatial analysis methods will be:

1. essentially descriptive;
2. essentially exploratory;
3. probably not inferential in a traditional spatial hypothesis testing sense; and
4. that they may have to be invented from scratch although often the basic techniques already exist.

It is apparent that some suitable methods, such as some spatial modelling techniques, already exist and that these can be readily ported into GIS environments without too much difficulty. The principal problem is that of discriminating between those that are useful, those methods that might usefully be reconditioned or refurbished for spatial data applications, and those that are not useful.

2.5.2 Learning to live with exploratory spatial data analysis and spatial modelling

One major awareness problem concerns the perhaps surprising lack of appreciation of the GIS world's needs for spatial analysis technology. They are not necessarily the same as found in other areas where quantitative methods have traditionally been

Table 2.4 Technical specification

High degree of automation
Self-checking
Constantly seeking to improve performance
Capable of real-time operation
Can be presented as a black-box but is understandable
Contains mechanisms for user interaction and incorporating human knowledge
Can handle localised spatial heterogeneities
Not hindered by lack of knowledge of what, or when to look for
Handles rather than ignores spatial data characteristics
Can respond to the unexpected in intelligent ways
Can generate new hypotheses as an output from the analysis process

applied. The needs and views of GIS can be different and the data are different.

Table 2.4 outlines some of the basic design principles that have been deemed important (Openshaw, 1995). Clearly, not all GIS users will agree with them and probably even fewer statisticians, but they do at least draw attention to the need to think about the underlying design aspects; in particular who the end-users are, their levels of skills and what they see GIS as providing them with. The typical GIS end-user should not be assumed to be an academic mainly interested in high-level research activities that need extensive research training. Equally, designers of GIS relevant spatial analysis technology should not be mesmerised by a search for a 'holy grail' of soundly based statistically theory-rich methods: such things in general do not yet exist for use with spatial data generated by GIS. The real problem is that users do not know what patterns or relationships exist in the data. They do not know where to look, when to look, or even what to look for. GIS databases are multivariate data rich. There are lots of them and it is likely that whatever prior knowledge the user has will either be wrong (maybe only partially) or be inappropriate (due to the unexpected presence of much more distinctive but initially unknown order).

The goal is easily stated: a search for spatial order and regularities that are either recurrent elsewhere, or of purely local importance. The problem is that as Openshaw (1994a, d; 1995) explains, we are blind to the spatial patterns and processes that exist in the 'geocyberspaces' generated by GIS technologies. We have probably only ever found a minuscule fraction of the regularities that exist and worse still what we have found so far might well be artefacts created by the arbitrary selections and restrictions imposed on the data just so that they can be analysed. There is a desperate need for new smart spatial data explorers and modellers that can function without being told, in advance, what to look for. Artificial intelligence (AI) technology holds some promise and there are a number of potentially interesting ways forward based on neurocomputing and artificial life, as well as other forms of machine learning (Openshaw, 1992). However, in many ways these methods still need to be developed. Here it is more appropriate to develop a more general specification of the sort of GIS relevant technology that meets the design criteria that have been discussed.

2.5.3 Handling the human knowledge interface

A key need is the ability to create means of incorporating human knowledge, human skills, and human intuition into spatial analysis. No automated spatial analysis technology will develop far by itself, but equally it does not make much sense to cripple the new methods by expecting all the critical and hard decisions to be made by humans. As we have suggested earlier, model specification or *a priori* hypothesis is not easy in a GIS context. There is need for a conceptual view of exploratory analysis and modelling in GIS that can combine the strengths of both the human being and the power of AI-based machine learning technology, whilst simultaneously attempting to avoid their weaknesses. Figure 2.1 outlines in a general schematic fashion how this might be operated. It is independent of the exploratory analysis and modelling technology being used; precisely what is incorporated into the 'Spatial Analysis' or 'Exploratory Analysis and Modelling' boxes is irrelevant. Far more important is the notion of trying to create a spatial analysis tool or model builder that combines symbiotically both human and machine abilities, with in-built feedback and learning mechanisms which in time can be formalised and incorporated into intelligent software systems, but this is still at least 5–20 years away.

Implicit in Figure 2.1 is the assumption that the initial human inputs should be broad-brush rather than specific; they should reflect general goals rather than precise specification; and that, rather than start by defining the variables to be used, the universe of potentially relevant variables should be tested. It is also important to provide a means whereby the end-user can visualise what is going on. Animation of the spatial analysis process is one possibility here. Openshaw and Perrée (1996) offer a highly visual approach to spatial analysis that is designed to be almost instantly understandable by offering the user four different views in the form of computer movies of an intelligent exploratory spatial analysis procedure operating on data.

2.5.4 The SAT/1 proposal

It is in this context also that the recent Spatial Analysis Toolkit for GIS proposal (SAT/1) should be placed. The idea here is to develop a set of GIS independent, generic or application independent spatial analysis tools that conform to the principles outlined in Section 2.2 and which would initially be interfaced to Arc/INFO (although the interface design is the only non-portable part). Table 2.5 lists the initial procedures that have been suggested as potentially suitable but others could be added.

SAT/1 is designed to be 'loosely coupled' for the following reasons:

Table 2.5 Suggested SAT/1 tools

Diggles point density method
A simple GAM
Besag–Newell method
Cartograms
Kernel estimated surfaces
Stones method
Getis–Ord *g* statistics
Zone design methods
Regionalisation and classification
Zone ranking
Spatial regression modelling

1. Ease of creation.
2. Independence of any particular GIS.
3. Separation of vendor software from SAT/1 consortium code.
4. None of the methods needs a high degree of geographical interaction.
5. It is always possible later to convert loosely coupled into closely coupled integration if needed.

Each SAT/1 procedure will have its own AML, but it could also be a stand-alone system in its own right. By these means the investment of time and effort in creating the toolkit is preserved and considerable system specific programming avoided. Clearly the SAT/1 approach could be extended beyond the initial consortium of Lancaster, Leeds and Newcastle Universities into an international project. Further details can be obtained from the authors.

2.5.5 Database pattern hunting creatures

The ultimate machine-oriented implementations of Figure 2.1 are to develop 'smart' pattern and relationship seeking tools to handle the analysis and modelling aspects. Openshaw (1994c, 1995) outlines one way of achieving this goal using ideas borrowed from Artificial Life (ALife); also see Langton (1989) and Langton *et al.* (1992). The basic idea is very simple and involves converting the spatial analysis and modelling tasks into a search problem to find the answer to the questions where is the pattern and what form of equation fits best? Genetic algorithms and genetic programming techniques can be used to explore the full complexity of the GIS database in a bottom-up manner.

The ALife forms of Openshaw (1995) and Openshaw and Perrée (1996) consist of many hyperspheres that move around the data trying to engulf unusual concentrations of points, under their own control. The degree of 'unusualness' is measured via a Monte Carlo significance test. These hypersphere creatures can be left permanently roaming in a never-ending search for pattern. The genetic algorithm will allow them to adapt to whatever local environments they discover in the database. There are, of course, other mechanisms for driving the search process and neural nets might equally be extremely effective in bottom-up pattern recognition. These ideas still need to be fully developed and are still sufficiently 'blue skies' research that they might form yet another area for international collaboration.

2.6 Where to now?

There are at least three ways forward. The first is to develop special single purpose generic tools and this has been the main route considered here. These could be embedded in a particular GIS or its map interface, or alternatively the GIS may be embedded within the method (for example Openshaw *et al.*, 1987). Developments in heterogeneous distributed computing suggest that networking specialist spatial analysis and modelling tools on a national, European and international scale is a viable proposition. It would not matter at all what or where the hardware was on which a particular highly specialised task was run. This strategy could be particularly useful for highly complex analysis tasks but, as hardware speeds improve, it will downsize onto personal workstations. A second strategy is to develop more generally useful GIS-independent generic spatial analysis tools that are subsequently interfaced to a particular GIS. Although developed in a different context, the NAG subroutine library is one example of this approach. At some future date this might also be appropriate for spatial models in general. GIS vendors or developers could gain access to the basic technologies and design whatever interfaces they considered necessary in whatever way was deemed appropriate. A third strategy is to import traditional methods into GIS, or vice versa, and disregard the wider objectives that are of such importance. This will happen and is already happening and whilst academically attractive and better than nothing, it is clearly not at all satisfactory when viewed from a GIS end-user's viewpoint.

From a GIS DATA perspective, it is *collaborative* aspects that matter most. This book is testimony to the international collaboration now under way on improving the levels of spatial analysis within GIS. We hope we have offered some ideas on one possible research strategy for the future.

REFERENCES

Anselin, C. (1989). What is special about spatial data?, in NCGIA Technical Report 89/4, pp. 1–15, Santa Barbara: National Center for Geographical Information and Analysis.

Batty, M. (1993). Using GIS in urban planning and policy making, in Fischer, M.M. and Nijkamp, P. (Eds) *Geographical Information Systems, Spatial Modelling and Policy Evaluation*, pp. 51–72, Berlin: Springer-Verlag.

Benoit, D. and Clarke, G.P. (1995). Assessing GIS for retail analysis and planning, Working Paper, School of Geography, Leeds, UK: University of Leeds.

Birkin, M., Clarke, G., Clarke, M. and Wilson, A.G. (1996). *Intelligent GIS, Geo Information*, Cambridge: Cambridge University Press.

Clarke, G.P. and Clarke, M. (1995). The benefits of customised spatial decision support systems, in Longley, P. and Clarke, G.P. (Eds) *GIS for Business and Service Planning*, Cambridge: Geo Information.

Cliff, A.D. and Ord, J.K. (1973). *Spatial Auto-correlation*, London: Pion.

Ebdon, D. (1977). *Statistics in Geography*, Oxford: Blackwell.

Fotheringham, S. and Rogerson, P.A. (1993). 'Letter to the editor', *Environment and Planning A*, 1367.

Fotheringham, S. and Rogerson, P.A. (1994). *Spatial Analysis and GIS*, London: Taylor & Francis.

Gregory, S. (1963). *Statistical Methods and the Geographer*, London: Longmans.

Griffith, D.A. and Amrhein, C.G. (1991). *Statistical Analysis for Geographers*, New Jersey: Prentice Hall.

Haining, R. (1992). *Spatial Data Analysis in the Social and Environmental Sciences*, Cambridge: Cambridge University Press.

Langton, C.G. (1989). *Artificial Life*, California: Addison Wesley.

Langton, C.G., Taylor, C., Farmer, T.D. and Rasmussen, S. (1992). *Artificial Life II*, California: Addison Wesley.

Martin, R.L. (1990). The role of spatial statistical processes in geographic modelling, in Griffith, D.A. (Ed.) *Spatial Statistics: Past, Present and Future*, pp.107–128, Institute of Mathematical Geographers, Ann Arbor.

Openshaw, S. (1989). Learning to live with errors in spatial databases, in Goodchild. M. and Gopal, S. (Eds) *The Accuracy of Spatial Databases*, pp. 264–76, London: Taylor & Francis.

Openshaw, S. (1991a). A spatial analysis research agenda, in Masser, I. and Blakemore, M. (Eds) *Handling Geographical Information: Methodology and Potential Applications*, pp. 18–37, London: Longmans.

OPENSHAW, S. (1991b). Developing appropriate spatial analysis methods for GIS, in Maguire, D.J., Goodchild, M.F. and Rhind, D.W. (Eds) *Geographical Information Systems: Principles and Applications*, pp. 389–402, London: Longmans.

OPENSHAW, S. (1992). Some suggestions concerning the development of AI tools for spatial analysis and modelling in GIS, *Annals of Regional Science*, **26**, 35–51.

OPENSHAW, S. (1994a). Computational human geography: exploring the geocyberspace, *Leeds Review*, **37**, 201–220.

OPENSHAW, S. (1994b). What is GISable spatial analysis, in *Proceedings of the Workshop on New Tools for Spatial Analysis*, 36–44, Luxembourg: Eurostat.

OPENSHAW, S. (1995). Developing automated and smart spatial pattern exploration tools for geographical information systems applications, *The Statistician*, **44**, 3–16.

OPENSHAW, S. and BRUNSDON, C. (1991). An introduction to spatial analysis in GIS, *International GIS Sourcebook*, pp. 401–5, GIS World, Collared.

OPENSHAW, S., CHARLTON, M., WYMER, C. and CRAFT, A.W. (1987). A mark I geographical analysis machine for the automated analysis of point data sets, *International Journal of Geographic Information Systems 1*, 335–58.

OPENSHAW, S. and FISCHER, M.M. (1996). A framework for research on spatial analysis relevant to geo-statistical information systems in Europe, *Geographical Systems* (forthcoming).

OPENSHAW, S. and PERRÉE, T. (1996). User-centred intelligent spatial analysis of point data. *Innovations in GIS3*, pp. 119–134, London: Taylor & Francis.

TAYLOR, P.J.(1977). *Quantitative Methods in Geography*, Boston: Houghton Mifflin Co.

CHAPTER THREE

Scripting and toolbox approaches to spatial analysis in a GIS context

R. S. BIVAND

3.1 Introduction

In moving towards better and more timely support for decisions in the environmental and socio-economic arenas, it is of relevance to examine the availability of spatial analysis tools in geographic information systems. An issue of balance arises in fitting the richness of analyses which may be performed to the skill and training of system users, and response time in executing analyses. Both of these barriers are now being eroded by faster hardware and more open software environments, leading to the widespread use of interpreted scripts in linking program modules of very differing provenance in a customised form to yield desired results. This in turn leads to opportunities for adding value to geographic data in training and customisation, and also to questions regarding the openness of data models used by co-operating programs, including the need for arbitrary data structures stored in a machine-independent manner. The chapter considers examples taken from within GIS, and from other systems for geographical data manipulation and presentation, including the GMT-system. It concludes that a new class of 'script'-coupled spatial analysis tools may be easier to create than either to elaborate a range of loose-coupled programs, or to convince GIS-houses of the commercial wisdom of incorporating spatial analytical tools into their already overcomplex products.

3.2 Turning artificial intelligence on its head

One of the goals of artificial intelligence in the 1970s was to model understanding of natural language and human behaviour. As Winograd and Flores (1986) point out, this goal was not met for a number of reasons, not least the conceptual complexity of the contexts invoked in natural language discourses. The aim of modelling understanding of natural language was to permit a machine running an artificial intelligence program to play the role of a human partner in such discourses. Partly this would tell us more about language, partly about cognitive processes, thus forming an experimental laboratory for the study of human behaviour. Decision support was an instrumental ingredient in these research projects, as indeed it remains today.

However, several of the observations made then are of use in confronting the challenges we meet in manipulating large data sets and structures, in that they permit us to impose some order on our surroundings. Minsky (1981) developed the concept of a frame system. These are data structures which represent stereotyped situations, with attached information about the use of the frame, about expected results, and about error states that may arise when expectations are not confirmed. The individual frame has numerous terminals or slots, which must be filled with specific instances or data, and which may be subject to conditions, possibly binding several slots together. The different frames making up a frame system will share information residing in the same terminals. The terminals will most often be filled with default values, constituting presuppositions about the instantiation not specifically grounded in the data.

Schank and Abelson (1977) describe these regular structures in the following way: 'We use specific knowledge to interpret and participate in events we have been through many times. Specific detailed knowledge about a situation allows us to do less processing and wondering about frequently experienced events'. In their description of scripts, they stress the large contextual component in understanding a sequence of events. If the participant were obliged to reason through each event from first principles, a great loss of efficiency would occur. In human behaviour, they maintain that scripts describing frequently repeated situations dramatically reduce the need for information processing, by permitting generalisation from past experience, providing a contextual key to the problems of natural language interpretation, provided that all participants share the same script.

In artificial intelligence, attempts were directed to programming machines to infer, or 'learn' the scripts that human participants had internalised through socialisation, experience and training. As indicated, it proved to be unrealistic to approach natural language in this general way. Turning these AI results on their heads is however a fruitful strategy: instead of wanting the computer to infer the script implied in a given situation, we can choose to encode the script in a suitable language for the computer to execute under our control. It is clear that the combination of a range of tools and scripts in a powerful command language, interpreted at run-time rather than compiled, is an efficient means of carrying out the analyses we need. The trade-off between monolithic program systems run in interactive mode and script-based toolkits occurs in relation to the frequency with which particular types of tasks have to be carried out. This is analogous to Minsky's frames, which regularise and structure stereotyped situations, or to Schank and Abelson's specific detailed knowledge. If tasks are seldom repeated, then the effort of writing scripts will be wasted, since once-only interactive sessions will be at least as efficient. If tasks are repeated sufficiently often, then gains can be made in using a command language to capture operator and analyst experience and knowledge in script form. Scripts may be seen as necessary in any case if an interactive session cannot provide satisfactory debugging or auditing compared with the command language of choice.

This chapter will argue the case for the construction of spatial analysis tools as cleanly interfacing modules, under the control of suitable command languages. While there have been major improvements in the quality of software engineering in geographical information systems, it would be unreasonable to believe that there was no room for further advance. In this context I have chosen to concentrate on UNIX system tools and languages, partly because they represent the mainstream in GIS platform operating system environments, partly because the software tool approach has characterised UNIX since its beginnings (Kernighan and Plauger, 1976; Bentley, 1986), and finally from personal preference. To elaborate on the second of these arguments, it is worth describing the results reported by Bentley. He had challenged Knuth and McIlroy, two of the most respected computer scientists and programmers of our times, to solve a problem by methods of their choice. The problem was to print the *k* most common words in a given text file, for given *k*, with their numbers of occurrences in decreasing frequency. Knuth wrote a lengthy Web program, combining TeX and Pascal code, inventing a new way of ordering stored words in the process. McIlroy wrote a shell script under UNIX, using four system utilities, all incidentally documented in Kernighan and Plauger (1976), of which at least one was originally programmed by McIlroy. As McIlroy comments: 'Very few people can obtain the virtuoso services of Knuth (or afford the equivalent person-weeks of lesser personnel) to attack problems such as Bentley's from the ground up. But old UNIX hands know instinctively how to solve this one in a jiffy' (Bentley, 1986, p. 480).

3.3 Shells and command languages

With the advent of user access to computers, no longer mediated by operators in white coats, it became important both to enhance operating system functionality and to document the user interface. The latter was accomplished by specifying the syntax and semantics of commands needed to run jobs, enabling users, assisted by expert operators, to punch either on cards, or using a teletype, the instructions required. These were most similar to magical incantations, and were radically improved with the introduction of visual display units, emulating teletype printers on a cathode-ray tube. It is uncertain whether there was any attempt to design these command languages; many have subsequently been hidden under very successful control applications, such as CICS. Many early microcomputers used a BASIC dialect as the user interface, functioning as a programmable command language as well as an interpreter for user source code. IBM PCs also included BASIC as a supplement to PC-DOS; CP-M microcomputers were equally dependent on the use of BASIC, as indeed were Apple II machines.

Both MS-DOS and MacOS implement an operating system model which separates the workings of the system from the user interface. In the former, the operating system kernel, device drivers, and file system are hidden behind the interface program: `command.com`. In the latter, it is the Finder GUI which performs the same function. In the MS-DOS case, the Windows GUI hides `command.com`, taking over its role. Windows NT substitutes a new kernel providing for a variety of personalities, including one which resembles the existing Windows GUI. Windows NT does not provide a command language interface permitting script execution, although third-party applications are available to provide this functionality. In all of these cases, we observe that the internal workings of the operating system are hidden from the user by an interactive command language shell, which may or may not permit the input from a file of a script of commands.

3.3.1 UNIX shells

The first UNIX shell was developed at Bell Labs, and modernised in late 1975 and early 1976 by Bourne (1978), becoming widely known with the release of the Seventh Edition in 1979. Bourne (1987) describes its function as follows:

> The shell is a command language that provides a user interface to the UNIX

> operating system. The shell executes commands that are read either from a terminal or from a file. Files containing commands may be created, allowing users to build their own commands. These newly defined commands have the same status as 'system' commands. In this way a new environment can be established reflecting the requirements or style of an individual or a group. The standard input and standard output are used by many commands as the default for reading and writing data. Commands that process data in this way are called filters. Pipes allow such processes to be linked together so that the output from one command is the input to the next. The shell provides a notation enabling pipes to be used with a minimum of effort (p. 5).

The syntax of the Bourne shell is based on Algol and Algol68, with characteristic control flow expressions: `if ...; then ...; fi`, `case ... esac`, or `for ...; do; ...; done`. Given the very wide range of programs written in this command language, it has been considered important to maintain strict compatibility with the earliest versions, while incorporating the most frequently used system utilities, such as echo and test as internal commands. Differences in detail exist between BSD and System V `/bin/sh` shell implementations without constituting major barriers to the portability of shell scripts. The current state of `/bin/sh` technology seems to be 1984/5, with efficiency changes subsequently related to optimisation of stable, mature code.

Many users have migrated to alternative shells providing a more congenial interactive environment. The Berkeley C shell, developed by Joy, explicitly supports the interactive user with a history mechanism to log commands executed, as well as more powerful job control mechanism for background and foreground processes. It is the preferred shell on BSD UNIX systems, while on System V derivatives this position is taken by the Korn shell (Bolsky and Korn, 1989). The Korn shell is compatible with the Bourne shell in that essentially all `/bin/sh` scripts will run under ksh. The latter provides a more convenient history mechanism than `csh`, permitting full command line editing and recall as well as the recording of all commands executed up to an arbitrary line limit. These features are offered using the `emacs` or `vi` editing modes, to suit the user's preference.

All of these shells define an environment consisting of pairs of variable names and string values. Some environment variables are set by default, while others are initialised for the user at login. Some are also inherited by child shells, while others are not. They all now permit the construction of shell functions (`sh`, `ksh`) or aliases (`csh`), which differ in some ways, but permit the clean implementation of frequently used command combinations. Within a given shell instance, all variables are global, and the provisions made for arithmetic or string operations on these variables leave a good deal to be desired.

3.3.2 Modern shell developments

Although the Bourne shell is acknowledged to have been ahead of its time, and indeed forms the basis for a very large number of applications in all UNIX dialects, it has shortcomings both with regard to syntax and semantics. As with much of the work underlying the development of UNIX itself, AT&T Bell Laboratories' computer scientists have been working on leaner and more efficient operating and windowing systems and tools. In the course of work on the Plan 9 operating system, which may be thought of as the antithesis of Windows NT, and which is not a microkernel system, Duff

(1990) wrote a new shell, called `rc`. For the simplest uses, `rc`'s syntax is similar to that of the Bourne shell. It differs in more advanced syntactic structures, and in circumventing the rescanning of command lines which made the correct entry of compound expressions difficult. The source of the syntax is C and Awk, rather than Algol. It also introduces lists as a type of variable in addition to character strings. At present the shell is still evolving, but its source code is available for those wishing to try it out.

Subsequently, the computer scientist who reimplemented `rc` for use as a UNIX shell, thus making it available to users without access to Plan 9, Byron Rakitzis, in collaboration with Paul Haahr, has been tempted to go further. They wrote a new shell, `es`, which again works like the Bourne shell for simple commands (Haahr and Rakitzis, 1993). `es` adopts the syntax of `rc`, while exploring new semantics borrowed from modern programming languages. Almost all the standard shell constructs, such as pipes, are translated into a uniform representation as calls to functions. Replacing primitive functions by user-defined versions is a key innovation in `es`. In addition, all variables are lexically scoped, meaning that the use of a variable name is local to its scope, a natural requirement for a language based on the use of functions. Rich return values are carried over from `rc`; while current UNIX utilities, applications, and shell scripts return a single number between 0 and 255, an `es` function can return a number, a string, a program fragment, or even a list which mixes such values. Since both `rc` and `es` are experimental, it might seem foolhardy to base production scripts on access to these languages. On the other hand, both are freely available on Internet, and do indicate the direction of current research in shells and command languages.

3.3.3 Other command languages

While shell scripts can be very powerful mechanisms for controlling utilities and applications, experience has shown that the combination of such scripts with programmable utilities can be even more effective and efficient. There is a continuum between interactive shells and programmable utilities, since both provide command language facilities to a certain extent. Examples here are the UNIX utilities `grep`, `sed`, and `awk`, and the freely available report language PERL. Some tasks are however best solved by the shell, some by a utility, and some may be embedded into an application. The `es` shell described in brief above is small enough to be embedded into an application, a role derived from the very successful Tcl/Tk languages developed by John Ousterhout. In this section, we will touch on the most important features of these alternatives.

3.3.3.1 AWK and UNIX utilities

Early UNIX utilities, such as `grp` and the editors `ed`, `ex`, and the stream editor `sed`, were based on finding regular expressions. The stream editor can be programmed using script files to conduct operations on identified lines of files (Dougherty, 1991). Regular expressions or patterns form a key part of the AWK language, implemented first as `awk`, and in newer versions as `nawk`, `gawk` (Aho *et al.*, 1988; Dougherty, 1991), and mawk (Brennan, 1992). AWK is used widely in massaging data files composed of both numerical and string data in many contexts (van Wyk, 1986).

An AWK program is a sequence of `pattern {action}` pairs and function definitions. Short programs are entered on the command line usually enclosed in single quotes to avoid shell interpretation. Longer programs can be read in from a file. Data input is read from the list of files on the command line or from the standard input when the list is empty. The input is broken into records as determined by the record separator variable. Initially records are synonymous with lines. Each record is compared with each `pattern`, and if it matches, the program text for `{action}` is executed. Patterns may be omitted, and may include the key words `BEGIN` and `END`, which are executed before and after the reading of the records in the input. Function definitions can be included, most usually in file input in longer programs; the functions resemble C functions, and can be used to prototype full-scale programs.

The following code snippet is an AWK function for distance calculation:

```
function hypot(x,y,  xabs,abs, w, z, v)
{
        xabs = x;
        if (x < 0) xabs = x * -1;
        yabs = y;
        if (y < 0) yabs = y * -1;
        if (xabs > yabs) w = xabs;
        else w = yabs;
        if (xabs < yabs) z = xabs;
        else z = yabs;
        if (z == 0.0) return(w);
        else {v = z/w;
              return(w*sqrt(1 + (v*v)));
        }
}
```

Local variables are declared at the right-hand end of the function statement by convention, the left-hand expressions are passed by value, arrays by reference, and return values are numeric or string depending on the context. The AWK language stems from Bell Laboratories, but is now accessible on many platforms, and under many operating systems.

3.3.3.2 *PERL (Practical Extraction and Report Language)*

The 'Practical Extraction and Report Language' was developed later than AWK and the UNIX utilities, and seeks to address the need for more power. It has three data types: scalars, arrays of scalars, and associative arrays of scalars. Normal arrays are indexed by number, associative arrays by string, as in AWK. PERL runs very close to the C system libraries, implementing many system management commands directly. As an interpreted language, it shares the advantages and disadvantages of AWK, with which it shares many features; it includes a debugger. It can also handle binary data, though it is optimised for pattern matching. Translators from AWK and `sed` are available. The language is described in detail in Wall and Schwartz (1991).

3.3.3.3 *Tcl/Tk*

The `es` shell mentioned above derived some of its design features from John Ousterhout's tool command language (1990, 1991). Tcl was designed as an embeddable command language offering uniform shell programming structures within applications, especially within a windowing environment. This has been complemented with a message passing system for applications implementing Tcl/Tk in the X11 tool kit Tk.

Both are based on the assumption that interactive programs should support the dynamic, runtime modification of various of their attributes, and that a programming language like the traditional shells is the most appropriate way of providing this functionality. Ousterhout (1990) explains:

> Unfortunately, few of today's interactive applications have the power of the shell or Emacs command languages. Where good command languages exist, they tend to be tied to specific programs. Each new interactive application requires a new command language to be developed. In most cases application programmers do not have the time or inclination to implement a general-purpose facility (particularly if the application itself is simple), so the resulting command languages tend to have insufficient power and clumsy syntax. (p. 1)

Tcl/Tk can be used as a shell programming language, but are best suited to providing the general-purpose command language facility within applications. Work is under way to provide Tcl/Tk functionality within PERL, thus permitting PERL programs to manipulate X11-based GUI essentially from the shell level. This in turn is similar to the manipulation of the X11/Motif environment at the shell level provided by `xgen`, a program associated with GRASS. Commenting on windowing environments, Ousterhout (1991) writes:

> Current windowing applications are forced by the lack of good communication to lump large amounts of functionality into a single application. Tk makes it possible to replace such monolithic applications with collections of smaller specialised applications that communicate with each other using Tcl commands. These smaller tools are often re-usable for other purposes, thereby resulting in more powerful windowing environments. (p. 1)

Pressure to modularise applications is still present 20 years after the development of the concept of software tools. The benefit of modularising is not only reuse, but extends to simpler debugging and maintenance, since quality testing of small modules can be undertaken much more efficiently than in monolithic applications.

3.4 Command languages in geographical data analysis

Command languages are used both by operating system shells and applications. They do not have to be programmable, nor do they have to support more complicated language structures, like flow control. In many modern applications, the command language is GUI-based, relying on sequences of mouse events to give expression to the wishes of the user, provided these are within the syntax and semantics of the language used. With the coming of GUI, it is possible to step from one task, for instance digitising site co-ordinates, to check the reference code needed to identify the current site in a data file. With cut-and-paste, the necessary information can then be taken from one on-screen application to another without further work, but also without any verification of the transfer, or any journaling.

Almost all commonly used programs for geographical data analysis and presentation have their own command languages, which in many cases seem to have grown without very much thought being given to design issues. This applies both to statistics programs, geographical information systems, desktop mapping systems, and, not least, one-off programs written to solve geographical analysis problems. In commercial terms in a static, or slowly changing, customer environment, this is even advantageous to software houses and their customers, since they can concentrate on training and application customisation for a user base which has, or will shortly have, product experience. It is, however, instructive to note that very many of the technical questions which are raised on discussion lists concern the movement of data from one format to another, or about how to express a statement in one application's command language in that of another application.

Many of the applications in question base their customisation on the writing of macro files in their command languages, an activity which can be quite profitable for dealers and consultants, but which often has little support for concepts now ingrained in program writing for traditional programming languages, like debugging. Were it possible to use generic command languages, supporting computer assisted software engineering (CASE) concepts, many of the concerns about script and macro quality which one harbours would be reduced, if not removed. This issue is being addressed in a new generation of products in the larger scale GIS market, particularly within larger utilities, by companies like Smallworld Systems Ltd, with the Magik environment, and APIC Systemes Groupe. There is however a divergence between the CASE interest at the multi-user, production grade application level, and the experimental, often prototypical work being undertaken in analysis, with apparently less commercial leverage. This lack of leverage may in fact be a good reason to try to base command language choice more on openness, and on the access to modern software technology afforded by 'copyleft' and other non-commercial licensing schemes. It may also mean that we ought to avoid writing one-off command languages, and rather open up program modules to run from the command line, thus placing them under the control of the shell.

3.4.1 Geographic Resource Analysis Support System (GRASS)

GRASS is designed within a functional level scheme: the specialised interface level is reserved for heavily customised applications, and the programming and library levels are reserved for C programmers wishing to add new functionality through either new programs or modifications or additions to the existing libraries. The most used level is the command level:

> Using the user's login shell, GRASS commands are made available through modification of the PATH variable. Help and on-line manual commands are available.
>
> In version 2.0, GRASS programs included both user interface and program function capabilities and were highly interactive. GRASS 3.0 introduced complementary command-line versions of these functions in which the information required by the program was provided by the user on the command line or in the standard input stream (with no prompting). This provided the advanced user greater flexibility and the system analyst a high-level GIS programming language in concert with other UNIX utilities. However, this resulted in a doubling of the number of commands: one for the interactive form, another for the command-line form.
>
> In GRASS 4.1 the interactive and command-line versions of a program have been 'merged' into a single program (as far as the user is concerned). A standard command-line interface has been developed to complement the existing interactive interface, and an attempt has been made to standardise the command names (Shapiro *et al.*, 1993, p. 7).

The GRASS applications environment is entered by setting a number of environment variables, including PATH, and the prompt, which typically changes to GRASS as an indication that the user is able to access the necessary programs and data sets. Interactive use of the programs now assumes that a program name typed in from the UNIX shell without arguments is to be run in a dialogue with the user, through new commands, menus, or user responses to questions. If however the program is called with arguments, it will be executed, provided that necessary conditions are met, and control returned to the shell. A third way to execute programs is to redirect standard input from a file of responses to program requirements. For the `r.mapcalc` command, which uses mathematical syntax to perform analyses on existing map layers within GRASS, this means that three equivalent routes are available for conducting the following analysis.

```
1 Interactive:
GRASS 4.1 > r.mapcalc
MAPCALC >
trans.avail=transport.misc==2 || roads==1 || railroads==1
MAPCALC >
GRASS 4.1 >

2 Shell:
GRASS 4.1 > r.mapcalc trans.avail = 'transport.misc==2 ||
                                   roads==1 || railroads==1'
GRASS 4.1 >

3 Input redirection:
GRASS 4.1 > cat input.file
trans.avail = transport.misc==2 || roads==1 || railroads==1
GRASS 4.1 > r.mapcalc < input.file
GRASS 4.1 >
```

The example creates a zero/one map of transport availability from three input maps where any of the conditions is true (Larson *et al.*, 1991).

When using the interactive or redirected input interfaces, other tools and utilities are not necessarily available from the program's own prompt. It is here that GRASS shell scripts come in, making it possible to customise from the range of programs and utilities accessible from the UNIX prompt, here GRASS 4.1. It is not difficult to add to the skeletal script presented in Shapiro *et al.* (1993, p. 253):

```
#!/bin/sh
if test "$GISRC" = ""
then
        echo "Sorry, you are not running GRASS" >2
        exit 1
fi
eval 'g.gisenv'
:${GISBASE?} ${GISDBASE?} ${LOCATION_NAME?} ${MAPSET?}
```

which checks that the five most important GRASS environment variables are set. There is also a `g.ask` GRASS command permitting shell scripts to prompt the user for input.

Examples of the use of the shell for spatial statistics and presentation are contained in work by McCauley (1992, 1993), and McCauley and Engel (1993). From the easier automation of spatial autocorrelation for site data, appended to the script prologue above (1992, p. 6):

```
s.voronoi site="$1" vect="$1"
v.support "$1"
v.autocorr "$1"
exit 0
```

where "$1" is the file name, we can move to the use of `g.gnuplot` for contour displays (1993, p. 2):

```
g.region -p > /tmp/region
set n=`grep north /tmp/region | colrm 1 12`
set s=`grep south /tmp/region | colrm 1 12`
set e=`grep east /tmp/region | colrm 1 12`
set w=`grep west /tmp/region | colrm 1 12`
set nsr=`grep nsres /tmp/region | colrm 1 12`
set ewr=`grep ewres /tmp/region | colrm 1 12`
cat >! /tmp/awk$$ << EOF
$2 != row { print ""; row = $2 }
{ print s-nsr*$1, w+ewr*$2, $3 }
EOF
r.stats - 1x elevation | awk -f /tmp/awk$$ s=$s nsr=$nsr w=$w
                          ewr=$ewr >! elevation.dat (1993, p.2)
```

The script is written for `csh`. GRASS command `g.region` returns boundary information on the map under analysis, which is then transfered to shell variables by six `grep` and `colrm` steps. `Colrm` is a Berkeley UNIX addition, but these steps can easily be implemented using alternative, generic utilities, like `cut`. Next, a short AWK program is written to `/tmp/awk$$`, and executed on the output from `r.stats -1x`, which is run on the elevation cell file. This creates a data file with three columns, Y, X, Z, for input to the GNUPLOT scientific plotting program. This is of course not a finished script, just a code snippet illustrating the ease with which GRASS and UNIX commands may be used together. McCauley has released this as the GRASS `r.to.gnuplot` command:

```
#!/bin/sh
# r.to.gnuplot
# Author: James Darrell McCauley, Purdue University
# GRASS-GRID> r.to.gnuplot raster_file > data
# GRASS-GRID> d.mon sta=x0
# GRASS-GRID> g.gnuplot
# gnuplot> set parametric
# gnuplot> set contour base
# gnuplot> set nosurface
# gnuplot> set view 180,0
# gnuplot> splot 'data' notitle with lines
r.stats -1x $1 | awk '$2 != row{print "";row=$2}{print}'
```

3.4.2 Generic Mapping Tools System (GMT)

The GMT-SYSTEM is a public domain software package that can be used to manipulate and display two-dimensional (for example, time-series or (x,y) series) or three-dimensional data sets (for example, (x,y,z) grids). The processing and display routines within the GMT-SYSTEM are completely general and will handle any (x,y) or (x,y,z) data as input (Wessel and Smith, 1991). The programs are capable of a certain level of analysis, as is illustrated by the GMT 2.1.4 version of the GRASS `r.mapcalc` example above:

```
$ grdclip transport.misc -Gmisc.grd -A2/0 -B2/0
$ grdclip roads -Groads.grd -A1/0 -B1/0
$ grdclip railroads -Grailroads.grd -A1/0 -B1/0
$ grdmath misc.grd + roads.grd = tmp.grd
$ grdmath tmp.grd + railroads.grd = tmp1.grd
$ grdclip tmp1.grd -Gtrans.avail -A1/1
```

The additional commands are required because `grdmath` does not support logical operators, but on the other hand GMT has full floating-point numerical data, and the use of NaN (not a number) for missing values is permitted. The `grdclip` commands simply set the cell values to zero where the condition is not met: −A1/0 sets output cell values to zero if the input cell value is above 1, and −B1/0, if it is below 1. The fact that GMT can seem verbose is because it is designed to be used in shell scripts:

> The GMT-SYSTEM programs were designed to run under the UNIX operating system. UNIX is quite different from most earlier operating systems, and has proven to be a very productive tool for scientists and engineers worldwide. The UNIX philosophy is to let small, general-purpose programs work together in solving a problem, rather than designing large task-specific programs which are very time consuming to debug and maintain. We have written the GMT-SYSTEM in highly portable C as defined by Kernighan and Ritchie to be used with any hardware running UNIX and we have followed the modular design philosophy. For example, in order to plot a filtered marine bathymetry profile, we would first run one program to project the original (lon,lat,depth) data points onto a great circle, filter the data using another program, and use a third program to create the final plot. This modular approach brings several benefits:
>
> 1. only a few programs are needed,
> 2. each program is small and easy to update and maintain,
> 3. each step is independent of the previous and the data type so these programs can be used in a variety of applications, and
> 4. the programs can be chained together in Shell-scripts or with pipes and thereby be tailored to do user-specific tasks. (Wessel and Smith, 1993)

GMT 3.0 was made available in 1995, and release 3.1 is due in late 1996.

3.5 Conclusions

A number of issues deserve to be taken up in conclusion, before we round off by seeing whether standing AI on its head is reasonable. They are customisation, training, debugging, and journaling. In the context of the present and future market volume for spatial analysis software in a GIS context, one would be foolhardy to press software houses to build relatively experimental or prototypical analytical tools into monolithic applications. There will never be a need for all GIS users to access spatial analysis functions, since most use GIS as a spatial database application for land information systems and/or facilities management. It is arguable that GIS in themselves ought to be subject to CASE, which will often bring with it a strong modularisation, simply to ensure software performance quality. Further, partly because of the tender requirements of official procurement, most GIS adhere to open systems standards, including POSIX interface criteria. This means that even operating systems which are not UNIX will provide UNIX-like utilities and services, permitting applications to be run on a wide variety of platforms and network specifications. The wild card at the time of writing is undoubtedly Windows NT, which may cause a weakening in the current open systems consensus.

Spatial analysis in a GIS context has little to gain from increasing proprietary boundaries between operating systems and applications, because it inhabits an academically flavoured niche in the GIS market. This is a disadvantage if one had hoped to secure a comfortable second income from software sales, but not if one joins the mainstream of academic software development as represented by the volume of public domain and other non-commercial software available on the Internet. This however depends on adhesion to open systems standards, or to other standards accessible on the same copyright basis. Curiously, and perhaps of signal interest is the fact that Linux, a UNIX clone for Intel platforms, was developed under the GNU General Public License by interested persons and some companies in the time since Windows NT was due to be ready and its release. GRASS and GMT run fairly happily under Linux and X11.

The consequences of this market setting are that while links between say Arc/Info and SPSS, were they to be developed, would be promising, they do not perhaps exploit the strengths of the current software development environment. The use of shells, programmable tools, and the toolkit approach to program development leads to great ease in customising applications for users. Shell scripts and programmable tools like `awk` or `perl` use standard text files, which can be edited as required to provide special features or for instance menu items. It is not at all uncommon for AWK programs to write AWK programs; the same applies to shell programming. When a small utility needs to be written from scratch, this can be done using standard input and output formats to make data exchange between tools effective, and to allow the new tool to be used in new shell scripts.

Training is an awkward topic. Should spatial analysis in a GIS context be as apparently easy and intuitive as say Excel? There is a further question lurking behind the user-friendliness of GUI-based software: how many users know of or use how many of the functions they have (or should have) paid for? It has been said that steep learning curves may be bad for sales, but that they are an effective – also cost-effective – way of learning fast. Most spatial analysts, including users and not just developers, can be expected to have relevant higher education, so that there is no absolute need to impose an IQ< 90 limit on the way in which the software interface is engineered. Training will be necessary in any case to ensure that the users are in a position to benefit from the results of the analyses that they have carried out.

Debugging is possible in the Bourne shell, by executing the script with the -x option to echo to the screen the commands being executed, thus permitting erroneous substitutions to be found. The Korn shell has a debugging mechanism built in; this is described by

Rosenblatt (1993). AWK programs can be debugged by inserting print statements, or other modifications in the program itself; there does not seem to be an easy way of stepping through line-by-line (Dougherty, 1991). `Perl` has a built in debugging mechanism, but Tcl/Tk and the modern shells expect that the need for debugging should be reduced by the provision of better engineered and cleaner language structures. For example, most bugs in Bourne shell scripts stem from the fact that the command is rescanned, maybe even several times, with the consequent need for control characters, of meaning for the shell, to be escaped. The `rc` shell only scans its input once, removing this source of error.

Everyone can make mistakes, but not everybody knows what they did. If commands are journaled, for instance through a history mechanism like that offered by `csh`, `ksh` and as an addition to `rc`, one is at least in a position to trace back to the point at which the key data file became corrupted. Journals also permit audit trails to be followed, not only with regard to data files, but also scripts, programs, and applications, by using a source code or revision control system to archive the more valuable products of ongoing projects. Journaling does not necessarily mean that a given file can be reinstated; this is a system administration task, depending on the available hardware/software configurations. Plan 9 solves this by using a 300b G WORM jukebox for permanent daily back-ups subsequently available on-line.

Is it sensible to use well-structured command languages in interaction with computers when the problems being handled are of a considerable degree of complexity, as in spatial analysis in a GIS context? It seems unlikely that spatial analysts ought to lower themselves to the perceptual threshold of even the most potent workstation – the machine has to be programmed, the question is simply how it should be done. There were commercial arguments for monolithic solutions, but even these have paled under the onslaught of market and technological flux. There is a clear role to be played by shell scripts and programmable tools in providing timely solutions to the software needs of spatial analysis within GIS, and indeed in many cases for GIS as a whole. Well-crafted command languages are a necessity for the successful introduction and implementation of CASE methods, methods which can reduce time to market substantially, and which may decide the fate of GIS applications.

REFERENCES

AHO, A.V., KERNIGHAN, B.W. and WEINBERGER, P.J. (1988). *The AWK Programming Language*, Reading, MA: Addison-Wesley.

BENTLEY, J.L. (1986). A literate program, *Communications ACM*, **29**, pp. 471–83.

BOLSKY, M. and KORN, D. (1989). *The Korn Shell Command and Programming Language*, Englewood Cliffs, NJ: Prentice Hall.

BOURNE, S.R. (1978). The Unix shell, *Bell Sys. Tech. J.*, **57**(6), 1971–90.

BOURNE, S.R. (1987). *The UNIX System V Environment*, Reading, MA: Addison-Wesley.

BRENNAN, M., (1992). "mawk 1.1" Newsgroup: comp.sources.reviewed Volume 1, Issue 50.

DOUGHERTY, D. (1991). *Sed & Awk*. Sebastopol, CA: O'Reilly.

DUFF, T. (1990). Rc – a shell for Plan 9 and Unix systems. *UKUUG Conference Proceedings*, London, UK, pp. 21–33; also anonymous ftp: unix.hensa.ac.uk: /pub/uunet/pub/shells/rc.

HAAHR, P. and RAKITZIS, B. (1993). Es: a shell with higher-order functions. *USENIX Conference Proceedings*, San Diego, CA; also anonymous ftp: unix.hensa.ac.uk: /pub/uunet/pub/shells/es.

KERNIGHAN, B.W. and PIKE, R. (1984). *The UNIX Programming Environment*, Englewood Cliffs, NJ: Prentice Hall.

KERNIGHAN, B.W. and PLAUGER, P.J. (1976). *Software Tools*, Reading, MA: Addison-Wesley.

LARSON, M., SHAPIRO, M. and TWEDDALE, S. (1991). Performing map calculations on GRASS data: r.mapcalc program tutorial. U. S. Army Corps of Engineers, Construction Engineering Research Laboratory, Champaign, IL; also anonymous ftp: moon.cecer.army.mil: /grass/grass4.0/documents/mapcalc.ps.

MCCAULEY, J.D. (1992). Measuring spatial autocorrelation with GRASS. Department of Agricultural Engineering, Purdue University; also anonymous ftp: pasture.ecn.purdue.edu: /pub/mccauley/grass.

MCCAULEY, J.D. (1993). GRASS tutorial: g.gnuplot. Department of Agricultural Engineering, Purdue University; also anonymous ftp: pasture.ecn.purdue.edu: /pub/mccauley/grass.

MCCAULEY, J.D. and ENGEL, B. (1993). Spatial statistics and interpolation procedures for GRASS, *8th Annual GRASS GIS Users' Conference*, Reston, VA; also anonymous ftp: pasture.ecn.purdue.edu:/pub/mccauley/grass.

MINSKY, M. (1981). A framework for representing knowledge, in Haugeland, J. (Ed.) *Mind Design*, Montgomery, VT: Bradford Books.

OUSTERHOUT, J.K. (1990). Tcl: an embeddable command language, *USENIX Conference Proceedings*, pp. 133–46; also anonymous ftp: ftp.cs.berkeley.edu: /ucb/tcl/tclUsenix90.ps.

OUSTERHOUT, J.K. (1991). An X11 toolkit based on the Tcl language, *USENIX Conference Proceedings*, pp. 105–115; also anonymous ftp: ftp.cs.berkeley.edu: /ucb/tcl/tkUsenix91.ps.

ROSENBLATT, W. (1993). Debugging shell scripts with kshdb, *Unix World*, **10**(5), 74–9.

SCHANK, R.C. and ABELSON, R.P. (1977). *Scripts, Plans, Goals, and Understanding*, New York: John Wiley.

SHAPIRO, M. *et al.* (1993). GRASS 4.1 programmer's manual. U.S. Army Construction Engineering Research Laboratory, Champaign IL (anonymous ftp from: moon.cecer.army.mil).

WALL, L. and SCHWARTZ, R. (1991). *Programming Perl*. Sebastopol, CA: O'Reilly.

WESSEL, P. and SMITH, W.H.F. (1991). Free software helps map and display data, *EOS Trans. AGU*, **72**, 441, 445–6.

WESSEL, P. and SMITH, W.H.F. (1993). *Gmtsystem Manual Pages*. School of Ocean and Earth Science and Technology, University of Hawaii (anonymous ftp from: kiawe.soest.hawaii.edu: /pub/gmt).

WINOGRAD, T. and FLORES, F. (1986). *Understanding Computers and Cognition*, Reading, MA: Addison-Wesley.

VAN WYK, C.J. (1986). Awk as glue for programs, *Software – Practice and Experience*, **16**, 369–88.

CHAPTER FOUR

Designing a health needs GIS with spatial analysis capability

ROBERT HAINING

4.1 Introduction

This chapter is written towards the end of a three-year collaborative project with the Sheffield Health Authority (SHA), the aim of which has been to assess the utility of GIS for health needs assessment at the local authority level. In the course of the work a number of statistical spatial data analysis (SDA) techniques have been applied to small area health data. The aim here is to summarise those activities and show how, taken together, they offer the basis for undertaking a coherent programme of data analysis in this area thereby defining the analysis requirements for a health needs GIS. SDA can provide useful information for those concerned with identifying health needs for specific (small) geographic areas. Such information may also be useful in deciding how to organise the delivery of various services once there is a better understanding of the geography of needs.

An important aspect of health needs assessment in relation to geographic areas is the identification of areas of special need ('problem regions'). Such areas may justify receiving additional resources because of, for example, significantly high incidence rates of a disease, concentrations of some group with special needs (such as the elderly) or the low uptake of a screening programme. Once areas have been identified the focus shifts to the possibilities for and the objectives of a programme of geographically based resource targeting and how best to deliver those resources. Eventually an assessment is required of whether the goals of the programme have been met. The ability to address all these questions might be considered essential for a fully developed health needs GIS. (Indeed, it may not be until these later stages that the extended capability of a GIS is fully exploited, as opposed to the capability of a good mapping package.) We focus here on the identification of problem regions because this problem represents a natural domain for the application of SDA techniques and GIS since it requires the handling and analysis of large volumes of spatially referenced data from different sources.

Haining (1994) considered the general issues that need to be addressed if linkage between SDA and GIS is to take place. These were presented in the form of six questions and the paper discussed each, illustrating the arguments with examples from medical geography. The questions were as follows.

1. What types of data can be held in a GIS?
2. What classes of questions can be asked of such data?

3. What forms of SDA are available for tackling these questions?
4. What set of (individual) SDA tools are needed to support a coherent programme of SDA?
5. What are the fundamental operations needed to support these tools?
6. Can the existing functionality within GIS support these fundamental operations or is new functionality required?

Answers to these questions affect what forms of SDA can be undertaken within a current GIS. It was emphasised in Haining's paper that incorporating SDA into GIS is not simply a matter of incorporating individual techniques, but rather of providing a range of techniques that taken together enable the analyst to carry out a coherent programme of SDA. It is this last point which we explore here in relation to a health needs GIS.

This chapter describes the requirements for a customised GIS for defining problem regions as part of health needs assessment in a single health authority area. This means specifying the set of fundamental techniques required for such identification and determining whether GIS functionality can not only support them but also support them with sufficient ease and speed of operation.

In Section 4.2 we discuss database issues and in Section 4.3 we outline SDA techniques that are needed to underpin a health needs assessment GIS and comment briefly on the implications of this work for the geography of service delivery. Section 4.4 considers design issues that may inform future extensions to the current system.

4.2 Database issues for a health needs assessment GIS

Two important data sources underlie health needs assessment research. These data sources are health authority data recording the location of 'health events' (for example, patients, non-attenders at a screening programme, and so on) and census data recording socio-economic characteristics of the population. Health data, such as the Trent Cancer Registry that was used in the Sheffield project, has the patient address and the postcode; census data are currently available in the UK down to the level of the enumeration district (ED) a spatial unit that includes between about 170 and 210 households. In a city the size of Sheffield (0.5 million) there are just under 1200 EDs. There are several problems with these data sources. First, postcodes and EDs are not compatible areal units. Secondly, census data at the level of the ED is only collected every 10 years. For some parts of a city, where urban redevelopment has occurred or where there are high levels of in and out migration, analysis based on these statistics is only likely to be on firm ground, at best, during the period covered by the census. Methods for interpolating between census years and extrapolating beyond the last census are needed to underpin analyses of this kind and might well form part of an analysis capability for updating or extending the database. Thirdly, computerised health authority data may not be fully up to date and before 1981 it was not postcoded.

One method of linking health authority data and ED based census data is to obtain digitised addresses for the health events, using the PinPoint address code (PAC) which is accurate to 1 m and then allocating the health events to their corresponding EDs. This yields a high level of accuracy in the matching of health events to EDs. In the absence of the addresses, look-up tables identify which unit postcodes sit in which EDs (although any unit postcode may sit in more than one ED) and the central postcode directory (CPD) provides a digitised 'centroid' for any postcode which is the grid reference of the south west corner of the 100 m grid square in which the first house of the postcode falls. These data sources unlike PAC are free. Health events can then be allocated to EDs but obviously with less accuracy. The end product of these methods of linkage is a data set that matches (to different levels of accuracy) health events with a wide range of socio-economic characteristics relating to the EDs within which the events occur. Figure 4.1 shows the structure of the possible linkages on the left side while on the right are the corresponding data models (following Goodchild, 1992).

EDs are spatial units that are far too small for meaningful data analysis. 'Small populations naturally attract small numbers of events and this can pose problems in producing sufficient numbers for analysis, particularly where the phenomenon exhibits low incidence in the population' (Carstairs, 1981). ED populations rarely exceed 600 persons and it is important not to lose sight of the relatively large standard errors associated with ED level data. The effects of 'Barnardisation', applied to UK census data to meet confidentiality requirements, are most marked at this scale.

Wards, spatial units at the next scale up, are composed of nested groups of EDs. On average there are 20 EDs to a ward (in the case of Sheffield the average is closer to 40 since there are 29 wards in all). Whilst at this level small number problems are less serious, the price to be paid is greater socio-economic heterogeneity. Like the ED level, ward level population data are only available every 10 years at the time of the census so that it is only on these occasions that numerator and denominator data used in computing rates are strictly comparable. In the case of rare diseases, often handled by aggregating over years in order to reduce small number problems, this means centring the aggregation on the year of the census.

An appropriate spatial framework is one that combines the larger population numbers of a ward-based partition with the relative socio-economic homogeneity of an ED-based partition. As Carstairs (1981, p. 135) notes 'an approach which scores or ranks EDs on deprivation or other dimension and aggregates them into classes on that basis still offers the best prospect . . .' for constructing an appropriate spatial partition for spatial analysis. Carstairs' suggestion can be implemented as an exercise in region building which in turn is a particular type of classification problem (Grigg, 1967). A system of n spatial units (regions) is built from a system of k ($k>n$) smaller spatial units (such as EDs) according to some criterion of optimality such as minimising within region variation and maximising between region variation. Attribute variables associated with the k smaller areas are used as the basis of the classification. Since the EDs are themselves not homogeneous, the homogeneity of the final (aggregated) product should not be overstated.

Methods of classifying can be divided into those that are non-spatial (no information about the absolute or relative locations of the k spatial units is used in the classification procedure) and those that are spatial. Examples of the second group of classification methods include the minimal spanning tree method that uses the location of the spatial units in the classification procedure (Cliff *et al.*, 1975) and the method based on attribute spatial correlation developed by Oliver and Webster (1989). However there may be reasons to prefer a non-spatial classifier. Spatial classification procedures like the minimal spanning tree method may be very time consuming to implement as k increases. Any procedure that incorporates a measure of spatial proximity into the classification scheme must make a compromise between spatial compactness and within-area homogeneity. Spatial classification schemes may give undue emphasis to spatial compactness relative to intra-regional homogeneity. Both from an analysis and service delivery perspective spatial contiguity (of regions) is desirable but not necessarily spatial compactness. The final classification may create a set of regions of very different population sizes. This means that if each region is treated separately the reliability of estimated rates will vary across the set of regions. A scattering of small regions may survive so that further aggregation may be necessary to avoid small number problems.

Non-spatial classification methods only give rise to a regional partition 'by default'. That is, once each ED has been classified and the classes mapped, some adjacent EDs are usually of the same class so that some boundaries between EDs can be dissolved. The next stage is to build a less fragmented regional partition by merging EDs by some set of additional rules. This could be done for example by merging neighbouring EDs that happen to be of the adjacent class. Local knowledge of the areas can be employed in deciding which adjacent areas to merge and regions can be built up observing the need to ensure that all are of a minimum population size.

An early example of area delimitation for health needs assessment is provided by the Community Medicine Areas (CMAs) of the Greater Glasgow Health Board (Information Services Unit, 1981). Although no formal classification rule was used to create the 87 CMAs into which Glasgow was divided, the aim was to identify distinct socio-economic areas of the city. Haining *et al.* (1994) construct 48 'Townsend' deprivation regions for Sheffield using a more formal procedure. Townsend deprivation scores are computed for

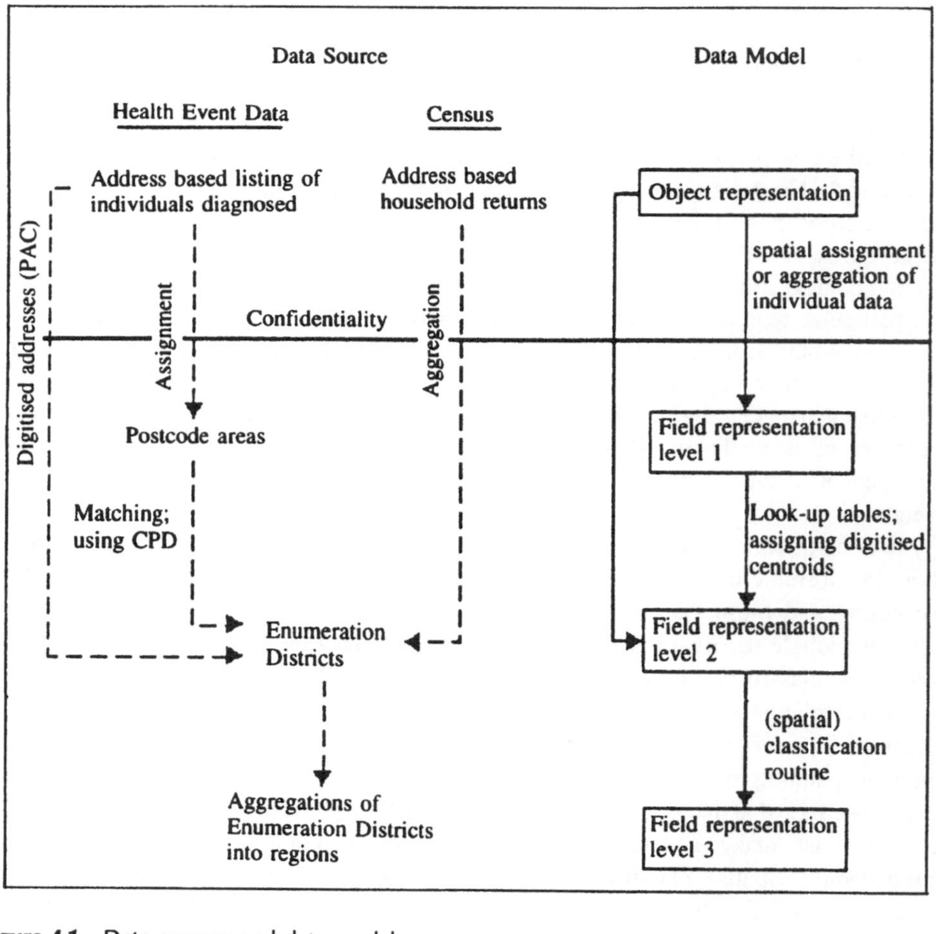

Figure 4.1 Data sources and data models.

each ED (Townsend *et al.*, 1988) and these scores are then grouped into four classes using an information based classification method (Johnston and Semple, 1983). This gives rise to the initial ('default') regionalisation. The second stage is to construct a less fragmented set of regions using two rules: an 'island' criterion and a 'like-neighbours' criterion. The 'island' criterion was to take small areas of one class that were entirely encircled by larger areas of the adjacent class and merge them into the larger areas. The 'like-neighbours' criterion involved the same merging (of a small area into a large area) when both were contiguous and of the adjacent class. Any large regions (generated at the first stage by the non-spatial classifier) are split and all regions are built up to have a population at risk of between 4000 and 7000. The effect of this merging of EDs at the second stage can be monitored by the same information loss statistic that is used to obtain the initial regionalisation. This type of work is similar in spirit, if not in detail, to social area analysis and factorial ecology (see for example Robson, 1969; Johnston, 1978).

The creation of an appropriate spatial framework is fundamental to the analytical work. First it is important that regions are all of a minimum size and do not vary greatly in size in order to ensure that estimated rates are as reliable as possible, given the scale of the study, and of comparable robustness across the set of regions. Secondly, regional homogeneity is important if links are to be established between disease rates and (in the above case) material deprivation. (For a review of regionalisation methods which meet these criteraia, see Wise *et al.*, 1996.) The spatial framework is perhaps most important in terms of any subsequent service delivery because it is through such a regional partition that any spatial variation in resource targeting might be implemented. Areas that are fragmented and difficult to define on the ground are not helpful from an administrative point of view. This appears to be a difference between an areal framework constructed for a service delivery system and one constructed for epidemiological research. In the latter case it is not clear that the construction of contiguous areas is either necessary or desirable since the purpose is to relate health events to aggregate properties of homogeneous areas however the latter are spatially constituted (see for example Reading *et al.*, 1991). In this case therefore it would be sufficient to use the output from a non-spatial classification routine (such as described above), accepting the classification of ED's generated, without proceeding to the second stage.

4.3 Spatial analysis techniques and application issues

A statistical approach to the identification of problem regions based on the Sheffield work is now described. It is important to point out that very few of the analysis operations described in this section were executed within the GIS. Also, the Sheffield study considered colorectal cancer which does not qualify as a rare disease for which other methodologies may be more appropriate (see for example Alexander *et al.*, 1991). The first stage of the analysis involved the computation of standardised rates for each subarea.

Socio-economic and demographic characteristics are available from the census. Citywide rates of the disease and age and gender specific rates of the disease are computed from the health event data. Problem regions are defined with respect to the occurrence of the disease in the city in question and not with respect to national rates. The only case for analysis based on national statistics would be where the city was bidding for additional resources to handle abnormally high rates relative to the rest of the country. We are assuming that the issue here is the distribution of resources within the city.

Let O_i denote the observed number of new cases of a disease in area i in a given period of time (one year). Let $E[x]_i$ denote the corresponding expected number of new cases given characteristic $[x]$ of the area. The maximum likelihood estimate of the relative risk in the ith area under the Poisson assumption (that each of a (large) number of individuals at risk has a small but equal probability of contracting the disease) is $\{O_i/E[x]_i\}$.

Approach 1. $E[\text{pop}]_i$ is the expected number of new cases of a disease given the population at risk in area i. Thus, if p denotes the proportion of new cases in the at-risk population of the city, $p = \{$observed number of new cases in the at-risk group in the city during time t/population of the city in the at risk group during time $t\}$ then for area i:

$$E[\text{pop}]_i = p\ (\text{pop}_i)$$

where pop_i denotes the population at risk in area i. The population at risk refers to the appropriate age cohort minus those already diagnosed as having the disease. The value p can be thought of as an estimate of the probability that any individual (selected at random from the population) is diagnosed as having the disease under the assumption that the disease occurs at random in the population.

Approach 2. $E[\text{popAG}]_i$ is the expected number of new cases of a disease adjusting for the age and gender composition of the population at risk in the area. High rates detected under approach 1 may be the consequence of a disproportionate number of a particular category of highly at-risk individuals (for example, the elderly). To adjust for this, separate rates are calculated for separate age cohorts distinguishing between male and female rates if appropriate. Thus let $p_F(k)$ denote the proportion of new female cases in the at-risk population in age cohort k. So for area i:

$$E[\text{popAG}]_i = \Sigma k\ p_F(k)\ (\text{pop}_F(k)_i) + \Sigma k\ p_M(k)\ (\text{pop}_M(k)_i)$$

where M denotes male and $\text{pop}_F(k)_i$ denotes the female (M = male) population at risk in cohort k in region i. The sum is over all cohorts. While this measures controls for the population characteristics of the areas, the rates $p_F(k)$ and $p_M(k)$ are based on yet smaller (sub)populations within the city. This may raise concerns about the robustness and reliability of these estimates and suggests the need to base values for $p_F(k)$ on time series of data that are as long as possible. Notwithstanding these remarks, approach 2 is to be preferred to 1 because age profiles can differ markedly across the areas of a city.

Areas can be rank ordered in terms of their relative risk scores but the question remains as to whether any should be classified as problem regions and if so, which? Significance tests are one possibility by identifying the probability of a count as large or larger than O_i under the Poisson assumption with parameter $E[x]_i$. Reporting maps of relative risks together with maps of significance is one possible way of presenting results. There is a problem with the latter in that significance is determined by sample size (Marshall, 1991). Testing for significance at the alpha percent level generates a corresponding percentage of problem regions. Maps of significance should not therefore be shown on their own. Attention has focused recently on improving relative risk estimation and empirical Bayes estimators have been used in the medical literature for this purpose.

Empirical Bayes estimators (also known as James–Stein estimators) adjust for variation in the at-risk population across the set of areas of the city by assuming a mixture model for the distribution of relative risks. The gamma mixture model adjusts each relative risk ($O_i/E[\text{x}]_i$) so that values where both O_i and $E[\text{x}]_i$, for example, are large remain almost unchanged while values where both are small are adjusted towards the mean of the set of relative risk values. Areas with small values of O_i will thus be adjusted towards the mean. The log-normal mixture model introduces a similar adjustment. This approach reduces the chance that areas will be flagged as problem regions simply because they have small populations (where the presence of one or two extra cases can have a disproportionate influence on the classification of the area). This framework is used by Clayton and Kaldor (1987), and Marshall (1991) reviews other applications. Each area is treated independently in the gamma-Poisson model thereby ignoring the spatial distribution of the areas. But, independence may not be a tenable assumption in the case of computing rates for small adjacent areas. Clayton and Kaldor (1987) show how the log-normal model can be modified so that log relative-risks are spatially correlated. Other approaches that introduce space into the estimation procedure are reviewed by Marshall (1991). These spatial methods tend to shrink each rate towards its neighbourhood average (rather than towards the overall mean in the case of the non-spatial estimators) but depend on what are often ad hoc definitions of 'neighbour' (Haining, 1993). Where socio-economic conditions vary between adjacent regions (as they may do by construction), adjusting to a nearest neighbour average may not be justified. Nearest neighbour adjustment may, however, be justifiable as a way of moderating the effects of locating event data by a low accuracy allocation procedure such as the Central Postcode Directory.

It may also be of interest to establish whether there is any association between (Bayes adjusted) relative risks and other characteristics of the population. One such characteristic is material deprivation (Townsend *et al.*, 1988). Computing separate rates by age, gender and material deprivation as described in approaches 1 and 2 above, will be impractical since numbers falling into the different categories will be too small. Haining *et al.* (1994) use the fact that the areas have been constructed to preserve as far as possible intra-region homogeneity in terms of material deprivation to then plot (log transformed) Bayes adjusted relative risk against material deprivation for the 48 regions and fit a bivariate regression model. Regression outliers from the model flag regions where given the age, gender, population size *and* deprivation level of the region, rates are high relative to the city as a whole.

Spatial analysis tools enable the analyst to describe and better understand the distribution of relative risk across a city. How information on disease incidence is used in deciding service delivery depends at least in part on the type of service proposed, for example, whether it is concerned with caring for those with the disease or trying to prevent new occurrences. Areas with very high rates even after allowing for the size, age and material condition of the population may suggest the need for a closer analysis of the area in question to identify if there are place-specific reasons. However, before any conclusions should be drawn regarding service delivery and certainly before any additional (fixed) resources are committed it will be important to establish whether any high rates are 'real', persistent and geographically stable. For each health event data value we have an attribute value (number of events), a timing and a location. It should be possible to explore the robustness of findings to the time period of study and to the spatial partition used. The first may imply looking at results over different time periods (years) and across different aggregations of years. The rules invoked in Haining *et al.* (1994) for constructing Townsend regions admit several equally plausible areal partitions, in which case the analyst may want to re-run the analysis to see to what extent results change with small changes in the areal partition. Also the analyst may want to explore the robustness of findings to the accuracy of the attribute values for any given spatial–temporal framework. This might involve computing, in the case of areas at or near the boundary of statistical significance, the absolute and/or percentage shift in the number of cases that would be sufficient to lead to a reclassification of the area as a problem region or not. Areas that change status as a result of just a small addition or subtraction to the number of cases need to be flagged.

Examining the temporal, spatial and attribute value robustness of results is an analysis task. Answers to the questions posed above may help to confirm or refute the existence of patterns in the data before committing any level of additional resources to particular places.

4.4 Looking ahead: designing future health needs GIS

The previous sections described the progress to date in designing a health needs GIS. In this section we consider the shortcomings of this early design and then discuss additional design criteria and how the system might be improved and extended in the future.

The early work can be thought of as comprising two phases. The first phase involved the construction of a sound database. Setting aside issues of data accuracy this has meant dealing with the problem of linking health data with socio-economic data for it is through such types of linkage that value is added to health service data. (This issue of linkage was not a trivial problem because the two principal data sets were not available on identical spatial frameworks. The problem was partly resolved because health event data were made available as point addresses.) What emerged from this phase was the view that any system should enable the user to construct meaningful areas in terms of which analysis could be carried out. On the one hand this meant rejecting pre-ordained spatial frameworks (such as local authority wards which are socially heterogeneous) and on the other it meant providing a facility that would enable the user to construct areas to his or her specification.

The fundamental building block for constructing new areal frameworks was provided by the enumeration district. The construction of such areas draws on objective rules and local expert knowledge of social areas within the city. The second phase involved the construction of tools that would enable the user to examine the distribution of health events and to explore counts in relation to the size and composition of the underlying population. This is a minimal requirement of any health needs GIS that purports to have analysis capability. High numbers of cases of a disease in certain geographical areas may simply reflect the distribution of population. Once this is allowed for, apparent clusters of events could be the result of concentrations of particular age groups in certain areas. This early system was designed to provide the user with a basic set of tools. Through the system it is possible to ask (and answer) questions such as: What is the geographic distribution of cases of a disease across the city? To what extent is that distribution a consequence of the distribution of population, or its age profile? Is there a link between the incidence of disease and certain socio-economic characteristics of the population such as deprivation? However, as a system with analysis capability it is still relatively crude – both in statistical terms and computing terms. Statistical crudeness stems largely from the range of analysis tools available. In computing terms the crudeness stems from the way the user is able to interrogate and explore the database. We develop these points now.

In two respects at least the current system is capable of statistical development that would add considerably to its utility and rigour. First, there is a large amount of literature in exploratory data analysis (EDA) including exploratory spatial data analysis (ESDA). These techniques which combine numerical, graphical and (in the case of ESDA) cartographical methods, have been developed to identify data properties, suggest hypotheses from data and identify data patterns. They also have a role to play in model development and model criticism. They represent a useful body of techniques for analysing and interrogating data sets including spatial data sets. Secondly, the field of spatial statistics has developed as a rigorous methodology for analysing spatial data, recognising the special problems that spatial data give rise to. Many of these problems revolve around the fact that spatial observations are not independent and it is necessary to model this dependence and allow for it in data analysis, especially inferential analysis. There are a number of techniques within both these areas of modern statistics that could be added to the system to its overall benefit. For example, regression modelling with spatial data raises a number of distinctive statistical issues including the presence of residual spatial correlation. Residual correlation is particularly likely to arise when fitting simple regression models since the residuals will contain information on important omitted variables that are themselves spatially correlated. An illustration of this was given in the previous section looking at the relationship between incidence rates and deprivation. The geographic patterns of a disease are likely to be the consequence of a wide range of factors, not just deprivation. For a review of problems and possible remedial approaches see for example Haining (1991). A wide range of EDA and ESDA and spatial statistical methods are reviewed in Cressie (1991) and Haining (1993) illustrated by several disease data examples. There is an extensive review of relevant methods for the analysis of disease data in Marshall (1991).

The introduction of ESDA and spatial statistical methods into software systems that handle spatial data is important. The former is important because EDA does not contain a sufficient breadth of techniques to explore spatial data and identify spatial patterns. The latter is important because spatial statistics are needed to bring rigour to the analysis of spatial data. There is a problem with the latter, however, which is that as a statistical methodology it is not widely known. The fact that it is also difficult may give rise to a conflict with a later design criterion to be discussed. We now turn to the second of the two shortcomings of the current system identified above.

In statistical computing terms the system is in some respects quite a traditional one and one which could make better use of advances in computing technology. For example, statistical analysis was once time consuming because it was a laborious task to compute descriptive data summaries (means and variances for example) or to fit models (regression models for example) whether the data set was large or small. The earliest computing revolution was characterised by placing in the hands of analysts, systems that would evaluate formulae for given data sets. Through various statistical packages it became possible to carry through a programme of statistical analysis in a fraction of the time previously required using paper and pencil or mechanical calculators. Whilst the new computing systems did enable some new forms of analysis to be undertaken their main benefit lay in speeding up the basic processes of computation. (Through time, software systems became more sophisticated enabling more thorough statistical analyses to be undertaken. The range of statistical techniques that could be implemented on computers also increased as software packages were modified and updated and as new systems came onto the market.) As a system for analysing a data set it is precisely these computational qualities that are demonstrated in the health needs GIS described above. The system computes numbers, for statistical formulae: it does so rapidly and returns them so they can be mapped.

Modern statistical computing is increasingly interactive and graphical. It is interactive (as opposed to batch) in a very important sense. Analysis is driven by looking at the data set from many different perspectives in many different ways and, crucially, the analysis (or view) at step *n* is dependent on the findings of earlier phases of analysis (or data interrogation). It is by interacting with the data through a wide range of numerical, graphical and cartographical summaries that the analyst may develop an understanding of data properties. This type of analysis is well illustrated by packages such as SPIDER and REGARD (Haslett *et al.*, 1991). The effectiveness of this type of E(S)DA depends on being able to look rapidly at the data in many different ways driven by a combination of acquired insight obtained from prior analysis or interrogation, intuition and, where it exists, hard information. Judged by these criteria the present system is static rather than dynamic, 'batch-like' rather than 'interactive' and computational rather than graphical in the way it allows the user to interrogate the database. It does not encourage the participation of users with expertise in medical issues unless they also have expertise in analysis issues because there are no facilities by which to analyse the database in ways that do not require some expert knowledge of statistical method.

The purpose of data analysis is to draw out useful information from raw data. The evaluation of statistical formulae and the fitting of statistical models is only one part of this process. The conduct of data analysis involves a number of different activities. Here we characterise the conduct of contemporary data analysis (DA) in terms of a number of defining attributes.

1. *DA is computer based but interactive/manual rather than batch/automatic.* Data analysis involves exploring data for important characteristics which often means querying the database rapidly. The answer to a question may have a critical bearing on the next question asked. The extent to which this is true means that the process often cannot be automated, often cannot be reduced to a series of predetermined commands that can be submitted in 'batch mode' so that the analyst can return later to digest the output.
2. *DA often requires that a number of individuals with different types of expertise are engaged simultaneously in a co-operative venture.* Good data analysis in any area of research has always necessitated co-operation between those with expertise in the substantive field of research and those with expertise in statistical analysis. There was presumably a limit as to how interactive such co-operation could be in the past simply because of technological limitations. Data analysis of any type (numerical, graphical or cartographical) used to be a slow process. It is arguable that today however, not only is interactive co-operation (at the computer terminal) possible it is more important than ever because of the size and complexity of modern data sets.
3. *DA is not just based on evaluating formulae and providing numerical summaries but is based on graphical (and cartographical) techniques.* The analyst wants to be able to look at data from many different perspectives in ways that will help develop insight and detect patterns. Visual methods are often very successful in stimulating ideas, helping to 'see' possible patterns that can be subject to more rigorous analysis later. Such methods are often simple to use and explain since they retain close contact with the raw data or involve only simple (or intuitive) data manipulations. Simple exploratory methods are likely to be important if a co-operative exploration of a large data set is to be undertaken because the specialist with expertise in the problem area may have a limited grasp of statistical theory. As suggested above, however, some areas of spatial statistics fail this simplicity test although much of the complexity of this field relates to the area of rigorous confirmatory data analysis.
4. *DA is concerned with robustness of findings.* For researchers carrying out data analysis

in the medical field, the spatial and temporal windows (the study areas) and the spatial and temporal partitions (how the study areas are divided up) become critical. There is often no natural window or partition for data analysis in the 'observational' sciences that deal with the world as it is — as opposed to the experimental sciences. For this reason any positive result (any detected problem region) needs to be looked at again and again to see if it persists when a different time period is taken or if a slightly different spatial partition is adopted. The exploration of data from different space–time perspectives is as important as more conventional forms of sensitivity analysis concerned with (for example) robustness to attribute error or data outliers.

4.5 Conclusions

A number of specific analysis functions are required to support a health needs GIS. Some of these, because of their complexity and highly specialised nature, may be better implemented outside the GIS and the results imported into the GIS (for example updating or interpolating census data in the database). Others may be better implemented through some form of coupling with the GIS (Haining and Wise, 1991).

The following techniques appear to be central to the analysis requirements of a health needs GIS.

1. A regionalisation module for constructing an appropriate spatial framework.
2. A rate estimation module for constructing different relative risk rates. These rates may be crude population-based rates; population-based rates adjusted for the age and gender profiles of areas; rates adjusted for the variable robustness of rates computed over large and small populations; rates that also recognise the distribution of areas across the city.
3. A statistical module for examining the relationship between area rates and various socio-economic characteristics of the areas. It will not be sufficient to utilise standard statistical packages that possess the facility to fit the (standard) regression model without modifying them to handle some of the special problems generated by spatial data. This last point means including the facility to test for spatial correlation in regression residuals and fit simple spatial forms of the regression model as a minimum requirement.
4. A 'robustness' module that allows the analyst to examine the sensitivity of results to small changes in the data, particularly the number of cases allocated to any area.

Finally, GIS offer display capabilities that may assist the identification of areas with special properties. An important question is the extent to which interactive capability can be developed for these systems that will allow a rapid transition between statistical properties of the attribute data and the location of these events on the map (and back again) as developed, in a different computing environment, by Haslett and co-workers (1991) at Trinity College Dublin. This raises a further set of issues that go beyond the capability of current GIS to support spatial analysis techniques and concern the 'nimbleness' of current GIS when faced with these demands.

Since this Chapter was written, the author has been involved (with Stephen Wise) in the development of a software system for the spatial analysis of health data linked to a GIS (SAGE) with the aid of an ESRC research grant. For further details, see Haining *et al.* (1996). The system builds on the experiences reported in this chapter.

Acknowledgement

Thanks to Stephen Wise and Marcus Blake for many helpful discussions on the subject matter of this paper and to members of the GISDATA workshop. All errors remain the responsibility of the author.

REFERENCES

ALEXANDER, F.E., RICKETTS, T.J., WILLIAMS, J. and CARTWRIGHT, R.A. (1991). Methods of mapping and identifying small clusters of rare diseases with applications to geographical epidemiology, *Geographical Analysis*, **23**, 158–73.

CARSTAIRS, V. (1981). Small area analysis and health service research, *Community Medicine*, **3**(2), 131–9.

CLAYTON, D. and KALDOR, J. (1987). Empirical Bayes estimates of age standardised relative risks for use in disease mapping, *Biometrics*, **43**, 671–81.

CLIFF, A., HAGGETT, P., ORD, J.K., BASSETT, K. and DAVIES, R. (1975). *Elements of Spatial Structure*, Cambridge: Cambridge University Press.

CRESSIE, N. (1991). *Statistics for Spatial Data*, New York: John Wiley.

GOODCHILD, M.F. (1992). Geographical data modelling, *Computers and Geosciences*, **18**, 401–8.

GRIGG, D. (1965). The logic of regional systems, *Annals Assoc. American Geographers*, **55**, 465–91.

HAINING, R.P. (1991). Models in human geography: problems in specifying, estimating and validating models for spatial data, *Spatial Statistics: Past, Present and Future*, in Griffith, D.A., pp. 83–102, Ann Arbor, Michigan: Michigan Document Services.

HAINING, R.P. (1993). *Spatial Data Analysis in the Social and Environmental Sciences*, Cambridge: Cambridge University Press.

HAINING, R.P. (1994). Designing spatial data analysis modules for Geographical Information Systems, in Fotheringham, S. and Rogerson, P. (eds) *Spatial Analysis and GIS*. London: Taylor & Francis.

HAINING, R.P., MA, J. and WISE, S.M. (1996). Design of a software system for interactive spatial statistical analysis linked to a GIS, *Computational Statistics*, (in press).

HAINING, R.P. and WISE, S.M. (1991). *GIS and Spatial Data Analysis: Report on the Sheffield Workshop*, Regional Research Laboratory Discussion Paper, 11. Department of Town and Regional Planning, University of Sheffield, 43 pp.

HAINING, R.P., WISE, S.M. and BLAKE, M. (1994). Constructing regions for small area analysis: health service delivery and colorectal cancer *J. Public Health Medicine*, **16**, 429–38.

HASLETT, J., BRADLEY, R., CRAIG, P.S., WILLS, G. and UNWIN, A.R. (1991). Dynamic graphics for exploring spatial data with application to locating global and local anomalies, *American Statistician*, **45**, 234–42.

INFORMATION SERVICES UNIT (1981). *Census 1981, Maps for Community Medicine Areas*, Glasgow: Greater Glasgow Health Board.

JOHNSTON, R.J.J. (1978). Residential area characteristics: research methods for identifying urban subareas: social area analysis and factoral ecology, in Herbert, D.T. and Johnston, R.J. *Social Areas in Cities*, pp. 175–217, John Wiley.

JOHNSTON, R. and SEMPLE, K. (1983). *Classification using Information Statistics. Concepts and Techniques in Modern Geography*, Norwich: GeoBooks.

MARSHALL, R.J. (1991). A review of methods for the statistical analysis of spatial patterns of disease, *J. Royal Statistical Society, Ser A*, **154**, 421–41.

OLIVER, M.A. and WEBSTER, R. (1989). A geostatistical basis for spatial weighting in multivariate classification, *Mathematical Geology*, **21**, 15–35.

READING, R., OPENSHAW, S. and JARVIS, S. (1991). Measuring child health inequalities using aggregations of enumeration districts, *J. Public Health Medicine*, **12**(2): 160–97.

ROBSON, B.T. (1969). *Urban Analysis*, Cambridge: Cambridge University Press.

TOWNSEND, P., PHILLIMORE, P. and BEATTIE, A. (1988). *Health and Deprivation: Inequality and the North*, London: Croom Helm.

WISE, S.M., MA, J. and HAINING, R.P. (1996). Regionalisation tools for the exploratory spatial analysis of health data, in Fischer, M. and Getis, A. (Eds) *Recent Developments in Spatial Analysis – Spatial statistics, behavioural modelling and neurocomputing*, Berlin: Springer (in press).

CHAPTER FIVE

Geostatistics, rare disease and the environment

M. A. OLIVER

5.1 Introduction

During the past 15 years or so GIS have developed and their application has expanded considerably. However, their progress has been curtailed because they have not incorporated more specific and advanced forms of spatial analysis. One way forward could be by linking these systems with methods of Exploratory Spatial Data Analysis (ESDA) which enable the complexity of spatial variation to be described using graphical, cartographical and statistical tools. Geostatistical methods could play a major role in the latter. Geostatistics was originally developed to estimate the concentrations of ores and recoverable reserves in mining, but it is now firmly established in the earth sciences for analysing spatial data: in for example groundwater studies, soil science, geology and ecology. It embraces a large suite of analytical techniques, such as the many forms of kriging, variography, simulation, coregionalisation, multivariate and fuzzy geostatistics, the design of optimal sampling schemes, geostatistically constrained multivariate classification, and so on (Armstrong, 1989; Soares, 1993), all underpinned by a coherent and consistent theory.

Interpolation is an important feature of GIS because geographically referenced data are often fragmentary. In many instances investigators want to know how their data vary spatially, and this usually means displaying the information as isarithmic or choropleth maps. Others may wish to infer values at particular points between the observations. Both situations require some method of interpolation to estimate values at unobserved locations. There are many methods of interpolation and mapping, but kriging has been shown to outperform other methods by Laslett *et al.* (1987). Recently, Oliver and Webster (1990) suggested that GIS could be improved by incorporating kriging as the principal method of interpolation. It is already an option in ARC/INFO. Kriging produces optimal unbiased estimates and an estimation variance associated with each, the kriging variance. Curran (1988) and Atkinson *et al.* (1990) have shown that information, such as data from remote sensing, which may not be fragmentary, can also be analysed geostatistically. Very large sets of data of this kind are increasingly available and require analysis.

In existing GIS, such as ARC/INFO, and in mapping packages, such as UNIRAS, kriging is an automatic procedure. This is undesirable because kriging involves stages that should remain separate and distinct since they involve making decisions and provide additional information about the data. For instance, the variogram and its model can give considerable insight into the data at the exploratory stage and they are also essential for kriging. Model fitting is by no means automatic; it needs experience to obtain satisfactory results. Furthermore, any geostatistical procedure requires some preliminary data analysis which should also form part of any spatial analysis system.

To progress, GIS need to have explicit links to other software, such as statistical packages and other spatial analytical routines. Geostatistical software is now more readily available, for example GSLIB (Deutsch and Journel, 1992) and ISATIS (1993). Existing GIS could have an interface with such libraries or packages and this would enhance their spatial analytical capabilities. The integration of these technologies will provide a powerful set of tools, especially for environmental analysis which involves many properties and interrelated systems. The need to understand the complex spatial patterns, and relations between properties and different systems in the environment, is likely to provide the main driving force in furthering spatial analysis in general as the needs in mining did with geostatistics.

The aim of this chapter is to suggest that geostatistics, and in particular variogram analysis and kriging, are suitable tools for ESDA and could provide some of the analytical needs of GIS. Rather than review the many geostatistical techniques that could be linked with GIS I shall describe two case studies; binomial kriging of a rare disease and disjunctive kriging for environmental management. They illustrate the analytical power of geostatistics and the role it could play in an advanced spatial analysis system.

5.2 Exploring spatial data geostatistically

5.2.1 Examination and statistical summary of data

An examination of the data for erroneous values which should be corrected or removed, and for outliers which should also be removed, should be the first stage of any analysis. Summary statistics, such as the mean, variance, and skewness describe the distribution of the data and indicate whether there is a need to transform it to normality. For data that cannot be transformed easily non-parametric geostatistics is available which is distribution free (Journel, 1983). The histogram also provides insight into our data: Haslett *et al.* (1990) and Unwin *et al.* (1990) describe how it can be explored using dynamic graphics. Having inspected the data and displayed them simply as graphs and dot or area maps we can proceed to the next stage.

5.2.2 Spatial variation and regionalised variable theory

Variation in the environment is common and to describe and map it has been an important task for geographers. Recently, the tools for doing this have been augmented by geostatistics. The spatial variation of most properties is so complex that it defies simple mathematical description. Regionalised variable theory treats spatial properties instead as the outcomes of random processes, and the particular values of a variable constitute just one realisation of such a process. However, their variation is not generally unstructured, it is almost always spatially dependent at some scale, that is, points within a given distance apart depend on one another in a statistical sense. This structure may be overlain by more or less erratic local variation, or 'noise' at the working scale. The whole can be described by the variogram which summarises the variation succinctly. Thus regionalised variable theory (RVT), which underpins geostatistics, treats spatial data as continuous spatially dependent variables, based on a stochastic model of spatial variation.

5.2.3 Geostatistical theory

In general the value of the random variable, $Z(\mathbf{x})$, where $\mathbf{x}$ denotes the co-ordinates, x_1 and x_2, for two dimensions is regarded as the sum of two terms:

$$Z(\mathbf{x}) = m(\mathbf{x}) + \epsilon(\mathbf{x}), \tag{5.1}$$

where $m(\mathbf{x})$ is some function of $\mathbf{x}$, known as the *drift* or trend, and $\epsilon(\mathbf{x})$ is a random variable with variance defined by

$$\mathrm{var}[\epsilon(\mathbf{x}) - \epsilon(\mathbf{x}+\mathbf{h})] = \mathrm{E}\{[\epsilon(\mathbf{x}) - \epsilon(\mathbf{x}+\mathbf{h})]^2\} = 2\gamma(h), \tag{5.2}$$

where $\mathbf{h}$ is a vector, the lag, that separates the two places $\mathbf{x}$ and $\mathbf{x+h}$ in both distance and direction. The symbol E denotes the expectation. In many instances $m(\mathbf{x})$ is constant, at least locally, and can be replaced by μ_v, so that Eq. (5.1) can be written as

$$Z(\mathbf{x}) = \mu_v + \epsilon(\mathbf{x}). \tag{5.3}$$

In these circumstances Eq. (5.2) is equivalent to

$$\mathrm{var}[Z(\mathbf{x}) - Z(\mathbf{x}+\mathbf{h})] = \mathrm{E}\{[Z(\mathbf{x}) - Z(\mathbf{x}+\mathbf{h})]^2\} = 2\gamma(\mathbf{h}). \tag{5.4}$$

The assumptions that μ_v, the local mean, and the variance of the differences for a given lag are constant constitute Matheron's (1965) *intrinsic hypothesis*. The quantity $\gamma\mathbf{h}$ is the semivariance, and the function that relates γ to $\mathbf{h}$ is the variogram. Where the intrinsic hypothesis holds, the variogram contains all the information about the spatial variation in Z.

Where the mean and variance remain constant over a whole region the semivariance is equivalent to the autocovariance. The covariance at lag $\mathbf{h}$ is

$$C(\mathbf{h}) = \mathrm{E}\{[Z(x) - \mu][Z(\mathbf{x}+\mathbf{h}) - \mu]\} = \mathrm{E}[Z(\mathbf{x}).Z(\mathbf{x}+\mathbf{h})] - \mu^2, \tag{5.5}$$

where μ is the mean over the region, i.e. $\mathrm{E}[Z(\mathbf{x})]$. The semivariance is related to it simply by

$$\gamma(\mathbf{h}) = C(0) - C(\mathbf{h}), \tag{5.6}$$

where $C(0)$ is the covariance at zero lag or the *a priori* variance of the process. Where the mean changes without limit the covariance does not exist. The variogram always exists, and this makes it more generally useful.

5.2.4 The variogram

The variogram is central to geostatistics and essential for most other geostatistical analyses. It compares the similarity between pairs of points a given distance and direction apart (the lag), and expresses mathematically the average rate of change of a property with separating distance. If the intrinsic hypothesis holds then the variogram can be used throughout our region of interest.

The experimental variogram can be estimated from sample data in one, two or three dimensions. It should also be estimated in at least three directions to examine the variation for directional dependence. The usual computing formula for the variogram is

$$\hat{\gamma}(\mathbf{h}) = \frac{1}{2M(\mathbf{h})} \sum_{i=1}^{M(\mathbf{h})} \{z(\mathbf{x}_i) - z(\mathbf{x}_i + \mathbf{h})\}^2, \qquad (5.7)$$

where $\hat{\gamma}(\mathbf{h})$ is the estimated semivariance at lag $\mathbf{h}$, $z(\mathbf{x}_i)$ and $z(\mathbf{x}_i{+}\mathbf{h})$ are the observed values at $\mathbf{x}_i$ and $\mathbf{x}_i{+}\mathbf{h}$ separated by $\mathbf{h}$, of which there are $M(\mathbf{h})$ pairs.

Since the true variogram is continuous we need to fit a mathematical function to the discrete values of our experimental variogram. The model must have the right shape and fit the experimental values well, but must also be authorised so that it cannot return negative variances (Journel and Huijbregts, 1978; Webster and Oliver, 1990). Some practitioners still fit models by eye. I prefer to fit by weighted least squares, with the weights in proportion to the number of paired comparisons in the estimates, using Ross' (1987) Maximum Likelihood Program (MLP). Since the parameters of these models are essential for kriging and for most other geostatistical procedures the fitting should be good.

Interpreting the variogram is also important to gain insight into the structure of the variation. In most instances the variance increases with increasing separating distance as in Figure 5.1(b). This corresponds with more or less strong correlation or spatial dependence at the shortest distances which weakens as the separation increases. Variograms often flatten when they reach a variance known as the 'sill' variance: they are bounded (Figure 5.1(b)). Bounded variograms suggest that there are transition features, patches or zones of different kinds of rock or soil for instance, whereas unbounded ones suggest continuous change over a region. The distance at which the sill is reached, the 'range', marks the limit of spatial dependence. The variogram often has a positive intercept on the ordinate, 'the nugget variance'. This is the part of the variation that we cannot predict. Much of it derives from spatially dependent variation within the smallest sampling interval, somewhat less from measurement error and purely random variation. A completely flat variogram – 'pure nugget' – means that there is no spatial dependence in the data and that interpolation should not be attempted. In this situation the sampling has missed the scale of spatial variation completely.

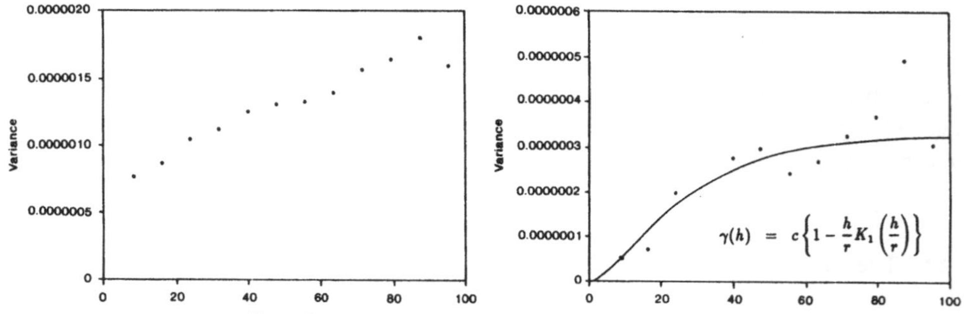

Figure 5.1 Variograms of childhood cancer in the West Midlands: (a) the experimental variogram of incidence $\hat{\gamma}_F(h)$; (b) the experimental variogram of the risk $\hat{\gamma}_R(h)$. The fitted model is Whittle's elementary correlation where c is the sill, r is a distance parameter and K_1 is the modified Bessel function of the second kind.

Hence the variogram can be used to assess the quality of existing data and to design further sampling at a suitable intensity for any particular purpose. The nature and scale of variation of several properties within a region can be compared using the variogram. This may suggest coregionalisation among some properties and influence how the existing data should be used. Hence the variogram is a valuable tool in its own right for ESDA. One of its most important uses is for optimal interpolation or kriging.

5.2.5 Kriging

The observations that describe many geographical properties are often fragmentary. To determine their spatial distribution we need to estimate their values at unsampled points. Kriging is the geostatistical method of estimating values locally. It provides optimal unbiased estimates with known estimation variances based on a model of the spatial variation. The term 'kriging' embraces a set of methods: simple kriging is with known mean, while ordinary kriging, with unknown mean, is the form most commonly used. Co-kriging is the logical extension of ordinary kriging to situations where two or more variables are spatially interdependent and the one whose values are to be estimated is often not sampled as intensively as the other with which it is correlated (Myers, 1982). For non-stationary variation, that is, where there is drift (trend), universal kriging (Matheron, 1973; Olea, 1975) or kriging with intrinsic random functions of order k (IRF-k) should be used. Disjunctive kriging, developed by Matheron (1976), is a more advanced tool based on indicator functions which could be used for environmental decision making (see Section 5.4). Probability and indicator kriging are non-parametric forms of kriging and are also based on indicator functions (Journel, 1983). Bayesian kriging (Abrahamsen, 1993) is a fairly recent development. It uses existing information, that is, 'priors', to improve the estimates of the variable of interest.

Kriging is essentially a method of estimation by local weighted averaging, but one in which the data carry different weights according to their positions both in relation to the unknown point and to one another. The kriging weights are obtained from the variogram or, equivalently, the spatial covariance function. Estimates can be made for points or over blocks of land, water and so on. Block estimates are more often what we want in practice (Oliver and Webster, 1990). In ordinary kriging the estimate of our variable Z over a block B is given by:

$$\hat{Z}(B) = \sum_{i=1}^{n} \lambda_i z(\mathbf{x}_i), \tag{5.8}$$

where λ_i are the weights which sum to one to avoid bias, and subject to this are chosen to minimise the estimation variance:

$$\begin{aligned} \sigma^2(B) &= \mathrm{E}\{[\hat{Z}(B) - Z(B)]^2\} \\ &= 2\sum_{i=1}^{n} \lambda_i \bar{\gamma}(\mathbf{x}_i, B) - \sum_{i=1}^{n}\sum_{j=1}^{n} \lambda_i \lambda_j \gamma(\mathbf{x}_i - \mathbf{x}_j) - \bar{\gamma}(B, B), \end{aligned} \tag{5.9}$$

where $\hat{Z}(B)$ is the estimate and $Z(B)$ is the true but unknown value, the quantities $\gamma(\mathbf{x}_i - \mathbf{x}_j)$ and $\bar{\gamma}(\mathbf{x}_i, B)$ are the semivariance between the data points $\mathbf{x}_i$ and $\mathbf{x}_j$, and the average semivariance between the data point $\mathbf{x}_i$ and the block B for which the estimate is required, respectively, and $\gamma(B, B)$ is the within-block variance. They are obtained from the variogram.

The estimation variance is minimised when

$$\sum_{i=1}^{n} \lambda_i \gamma(\mathbf{x}_i - \mathbf{x}_j) + \psi = \overline{\gamma}(\mathbf{x}_j, B) \quad \forall j \tag{5.10}$$

where ψ is a Lagrange multiplier to achieve minimisation.

These are the kriging equations, and their solution provides the weights to be inserted into Eq. (5.8) for estimating $Z(B)$. An estimate of the estimation variance is also obtained:

$$\hat{\sigma}^2(B) = \sum_{i=1}^{n} \lambda_i \overline{\gamma}(\mathbf{x}_i, B) + \psi - \overline{\gamma}(B, B). \tag{5.11}$$

For environmental properties, such as elements in the soil, rainfall, temperature, hydrology, and so on, kriging has been used mainly for interpolation and mapping to display and interpret their patterns of variation within a region. The kriging variances can be estimated at the same time and mapped similarly; they provide some information about how reliable the estimates are. Such maps can also be used to determine whether there are parts of a region that need more samples to improve the precision of the estimates.

5.2.6 Summary

Variography and kriging are fundamental tools in geostatistics, and they could also form an important component of ESDA. They could be separate from the GIS itself, provided that they can be accessed readily through an explicit interface from a library of spatial analytical routines including geostatistical ones. The whole would form an integrated spatial analysis and GIS structure. Kriging would improve the precision of interpolation in GIS (Oliver and Webster, 1990). This means that the quality of secondary data structures, and the results of mapping and of overlay in GIS would be enhanced. There are many other geostatistical techniques that would add to the analytical capabilities of GIS if they were linked, such as simulation (Deutsch and Journel, 1992), geostatistically constrained multivariate classification (Oliver and Webster, 1989), multivariate geostatistics (Wackernagel *et al.*, 1989), and space–time geostatistics (Goovaerts and Sonnet, 1993). Analysing and mapping the pattern of spatial variation are important in geographical analysis, and geostatistics is a suitable technology for this as the next two sections show.

5.3 Exploring the spatial pattern of rare disease

This case study shows how geostatistics can be used for ESDA in geographical epidemiology. Oliver *et al.* (1993a) modified standard geostatistical technique following the suggestions of Daly (1991) and Lajaunie (1991) to search for spatial pattern in rare disease. Geographical epidemiology is concerned with describing the spatial patterns of disease, and with comparing them with the spatial distributions of environmental factors to suggest possible causes. Historically, geographical studies of diseases have provided important clues to their aetiology. For example, Snow's (1855) observations on the distribution of cholera enabled him to identify its cause and led to its eventual control in the Western world. Similarly Burkitt's (Burkitt and Wright, 1970) painstaking mapping of the distribution of a lymphoma (Burkitt's lymphoma) enabled him to suggest a link with malaria. The aim is to formulate hypotheses about the aetiology of diseases by taking into account the spatial variation in environmental factors. The apparent geographical variation of disease must be interpreted with caution: other factors could contribute to the variation in the recorded frequency apart from environmental ones. This has led to statistical analyses of the spatial distributions of diseases to try to identify real differences in incidence from place to place. Marshall (1991) reviews many of these methods.

Much recent research and discussion on the geographical distribution of disease stems from the apparent concentration of cases of childhood leukaemia near to nuclear installations. As a consequence, much of the analysis has concentrated on the search for clusters in the spatial pattern. There have been two principal approaches. One is based on cell counts and includes Poisson probability mapping (Choynowski, 1959) and the Potthoff and Whittinghill (1966) test, which is a score based on the observed and expected case counts in a given area. In such methods the results depend on the arbitrary choice of cell size. Large cells cause too much smoothing; small cells emphasise local fluctuation. A further weakness is that the counts in neighbouring cells play no part. The other approach is based on the distances between cases, and it goes some way to avoiding these disadvantages. It includes the nearest-neighbour analysis of Cuzick and Edwards (1990) and Besag and Newell (1991), Knox's (1964) space–time clustering which examines pairs of cases within small distances and short times of each other and compares them with random expectation, Diggle's (1990) point process modelling, and the 'geographical analysis machine' of Openshaw *et al.* (1988). Each method has its shortcomings, and it is fair to say that none has yet provided conclusive results.

In the absence of point sources there is no evidence to suggest that the disease has any natural tendency to cluster. Therefore, it seemed to us that mapping the underlying risk of a child's developing cancer was of prime concern, rather than mapping the incidence of the disease itself or searching for clusters. Other workers have come recently to this conclusion. Clayton and Kaldor (1987), Besag *et al.* (1991) and Bernardinelli and Montomoli (1992) have used Bayes estimation to map the risk of various malignancies, including lip, thyroid and breast cancers, respectively. Bithell (1990) gets closest to our approach in using kernel estimation to determine the risk of childhood cancer in Cumbria. His method is similar to kriging, but it weights data arbitrarily, whereas kriging relies on a model of the spatial correlation derived from the data for estimating values. Carrat and Valleron (1992) recently used kriging for straightforward interpolation to map an influenza-like epidemic in France. Our model of the risk is based on a non-contagious disease, however, and to analyse a contagious one would require further development of theory.

5.3.1 The data

The West Midlands Health Authority Region (WMHAR) covers about 25 000 km^2. It includes rural, urban, industrial and suburban environments. During 1980 to 1984 inclusive there were 605 cases of cancer among the 1.13 million children under 15 years of age living there. The cases are distributed among 345 of the total 840 electoral wards (the areas for which population is recorded). When the cases were mapped their density varied enormously: it is much greater in the towns than in the rural areas, if only because many more children live there. However, this apparent pattern in the cases may not reflect the true pattern in the underlying risk. Our aim was, therefore, to explore the data for spatial autocorrelation with a view to identifying meaningful pattern in it.

5.3.2 Theory and analysis

We assume that there is an underlying risk, $R(x)$, of a child's developing cancer and to which all children are exposed, this is our regionalised variable. As a first step to estimating the risk locally we calculate a local 'frequency', $F(\mathbf{x}_i)$, that is, the ratio of the number of cases within each ward, $L(\mathbf{x}_i)$, to the number of children living there, $n(\mathbf{x}_i)$: $F(\mathbf{x}_i) = L(\mathbf{x}_i)/n(\mathbf{x}_i)$, where $\mathbf{x}_i$, $i = 1, 2, \ldots$, denote the centroids of the wards. We also assume that the different cases occur independently, so that R is the only possible source of correlation among them. Hence the observed frequencies, $F(\mathbf{x}_i)$, are drawn from a binomial distribution that depends on the risk, $R(\mathbf{x}_i)$, and the number of children, $n(\mathbf{x}_i)$.

More than half of the wards, 495, had no childhood cancer diagnosed, and few wards had more than one case each. The statistical distribution of the frequencies was very skewed as one would expect for a rare disease. However, the counts were poorly represented by the Poisson distribution, but the observed frequencies follow the binomial distribution well. We did not transform these data because we wanted to retain the distribution.

5.3.2.1 *Estimating the risk variogram*

To estimate the risk variogram we first compute an experimental variogram of the frequencies, $\hat{\gamma}_F$, with the usual formula, Eq. (5.7). This variogram embodies error arising from the binomial character of the frequencies and the variation in the numbers children over the region (Figure 5.1a).

For a given risk, R, the expectation and variance of $F(\mathbf{x}_i)$ are

$$\mathrm{E}[F(\mathbf{x}_i)|R] = R(\mathbf{x}_i) \tag{5.12}$$

and

$$\mathrm{var}[F(\mathbf{x}_i)|R] = \frac{R(\mathbf{x}_i)\{1 - R(\mathbf{x}_i)\}}{n(\mathbf{x}_i)} \tag{5.13}$$

We shall also need the expected product of two frequencies:

$$\mathrm{E}[\{F(\mathbf{x}_i)F(\mathbf{x}_j)\}|R] = R(\mathbf{x}_i)R(\mathbf{x}_j). \tag{5.14}$$

The relations in these three equations lead to the conditional squared differences of the frequencies:

$$\mathrm{E}[\{F(\mathbf{x}_i) - F(\mathbf{x}_i + \mathbf{h})\}^2/R] = \{R(\mathbf{x}_i) - R(\mathbf{x}_i + \mathbf{h})\}^2 + \frac{1}{n(\mathbf{x}_i)}R(\mathbf{x}_i)\} + \frac{1}{n(\mathbf{x}_i + h)}R(\mathbf{x}_i + \mathbf{h})\{1 - R(\mathbf{x}_i + \mathbf{h})\}, \tag{5.15}$$

in which $n(\mathbf{x}_i)$ and $n(\mathbf{x}_i+\mathbf{h})$ are the numbers of children in the wards centred at $\mathbf{x}_i$ and $\mathbf{x}_i+\mathbf{h}$ respectively. This leads to an expected value of the squared difference:

$$\mathrm{E}[\{F(\mathbf{x}_i) - F(\mathbf{x}_i + \mathbf{h})\}^2] = 2\gamma_R(\mathbf{h}) + \{\mu(1-\mu) - \sigma_R^2\}\left\{\frac{n(\mathbf{x}_i) + n(\mathbf{x}_i + \mathbf{h})}{n(\mathbf{x}_i).n(\mathbf{x}_i + \mathbf{h})}\right\}, \quad (5.16)$$

where $\gamma_R(\mathbf{h}) = \frac{1}{2}\mathrm{E}[\{R(\mathbf{x}_i) - R(\mathbf{x}_i + \mathbf{h})\}^2]$, σ_R^2 is the variance of the underlying risk, and μ is the mean risk. The variances obtained depend on the spatial location, because the population of children, $n(\mathbf{x}_i)$, varies from ward to ward, and so Eqs (5.15) and (5.16) do not define a variogram in the strict sense. Nevertheless, their averages over the whole area are correctly estimated by the empirical variogram of the frequencies obtained using Eq. (5.16). We may therefore replace μ, σ_R^2, and the semivariances of F and R by their estimates, $\overline{F}$, $\hat{\sigma}_R^2$, and $\hat{\gamma}_F(\mathbf{h})$ and $\hat{\gamma}_R(\mathbf{h})$ respectively, to obtain the following relation between the estimated variogram of the risk and the empirical variogram of the frequencies:

$$\hat{\gamma}_R(\mathbf{h}) = \hat{\gamma}_F(\mathbf{h}) - \frac{1}{2}\{\overline{F}(1-\overline{F}) - \hat{\sigma}_R^2\}\overline{\left\{\frac{n(\mathbf{x}_i) + n(\mathbf{x}_i + \mathbf{h})}{n(\mathbf{x}_i).n(\mathbf{x}_i + \mathbf{h})}\right\}}, \quad (5.17)$$

in which the quantity beneath the bar is the average over all pairs of wards involved in calculating $\hat{\gamma}_F(\mathbf{h})$. The mean of the data, $\overline{F}$, estimates μ without bias, but we have no value of $\hat{\sigma}_R^2$ to replace σ_R^2 in Eq. (5.16). However, with small $\overline{F}$ and large $n(\mathbf{x}_i)$ and $n(\mathbf{x}_i{+}\mathbf{h})$ it can be estimated from the sill of a fitted model. Then by iteration its estimate and the fitted model can be refined (Oliver *et al.*, 1993a).

We calculated semivariances, $\hat{\gamma}_R(h)$ where $h = |\mathbf{h}|$ for isotropic variation, from the incidences in this way, and the results are shown by the set of points in Figure 5.16(b). The solid line is the fitted model, Whittle's (1954) *elementary correlation*. In its isotropic form the function is:

$$\gamma(h) = c\left\{1 - \frac{h}{r}K_1\left(\frac{h}{r}\right)\right\}, \quad (5.18)$$

where c is the sill estimating the *a priori* variance of the process, r is a distance parameter, and K_1 is the modified Bessel function of the second kind.

5.3.2.2 *Binomial kriging*

To estimate the risk, R, of a child's developing cancer at points $\mathbf{x}_0$ throughout the region, and to map it, we use the frequencies, $F(\mathbf{x}_i)$, and the risk variogram. Our procedure is effectively an extension of ordinary co-kriging (Oliver *et al.*, 1993a). Assuming the mean to be unknown, the risk at a place, $\mathbf{x}_0$, is

$$\hat{R}(\mathbf{x}_0) = \sum_{i=1}^{n} \lambda_i F(\mathbf{x}_i), \quad (5.19)$$

where n is the number of data and λ_i are weights. It involves solving the kriging system

$$\sum_{i=1}^{n} \lambda_i C_F(\mathbf{x}_i, \mathbf{x}_j) + \psi = C_{FR}(\mathbf{x}_0, \mathbf{x}_j) \quad \forall j,$$

$$\sum_{i=1}^{n} \lambda_i = 1, \quad (5.20)$$

where ψ is a Lagrange multiplier, the $C_F(\mathbf{x}_i, \mathbf{x}_j)$ are the covariances of the frequencies, and the $C_{FR}(\mathbf{x}_0, \mathbf{x}_j)$ are the covariances between the frequency and the risk. The covariances of

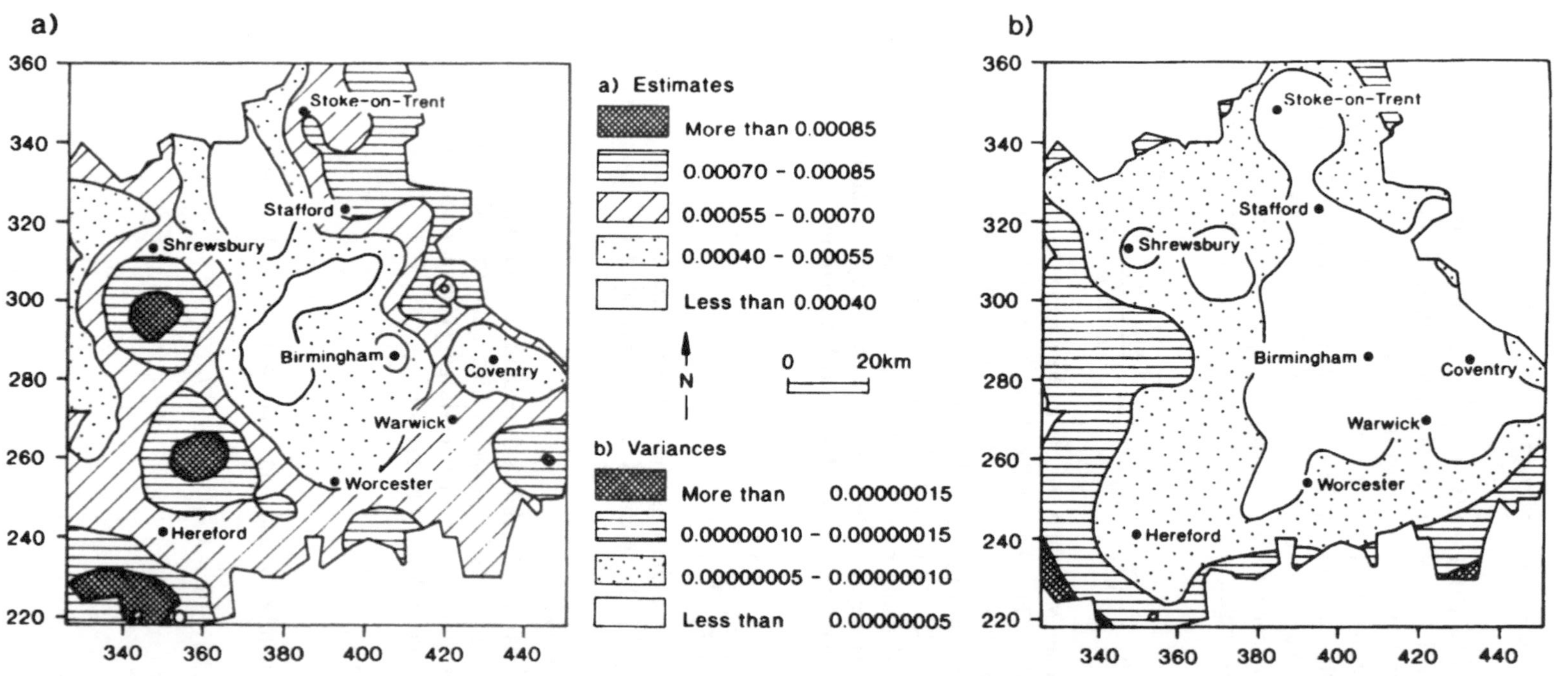

Figure 5.2 (a) Map of the estimated risk of childhood cancer in the West Midlands of England by binomial kriging; (b) map of the associated estimation variances.

the risk C_R are obtained from the risk variogram and from these the other covariances are derived (see Oliver *et al.*, 1993a). The risk of a child's developing cancer and the estimation variances were estimated by solving the kriging Eqs (5.20) at 2 km intervals on a square grid throughout the region. The kriging neighbourhood that we normally use for soil data, for example, contains 16 to 25 points. In this adaptation of kriging we found (Oliver *et al.*, 1993b) that more distant data carry sufficient weight to be included in the equation and we set n to 100. The estimates of the underlying risk and the estimation variances were then 'contoured' to produce isarithmic maps, Figure 5.2(a) and (b).

5.3.3 Results and discussion

Both the variogram of frequency, $\hat{\gamma}_F(h)$, and that of the risk, $\hat{\gamma}_R(h)$, are bounded. This suggests that the risk of a child's developing cancer has a coarse patchy distribution; that is, wards with large risk in general occur near to others with large risk, and similarly those with small risk occur close to others where it is small. The distance parameter of the risk variogram suggests that the pattern or autocorrelation in the risk extends to approximately 50 km. The semivariances of the frequencies are much larger than those of the risk, Figure 5.1(a), and the projection of the empirical values to the ordinate has an appreciable nugget variance. The latter is the principal difference between the variograms, it represents the error in estimating the risk from few cases among the finite population. There is no nugget variance in the risk variogram. This is what we should expect since our wards, which form the support of our sample, are contiguous.

The kriged map of the risk, Figure 5.2(a), supports the interpretation of the variogram: the risk has a patchy distribution. The rural areas and the fringes of the conurbation appear to have a large risk, whereas in general it is small in the urban areas. This contrasts markedly with the impression gained from the distribution of cases. Figure 5.2(b) displays the estimation variances which are greatest at the edges of the region. They are larger for the rural areas where the data are most sparse, than for the urban areas where the wards are smallest and where the data are most dense.

The pattern in the risk of childhood cancer for the WMHAR is like that observed elsewhere in the UK by, for example, Alexander *et al.* (1990) who found that the risk of children's developing lymphoblastic leukaemia was largest in wards farthest from large urban centres, and Greaves (1988) who observed more childhood lymphoblastic leukaemia in rural areas than in towns.

5.3.4 Summary

The epidemiology of a non-contagious disease can now be placed in a geostatistical setting. This study shows how ESDA of case records can be used to detect underlying spatial pattern in a rare disease, and to estimate the risk of developing it. As with all maps made by kriging it is the best map possible, in the sense that it displays unbiased estimates with least error after taking account of the autocorrelation present. Many other diseases could be analysed in this way, such as diabetes, spina bifida, coronary disease, arthritis and so on.

It now seems worth searching for possible environmental causes where the risk is large, and I plan to do that. Environmental properties can be estimated and mapped by ordinary kriging as described in Section 5.2.5. This will provide optimal estimates and enable visual comparisons between the distributions of properties as well as with the disease. The analysis will also be extended to the 20 years of available data, to subsets of particular forms of cancer, and to the four 5-year periods. Optimal estimates of the risk of disease for different time periods and different diagnoses could then be examined further with the overlay and data manipulation facilities in GIS. This would improve the likelihood of detecting possible temporal changes or stability in the patterns of the risk, patterns in subsets of the disease, and links with environmental properties. With such an integrated facility it should also be feasible to map and compare a range of levels of risk of developing the disease and of environmental concentrations in searching for common patterns that might provide clues about the aetiology of childhood cancer. The aim of such analyses is to provide epidemiologists with the kind of spatial information and advice that will enable them to focus attention on certain areas for further investigation.

In other situations where land or environmental managers need to make decisions based on estimated values ordinary kriging may not provide the solution, and a more advanced form, namely disjunctive kriging, is required.

5.4 Disjunctive kriging for environmental management

5.4.1 Introduction

There is increasing awareness that the environment can be improved by better management of resources and of human activities. To protect the environment from pollutants, governmental agencies are setting statutory or advisory maxima for the concentrations of certain substances in the soil, water supply, and the air. To apply such statutes they must be able to estimate reliably the local concentrations in areas from sample data and the risks that the true values exceed the thresholds.

Ordinary kriging provides optimal estimates, but not the risk of toxicity or deficiency associated with a given threshold value. Estimates always embody some error because they derive from sample information, and decisions based on them may be more or less sound in consequence. Decisions are easy where the estimated values are much less than or much greater than the critical level. The problem arises when the values are close to the threshold: the estimate might exceed the threshold even though the true value is less. Disjunctive kriging was developed by Matheron (1976) as a decision-making tool for miners faced with this problem. However, it is valuable wherever we are forced into decisions by the values of a property that varies spatially. Disjunctive kriging provides estimates of the property and of the risks of taking the estimate at its face value, that is, the probability that the true value exceeds or falls short of the threshold. Isarithmic maps of the probabilities enable us to identify those areas at greatest risk and in most need of attention. Yates and Yates (1988) showed how it could be used for controlling a bacterial pollutant, and Webster and Oliver (1989) and Wood *et al.* (1990) have explored its potential in agriculture.

5.4.2 Theory

Rivoirard (1994) has given a clear and detailed description of the complex theory; I shall only summarise the main points here. Essentially, disjunctive kriging involves defining a new disjoint variable or indicator function from the original continuous variable such that values equal to or exceeding a pre-defined threshold take the value 1, and 0 otherwise. If the original variable is denoted by $Z(\mathbf{x})$ and the threshold as z_c, then the indicator function is

$$\mathbf{1}[Z(\mathbf{x}) \geq z_c]. \tag{5.21}$$

At any one place this indicator is either 1 or 0; it cannot be both. At the sampling points it is known, elsewhere it has to be estimated. The estimate is the probability, given the data, that the indicator is 1, that is, the probability that z_c, the threshold, is exceeded.

To estimate the probabilities requires stronger assumptions than those for linear kriging. It is assumed that the values of the properties are the outcome of a second-order stationary process, and that the bivariate probability distribution is known and that it too is stationary throughout the region of interest. In Gaussian disjunctive kriging, the most common form for analysing environmental properties, the data are transformed to a standard normal distribution using a linear combination of Hermite polynomials. Hence, it is assumed that the bivariate distribution is also normal. The transformation has the form:

$$Z(\mathbf{x}) = \phi\{Y(\mathbf{x})\} = \sum_{k=0}^{\infty} Q_k H_k\{Y(\mathbf{x})\}, \tag{5.22}$$

where $Y(\mathbf{x})$ is the transformed value, ϕ is a linear combination of Hermite polynomials, H_k is an infinite series of Hermite polynomials and Q_k are coefficients evaluated by Hermite integration.

The disjunctively kriged estimates are linear combinations of the estimates of the Hermite polynomials of the transformed sample values, $H_k\{Y(\mathbf{x}_0\}$, given by

$$\hat{Z}_{DK}(\mathbf{x}_0) = \sum_{k=0}^{\infty} Q_k \hat{H}_k\{Y(\mathbf{x}_0)\}, \tag{5.23}$$

The Hermite polynomials are estimated by

$$\hat{H}_k\{Y(\mathbf{x}_0)\} = \sum_{i=1}^{n} \lambda_{ik} H_k\{Y(\mathbf{x}_i)\}, \tag{5.24}$$

where $\hat{H}_k\{Y(\mathbf{x}_0)\}$ is the estimated value of the kth Hermite polynomial at the site $\mathbf{x}_0$, n is the number of points in the neighbourhood, and λ_{ik} are the weights as in ordinary kriging and, are derived by solving these kriging equations (Rivoirard, 1991)

$$\sum_{i=1}^{n} \lambda_{ik} \mathrm{Cov}[H_k\{Y(\mathbf{x}_j)\}, H_k\{Y(\mathbf{x}_i)\}] = \mathrm{Cov}[H_k\{Y(\mathbf{x}_j)\}, H_k\{Y(\mathbf{x}_0)\}] \quad \forall j. \tag{5.25}$$

In disjunctive kriging the weights have to be found k times per estimate instead of once as in ordinary kriging, and the kriging equations have to be solved k times per estimate also.

To estimate the conditional probabilities that the true value exceeds the threshold we krige the indicator $\mathbf{1}[Z(\mathbf{x}) \geq; z_c]$. Based on the transformation the indicators are

$$\mathbf{1}[Z(\mathbf{x}) \geq z_c] = \mathbf{1}[Y(\mathbf{x}) \geq y_c]. \tag{5.26}$$

The indicator function is written in terms of the Hermite polynomials with respect to the observed values in the neighbourhood. Its disjunctive kriging estimate is:

$$\hat{\mathbf{1}}^{DK}[y(\mathbf{x}_0) \geq y_c] = 1 - G(y_c) - \sum_{k=1}^{L} \frac{1}{\sqrt{k}} H_{k-1}(y_c) g(y_c) \hat{H}_k^K\{y(\mathbf{x}_0)\} \tag{5.27}$$

where $G(y_c)$ is the cumulative probability distribution, $g(y_c)$ is the probability density function, and L is the number of Hermite polynomials used, usually no more than ten.

5.4.3 Soil salinity in Bet Shean

The problem of soil salinity in the Bet Shean Valley, Israel (Wood *et al.*, 1990) shows how disjunctive kriging can be used for dealing with excess concentrations in the environment. Sustainable agriculture under irrigation depends on controlling salinity. Salt concentration is usually measured in terms of the electrical conductivity (EC) of the soil solution: it is greater the larger the salt concentration. An EC of 4 mS cm^{-1} in the soil is usually regarded as critical, marking the onset of salinisation. The principal winter crops in Bet Shean, wheat and lucerne, yield less than their maximum where the EC exceeds this threshold value. Amelioration of the soil with gypsum, good quality irrigation water, improving drainage or a fallow year are costly remedies. The farmers need to know, therefore, the risks they are taking in not treating their soil. Wood measured the EC of 201 topsoil samples taken from an area of 2030 ha in November 1985 (Wood *et al.*, 1990).

5.4.3.1 *Analysis and results*

The measurements of electrical conductivity were strongly positively skewed and we transformed them to a standard normal distribution using Hermite polynomials. The sample variogram was computed on the transformed values, and an exponential model fitted to it. The model describes one form of second-order stationarity, and so this assumption underlying disjunctive kriging is satisfied. Figure 5.3 shows the sample semi-variances plotted as points, and the fitted model as the solid line. The electrical

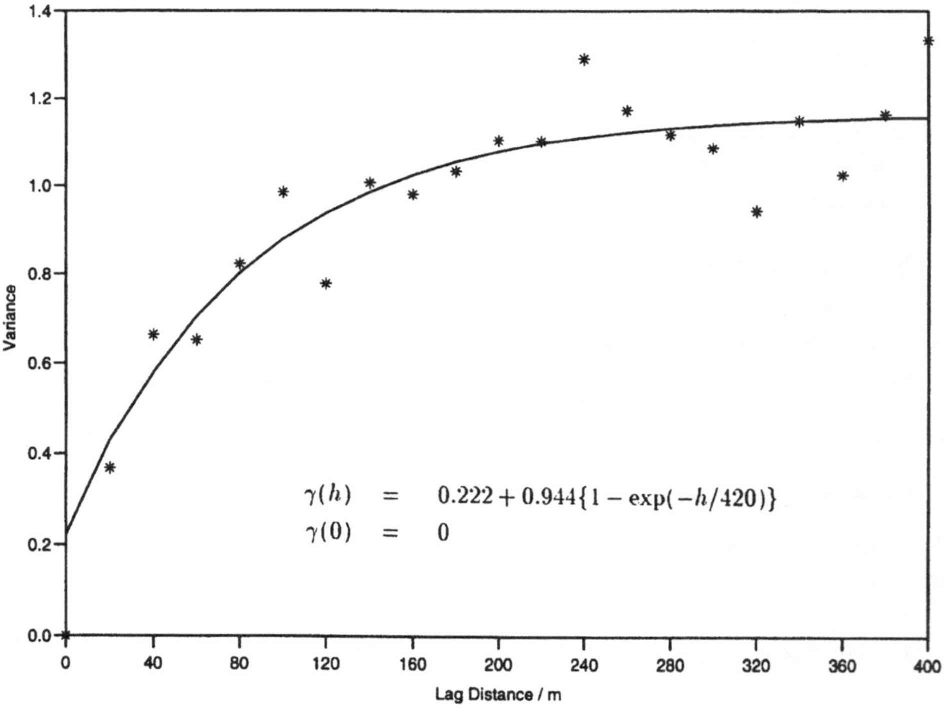

Figure 5.3 Variogram of electrical conductivity in Bet Shean after normalisation by Hermite transformation. The line is the fitted exponential model.

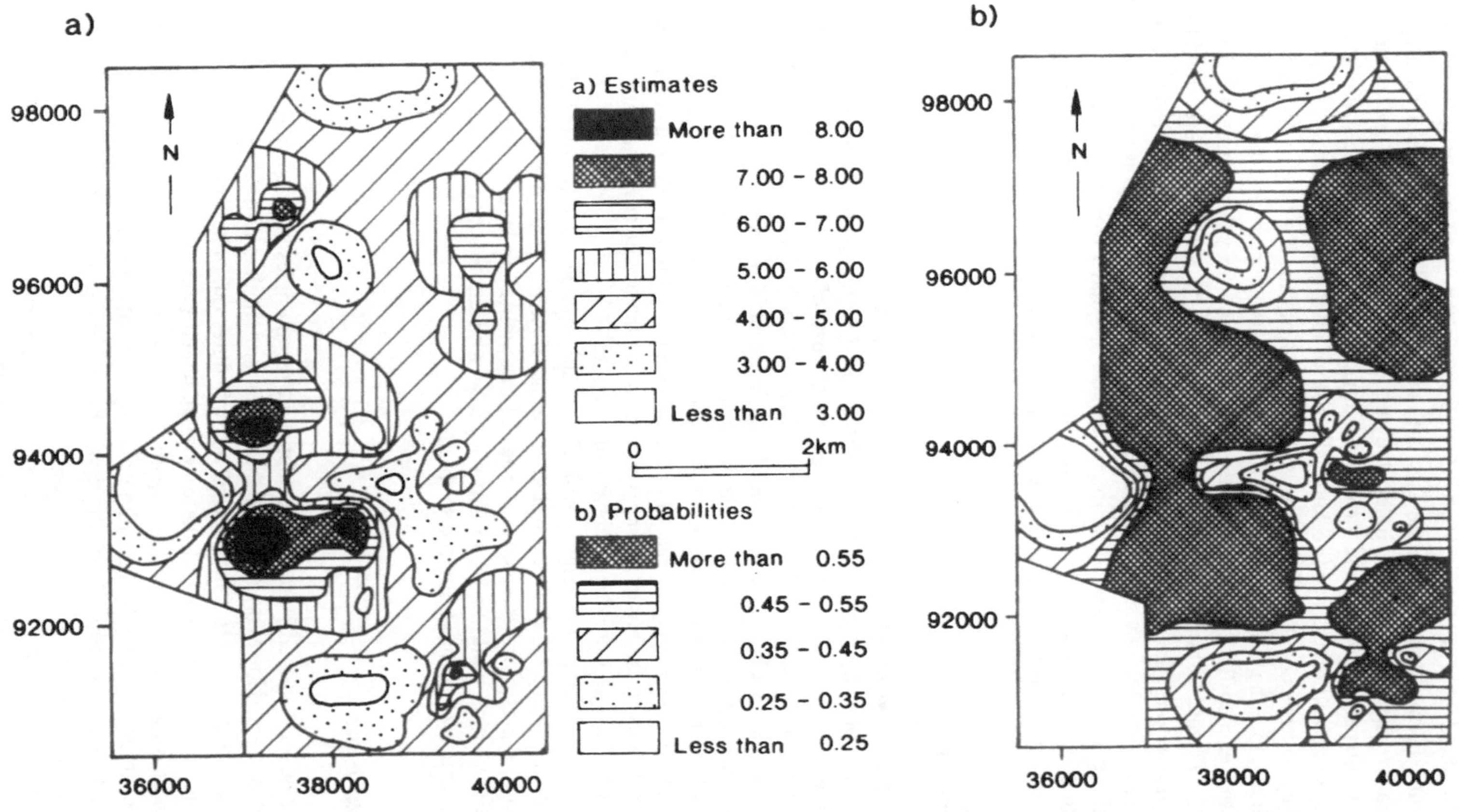

Figure 5.4 (a) Isarithimic map of electrical conductivity in mS cm $^{-1}$ for Bet Shean by disjunctive kriging; (b) map of conditional probability, the EC > 4 mS cm $^{-1}$.

conductivities and probabilities were estimated over blocks of 1 ha at intervals of 100 m on a square grid using the variogram model, the Hermite transformed values, and a critical threshold of $4\,\mathrm{mS\,cm^{-1}}$. Figure 5.4(a) shows the isarithmic map of electrical conductivity. In much of the area the estimates exceed $4\,\mathrm{mS\,cm^{-1}}$ and in these areas the farmers can expect losses of yield unless they treat their land. The more interesting point relates to those parts of the region where the estimated EC is approximately $4\,\mathrm{mS\,cm^{-1}}$ or less. In these areas farmers need to know the risk of taking these estimates at their face value. They covered about half of the area. Figure 5.4(b) shows the estimated conditional probabilities for the indicator $\mathbf{1}[EC(x) \geq 4]$ and they provide the answer. The probabilities exceed 0.4 over much of the area, only in a very small area are they less than 0.25. Farmers should be alerted to the risks of salinity where conditional probabilities are only 0.3. Thus the risks of salinity and of reduced yields are fairly large over a much greater area than the estimates suggested. The conditional probabilities identify those areas where there is an important risk of salinity even though the estimates suggest that the soil is not saline. Figure 5.4(b) provides a means of deciding where remedial action is most needed and where there is no need to act.

5.4.4 Summary

This example shows how the conditional probabilities can be used for assessing the risk of whether or not to ameliorate the land. The costs of doing so are considerable and the farmers must balance this against the potential loss of yield. In other agricultural situations we may be faced with the reverse situation where deficiencies in soil nutrients can lead to poor yields of crops and from livestock (Webster and Oliver, 1989). Disjunctive kriging should also provide environmental scientists with a useful decision-making tool, especially where there are sensitive issues that could result in litigation or damage to health. Assessing the risk of exceeding statutory limits of the concentrations of certain substances in the environment is now a major feature of environmental analysis, and disjunctive kriging provides a means of doing this.

A coupling of disjunctive kriging with GIS could provide a set of powerful tools for environmental management: this could be in terms of excess or deficiency of a property in relation to a prescribed threshold. Ranges of thresholds and levels of probability that might invoke action could be examined with ease with the two approaches providing a possible hierarchy for managing the problem.

5.5 Conclusions

Both geostatistics and GIS are well-established technologies that operate in different spheres at present. However, many of their aims are similar: to analyse, describe and interpret the nature of spatial variation of many features on, in or above the earth. Some kind of integration or coupling of these systems would benefit both groups of users: in GIS by increasing the scope of their spatial analysis, especially using the methods of ESDA, and in geostatistics by providing overlay and data manipulation facilities to aid the interpretation of results. Problems could be tackled from different aspects and so improve our understanding of reality.

The computing environment for this needs to be flexible and interactive wherever possible to allow GIS to be linked easily with a range of analytical procedures, such as geostatistics. In my view the integration of this technology should be such that the individual stages we work through in geostatistics at present would retain their distinct identity and not be embraced in a completely automatic procedure. We need to maintain the flexibility we have at present so that users are aware of the decisions they need to make at each stage, and also of any limitations that are imposed by the nature and quality of their data. The new software packages ISATIS of the Ecole de Mines de Paris and GSLIB (Deutsch and Journel, 1992) go some way to achieving this overall goal for geostatistics, and it would seem feasible to take it a stage further with an explicit interface to a GIS.

Geographical analysts in the broad sense must be ready to accept the initiative to improve and protect the environment. Geostatistics embraces a range of techniques suitable for analysing environmental data as well as for decision making. An advanced spatial analysis system (ASAS) linking geostatistical techniques with GIS could lead the way. Once a link is established it would become increasingly easy to add more routines to an existing library.

Acknowledgements

I thank the West Midlands Regional Children's Tumour Research Group and Dr G. Wood for the use of their data, and Professor R. Webster for software.

REFERENCES

Abrahamsen, P. (1993). Bayesian kriging for seismic depth conversion of a multi-layer reservoir, in Soares, A. (Ed.), *Geostatistics Tróia '92, Vol. 1*, pp. 385–98, Dordrecht: Kluwer Academic Publishers.

Alexander, F.E., Ricketts, T.J., McKinney, P.A. and Cartwright, R.A. (1990). Community lifestyle characteristics and risk of acute lymphoblastic leukaemia in children, *The Lancet*, **336**, 1461–5.

Armstrong, M. (Ed.), (1989). *Geostatistics*, Dordrecht: Kluwer Academic Publishers.

Atkinson, P.M., Curran, P.J. and Webster, R. (1990). Sampling remotely sensed imagery for storage, retrieval and reconstruction, *Professional Geographer*, **42**, 345–53.

Bernardinelli, L. and Montomoli, C. (1992). Empirical Bayes versus fully Bayesian analysis of geographical variation in disease risk, *Statistics in Medicine*, **11**, 983–1007.

Besag, J. and Newell, J. (1991). The detection of clusters in rare diseases, *Journal of the Royal Statistical Society, Series A*, **154**, 143–55.

Besag, J., York, J. and Mollié, A. (1991). Bayesian image restoration with two applications in spatial statistics, *Annals of the Institute of Statistical Mathematics*, **43**, 1–59.

Bithell, J.F. (1990). An application of density estimation to geographical epidemiology, *Statistics in Medicine*, **9**, 691–701.

Burkitt, D.P. and Wright, H.D. (1970). *Burkitt's Lymphoma*, Edinburgh: Livingstone Press.

Carrat, F. and Valleron, A.-J. (1992). Epidemiologic mapping using the 'kriging' method: application to an influenza-like illness epidemic in France, *American Journal of Epidemiology*, **135**, 1293–1300.

Choynowski, M. (1959). Maps based on probability, *Journal of the American Statistical Association*, **54**, 385–8.

Clayton, D. and Kaldor, J. (1987). Empirical Bayes estimates of age-standardized relative risks for use in disease mapping, *Biometrics*, **43**, 671–81.

CURRAN, P.J. (1988). The semi-variogram in remote sensing: an introduction, *Remote Sensing of the Environment*, **24**, 493–507.

CUZICK, J. and EDWARDS R. (1990). Spatial clustering for inhomogeneous populations, *Journal of the Royal Statistical Society, Series B*, **52**, 73–104.

DALY, C. (1991). *Application de la géostatistique à quelques problémes de filtrage*. Doctoral thesis, Fontainebleau, Ecole des Mines de Paris.

DEUTSCH, C. and JOURNEL, A.G. (1992). *GSLIB: Geostatistical Software Library*, New York: Oxford University Press.

DIGGLE, P.J. (1990). A point process modelling approach to raised incidence of a rare phenomenon in the vicinity of a prespecified point, *Journal of the Royal Statistical Society, Series A*, **153**, 349–62.

GOOVAERTS, P. and SONNET, PH. (1993). Study of spatial and temporal variations of hydrogeochemical variables using factorial kriging analysis, in Soares, A. (Ed.), *Geostatistics Tróia '92*, pp. 745–56, Dordrecht: Kluwer Academic Publishers.

GREAVES, M.F. (1988). Speculations on the cause of childhood acute lymphoblastic leukaemia, *Leukaemia*, **2**, 120–5.

HASLETT, J., WILLS, G. and UNWIN, A. (1990). SPIDER – an interactive statistical tool for the analysis of spatially distributed data, *International Journal of Geographical Information Systems*, **4**, 285–96.

ISATIS (1993). *The Geostatistical Key*, Avon, France: Geovariances.

JOURNEL, A.G. (1983). Nonparametric estimation of spatial distributions, *Mathematical Geology*, **15**, 445–60.

JOURNEL, A.G. and HUIJBREGTS, C.J. (1978). *Mining Geostatistics*, London: Academic Press.

KNOX, E.G.(1964). The detection of space–time interactions, *Applied Statistics*, **13**, 25–9.

LAJAUNIE, C. (1991). *Local Risk Estimation for a Rare Noncontagious Disease based on Observed Frequencies*, Note N-36/91,G, Fontainebleau, Centre de Géostatistique, Ecole des Mines de Paris.

LASLETT, G.M., MCBRATNEY, A.B., PAHL, P.J. and HUTCHINSON, M.F. (1987). Comparison of several spatial prediction methods for soil pH, *Journal of Soil Science*, **38**, 325–41.

MARSHALL, R.J. (1991). A review of methods for the statistical analysis of spatial patterns of disease, *Journal of the Royal Statistical Society, Series A*, **154**, 421–41.

MATHERON, G. (1965). *Les Variables Régionalisées et leur Estimation*, Paris: Masson.

MATHERON, G. (1973). The intrinsic random functions and their applications, *Advances in Applied Probability*, **5**, 439–68.

MATHERON, G. (1976). A simple substitute for the conditional expectation: the disjunctive kriging, in Guarascia, M., David, M. and Huijbregts, C. (Eds), *Advanced Geostatistics in the Mining Industry*, pp. 221–36, Dordrecht: Reidel.

MYERS, D.E. (1982). Matrix formulation of co-kriging, *Mathematical Geology*, **14**, 249–57.

OLEA, R.A. (1975). *Optimal Mapping Techniques Using Regionalized Variable Theory*, Series on Spatial Analysis, No. 2., Lawrence: Kansas Geological Survey.

OLIVER, M.A., LAJAUNIE, C., WEBSTER, R. *et al.* (1993a). Estimating the risk of childhood cancer, in Soares, A. (Ed.), *Geostatistics Tróia '92, Vol. 2*, pp. 899–910, Dordrecht: Kluwer Academic Publishers.

OLIVER, M.A., LAJAUNIE, C., WEBSTER, R. *et al.* (1993b). Binomial kriging the risk of a rare disease. *Cahiers de Géostatistique*, Fascicule, 3, pp. 159–65, Ecole des Mines de Paris.

OLIVER, M.A. and WEBSTER, R. (1989). A geostatistical basis for spatial weighting in multivariate classification, *Mathematical Geology*, **21**, 15–35.

OLIVER, M.A. and WEBSTER, R. (1990). Kriging: a method of interpolation for geographical information systems, *International Journal of Geographical Information Systems*, **4**, 313–32.

OPENSHAW, S., CRAFT, A.W., CHARLTON, M. *et al.* (1988). Investigation of leukaemia clusters by use of geographical analysis machine, *The Lancet*, *i*, 272–3.

POTTHOFF, R.F. and WHITTINGHILL, M. (1966). Testing for homogeneity. II. The Poisson distribution, *Biometrika*, **53**, 183–90.

RIVOIRARD, J. (1994). *Disjunctive Kriging and Non-linear Geostatistics*, Oxford: Clarendon Press.

ROSS, G.J.S. (1987). *Maximum Likelihood Program*, Oxford: Numerical Algorithms Group.

SNOW, J. (1855). *On the Mode of Communication of Cholera*. 2nd Edition, London: Churchill.

SOARES, A. (Ed.), (1993). *Geostatistics Tróia '92*, Dordrecht: Kluwer Academic Publishers.

UNIRAS (1985). *Unimap Interactive Mapping System*, Lyngby: European Software Contractors A/S.

UNWIN, A.R., WILLS, G. and HASLETT, J. (1990). REGARD-Graphical analysis of regional data. *ASA Proceedings of the Section in Statistical Graphics*, pp. 36–41.

WACKERNAGEL, H., PETITGAS, P. and TOUFFAIT, Y. (1989). Overview of methods for coregionalization analysis, in Armstrong, M. (Ed.), *Geostatistics, Vol. 1*, pp. 409–20, Dordrecht: Kluwer Academic Publishers.

WEBSTER, R. and OLIVER, M.A. (1989). Disjunctive kriging in agriculture, in Armstrong, M. (Ed.), *Geostatistics, Vol. 1*, pp. 421–32, Dordrecht: Kluwer Academic Publishers.

WEBSTER, R. and OLIVER, M.A. (1990). *Statistical Methods for Soil and Land Resource Survey*. Oxford: Oxford University Press.

WHITTLE, P. (1954). On stationary processes in the plane, *Biometrika*, **41**, 434–49.

WOOD, G., OLIVER, M.A. and WEBSTER, R. (1990). Estimating soil salinity by disjunctive kriging, *Soil Use and Management*, **6**, 97–104.

YATES, S.R. and YATES, M.V. (1988). Disjunctive kriging as an approach to management decision making, *Soil Science Society of America Journal*, **52**, 1554–8.

CHAPTER SIX

Interactive spatial analysis of soil attribute patterns using exploratory data analysis (EDA) and GIS

J. L. GUNNINK and P. A. BURROUGH

6.1 Introduction

When preparing data such as point observations of soil attributes for geostatistical interpolation to map characteristic patterns and to determine areas where values may exceed required thresholds, it is useful first to examine the data for spatial homogeneity, stationarity and normality. Exploratory Data Analysis (EDA) is not only useful in the pre-interpolation phase, but it is also helpful for validating the results. The quality of kriging predictions is largely determined by the choice of variogram model, which can be seriously affected by a few extreme attribute values if these are closely located in space. Dynamic interaction with the data, in which several views of the data such as histograms, scattergrams, maps and variogram clouds can be examined simultaneously, permits the relations between extreme data values or trend residuals and geographical location to be easily seen. This study was carried out using the REGARD software package developed at Trinity College Dublin, with modifications made by us. The use of EDA for detecting spatial anomalies that can affect geostatistical interpolation is demonstrated using data on polluted floodplain soils in The Netherlands.

6.2 Interpolation errors

The interpolation of patterns of nutrients or pollutants in soil and sediments (Leenaers *et al.*, 1990) is an application of increasing importance in GIS. Samples collected at point locations are analysed in the laboratory for levels of nutrients or pollutants: these data are then converted to maps to help managers decide if, and where, levels are sufficient or excessive. In the case of excessive levels of pollutants, there could be legal requirements for expensive clean ups or removal of material.

Because the data are sampled at point locations, interpolated maps are always an approximation. Measurement errors and short range variation contribute to local uncertainty, which can sometimes be very large indeed (Burrough, 1993). Also, the method used for interpolation contributes to the uncertainties in the predicted, mapped values. Interpolation errors may be systematic, resulting from the use of an inappropriate method, or they may be unbiased, but essentially unknown because the interpolation method assumes that the spatial variation is smooth, continuous and follows a standard spatial model. A frequently used example of a standard model for interpolation is the inverse distance weighting method, in which interpolation weights are computed simply as a function of distance between sample sites and the site at which the prediction has to be made (Burrough, 1986).

Deterministic interpolation methods, like inverse distance, make predictions but contain no information about the quality of the maps produced. Faced with a map showing values that exceed critical thresholds, decision makers cannot judge if these are artefacts of the data or the interpolation method, or if they represent areas that are truly in excess of legal norms. Without knowing the reliability of the interpolations, decision makers must make judgements on the basis of insufficient information. Clearly, when interpolated levels are in the region of legal limits, it could be useful to know what the chance is that a predicted level exceeds a given threshold, so that decisions about expensive clean up operations can be well-founded.

6.2.1 Geostatistical interpolation

Geostatistical methods of interpolation (known as kriging) are designed to overcome the problems of dealing with errors of interpolation (Journel and Huijbregts, 1978; Isaaks and Srivastava, 1989). The spatial variation of an attribute (called a *regionalised variable Z*) is modelled in terms of three components:

- a structural component with a constant mean or a constant trend
- a random, spatially correlated component, known as the variation of the regionalised variable
- random noise or a residual error term.

Let x be a position in one, two or three dimensions. Then the value of a random variable Z at x is given by $Z(x)$:

$$Z(x) = \mathrm{m}(x) + \epsilon'(\mathrm{x}) + \epsilon'' \qquad (6.1)$$

where $\mathrm{m}(x)$ is a deterministic function describing the 'structural' component of Z at x, $\epsilon'(x)$ is the term denoting the stochastic locally varying, normally distributed but spatially dependent residuals from $\mathrm{m}(x)$ – the regionalised variable – and ϵ'' is a residual independent Gaussian noise term, having zero mean and variance σ^2.

The variation of $\epsilon'(x)$ is assumed either to be second-order stationarity, or to obey the *intrinsic hypothesis* (Journel and Huijbregts, 1978). In second-order stationarity, sample points located close to each other are more alike than points further apart. The degree of similarity, or *spatial covariation* which is measured by the mean variance of pair-differences, does not decrease indefinitely, but levels off at a distance that is characteristic for the kind of spatial pattern. Beyond that distance, known as the *range*, average pair-differences do not increase and the mean and variance of the data do not change. Second-order stationarity assumes therefore, that while local variation can exist, over larger areas the spatial variation is homogeneous.

In many real situations, however, the variance of pair-differences does not level out at a characteristic range, but continues to increase with the size of the area sampled. The *intrinsic hypothesis* of geostatistics relaxes the conditions of second-order stationarity so that the mean value of the attribute does not vary over an area, but only the pair-differences are normally distributed. Further, the intrinsic hypothesis assumes that the magnitude of the variance of pair-differences only depends on the distance between the sample points, and not on their absolute location.

Formally, the intrinsic hypothesis states that (Cressie, 1993; Myers, 1989; Isaaks and Shrivasatava 1989):

(1) the expected difference between values at any two places separated by distance **h** is zero:

$$E[Z(x) - Z(x + \mathbf{h})] = 0 \text{ and} \tag{6.2}$$

(2) the variance of differences depends on the distance **h** and is given by:

$$\text{var}[Z(x) - Z(x + \mathbf{h})] = E[\{\epsilon'(x) - \epsilon'(x + \mathbf{h})\}^2] = 2\gamma(\mathbf{h}) \tag{6.3}$$

where $\gamma(\mathbf{h})$ is known as the semivariance. Eqs (6.2) and (6.3) also hold for second-order stationarity. The two conditions, stationarity of difference and variance of differences, define the requirements for the intrinsic hypothesis of regionalised variable theory. This means that once structural effects have been accounted for, the remaining variation is homogeneous in its variation so that differences between sites are a merely a function of the distance between them.

If the conditions specified by the intrinsic hypothesis are fulfilled, the semivariance can be estimated from sample data:

$$\gamma(\mathbf{h}) = \frac{1}{2n} \sum_{i=1}^{n} \{z(\mathbf{x}_i) - z(\mathbf{x}_i + \mathbf{h})\}^2 \tag{6.4}$$

where n is the number of pairs of sample points of observations of the values of attribute z separated by distance **h**, known as the *lag*. A plot of $\gamma(\mathbf{h})$ against h is known as the *experimental variogram*. The experimental variogram is the first step towards a quantitative description of the regionalised variation. The variogram provides useful information for interpolation, optimising sampling and determining spatial patterns, but it is first necessary to fit a theoretical variogram model through the experimentally observed points. Only certain kinds of model variogram functions can be used: the most common forms are the *transitive* circular, spherical, exponential or Gaussian models (all of which reach a maximum variance at a certain range) and the *non-transitive* linear model, which increases without limit (Journel and Huijbregts, 1978). Figure 6.1 shows two of the most commonly used variogram models.

The form of the model variogram is critical for the quality of the interpolation and its shape near the origin is often sensitive to the presence of large and small data values in close spatial proximity. The model variogram is a result of spatial averaging so it obscures regional variations within the area sampled and differences caused by local anomalies. Knowledge of both regional differences and local anomalies could be useful for interpreting the data or for improving estimates of prediction errors. Often juxtaposition of spatial extremes cannot be seen in the histogram of data values. Also different regions may have different spatial patterns and hence different variograms. Therefore, it is essential to examine data for spatial contiguity, boundaries and short-range differences before modelling variograms.

Given that the spatially dependent random variations are not swamped by uncorrelated noise, the fitted variogram can be used to determine the weights γ_i needed for local

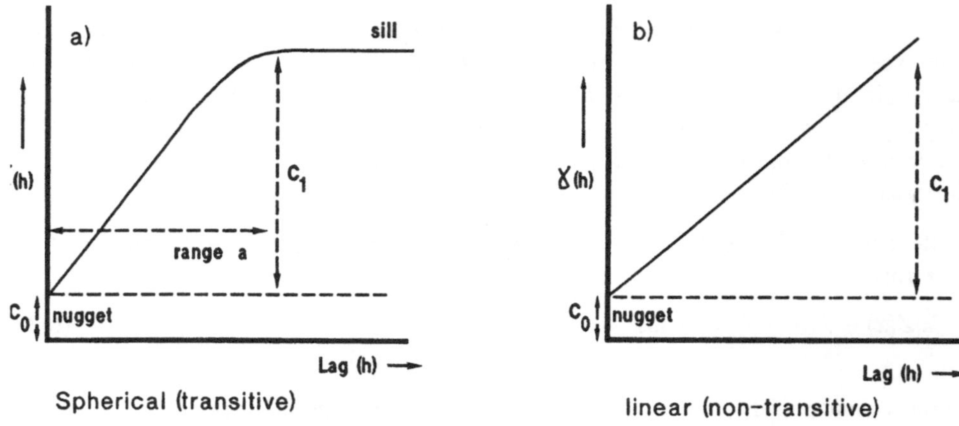

$$(h) = c_0 + \left\{ \frac{3h}{2a} - \tfrac{1}{2}(h/a)^3 \right\} \quad \text{for } 0 < h < a$$

$$(h) = c_0 + c_1 \quad \text{for } h > a$$

$$\gamma(h) = c_0 + bh$$

Figure 6.1 (a) Spherical and (b) linear variogram models used for interpolation.

interpolation. The procedure is similar to that used in weighted moving average interpolation except that the weights do not come from a deterministic spatial function, but from a geostatistical analysis of the way the pattern of the attribute Z varies. If the measured values of the regionalised variable Z are denoted by z_i, the value of z at unsampled locations is predicted by:

$$\hat{z}(\mathbf{x}_0) = \sum_{i=1}^{n} \lambda_i . z(\mathbf{x}_i) \tag{6.5}$$

with $\sum_{i=1}^{n} \lambda_i$ 1. The weights λ_i are chosen so that the estimate $\hat{z}(\mathbf{x}_0)$ is unbiased, and that the estimation variance σ_e^2 is less than for any other linear combination of the observed values.

The minimum variance of $\hat{z}(x_0)$ is given by:

$$\hat{\sigma}_e^2 = \sum_{i=1}^{n} \lambda_i \gamma(\mathbf{x}_i, \mathbf{x}_0) + \varphi \tag{6.6}$$

and is obtained when

$$\sum_{i=1}^{n} \lambda_i \gamma(\mathbf{x}_i, \mathbf{x}_j) + \varphi = \gamma(\mathbf{x}_j, \mathbf{x}_0) \text{ for all } j \tag{6.7}$$

The quantity $\gamma(\mathbf{x}_i, \mathbf{x}_j)$ is the semivariance of z between the sampling points $\mathbf{x}_i$ and $\mathbf{x}_j$; $\gamma(\mathbf{x}_i, \mathbf{x}_0)$ is the semivariance between the sampling point $\mathbf{x}_i$ and the unvisited point $\mathbf{x}_0$. Both these quantities are obtained from the fitted variogram and therefore are dependent on the way it was estimated and modelled, and on the underlying assumptions. The quantity φ is a Lagrange multiplier required for the minimisation. Kriging is an exact interpolator in the sense that when the equations given above are used, the interpolated values, or best local average, will coincide with the values at the data points. In mapping, values will be interpolated for points on a regular fine grid. The estimation error σ_e^2, can also be mapped to the grid to provide information about the reliability of the interpolated values over the area.

6.2.2 Summarising and testing the assumptions behind kriging

Although kriging is an optimal technique in the sense that all predictions are unbiased and have a minimum variance, the quality of the results depends on the degree with which the data match the assumptions of normality, stationarity and spatial homogeneity as embodied in the intrinsic hypothesis.

While assumptions of normality can easily be tested by conventional statistics, it is less easy to examine the data for departures from the intrinsic hypothesis. Indeed, because the data are used repeatedly to estimate semivariances over many lags, the setting up and testing of conventional hypotheses is complex, if not impossible. On the other hand, a visual examination of the spatial distribution of pair-differences could indicate departures from the intrinsic hypothesis that might provide new information. Such a visual examination could also be useful to see if the experimental variogram is being distorted by 'rogue' pairs of points – closely spaced data points having larger than usual pair-differences. The original data may not show up as outliers in a histogram, but only as a spatial deviation, or 'spatial outlier'.

Visual analysis of deviations from an ideal hypothesis might also be useful to see if an area contains more than one kind of spatial pattern. In many cases, the number of data points is so few (<100) that only one reliable experimental variogram can be computed for the whole area, irrespective of systematic differences in the kinds of patterns present. If two parts of the area had distinctly different patterns of variation, and therefore different 'true' variograms, a whole-area variogram would be an area-weighted average of both variograms, and therefore correct for neither. The average variogram would not compute optimal weights for interpolation because it would not correctly represent the patterns in the area. Consequently, it is sensible to test the assumption that the variation in the study area is homogeneous in the sense that there is only one kind of pattern present.

One way to test for non-stationarity in pattern over an area is to look for systematic differences in the size of the squared pair-differences $[z(x_i) - z(x_i + h)]^2$ with respect to location. This can be done by plotting each single value of $[z(x_i) - z(x_i + h)]^2$ against h to yield a figure called the *variogram cloud* (Chauvet, 1982). Ideally, the variogram cloud should vary as Chi-square at each lag, but if data from different patterns are mixed, then the cloud may show structure (Burrough, 1983). In particular, if two closely adjacent data points have large differences in attribute values, they can greatly affect the local form of the variogram cloud, even though neither value is a true outlier.

6.3 Exploratory data analysis and geostatistics

Using conventional statistical analysis it is not easy to see all at once if data are normally distributed, if the variance of pair-differences $[z(x_i) - z(x_i + h)]^2$ varies over an area, or if adjacent data points differ greatly in value. To do all these things requires the data to be displayed simultaneously in a map view, a histogram view and a scattergram view ($[z_i - z_{i+h}]^2$ versus h). These facilities are provided by the REGARD program, written for the Apple Macintosh by John Haslett and his co-workers (1990) at Trinity College Dublin to which we have added several modifications for computing objects such as the variogram cloud. The functionality of REGARD will become clear in the following section; here it is sufficient to state that the program allows multiple window viewing of complex data in which the views and the actions of the user can all be linked together.

REGARD is just one of a number of interactive, exploratory data analysis tools that are being developed for statistical packages and GIS (Walker and Moore, 1988; Cleveland and McGill, 1988; Velleman and Velleman, 1988; Cressie, 1993). Based on ideas by Tukey (1977), EDA and dynamic graphics offer new possibilities for exploring spatial relationships in the data, resulting in hypothesis generation, which then can be verified in a more formal way. A main problem with computerised EDA is that the printed page is a difficult medium for conveying the concepts of multi-window, dynamic, interactive graphics, because of the lack of interaction with the data. The figures presented in this chapter are snapshots from the computer screen and cannot show all the interactive aspects of the tool.

6.4 An example

6.4.1 Introduction

To demonstrate the use of EDA in geostatistics we use a data set consisting of 100 measurements of the concentration of zinc in sediments, deposited by the river Meuse in the southern part of The Netherlands, near the Belgian border. The area is about 5 km^2 and is almost totally inundated when the discharge of the river Meuse exceeds 1500–2000 m^3/s. Topsoil samples were collected at sites located by stratified random sampling, using information on elevation relative to the river and fluvial geomorphology, which describes the physical processes governing the deposition of sediment (Burrough *et al.*, 1996). The attributes measured are: easting and northing, zinc content (ppm), elevation with respect to the river, and the distance in metres from the sample site to the nearest point on the river bank.

Histogram analysis showed that zinc content and distance to the river were strongly log-normally distributed, so these attributes were transformed to their natural logarithms. All statistical analysis proceeded with the log-normally transformed zinc and distance data.

6.4.2 Preliminary analysis

To get an idea of the relationships in the data Figure 6.2 shows a map view of the data points and a histogram of the ln(Zn), together with scatterplots of lnZn versus relative elevation with respect to the river (RA_M) and lnZn versus distance to the river (lnDM). The river is shown as a broad black line in the map view.

In the histogram view, the observations with the higher zinc concentration have been selected with the mouse and the corresponding observations in the map view have been highlighted. These show a clear spatial pattern, being concentrated close to the river. The corresponding cases are also highlighted in the scatter plots, showing that the zinc content is not only correlated with distance to the river but is also correlated with relative elevation. Observations with large zinc content all occur at low elevation close to the river, indicating that both distance to the river and relative elevation are key variables for explaining the observed pattern of high zinc content.

The advantage of the linked window technique is obvious. Three different, linked views of the data – a map view, a histogram view and a scatterplot view – can be seen simultaneously: selecting observations in one view results in all corresponding cases in the other views being highlighted instantaneously.

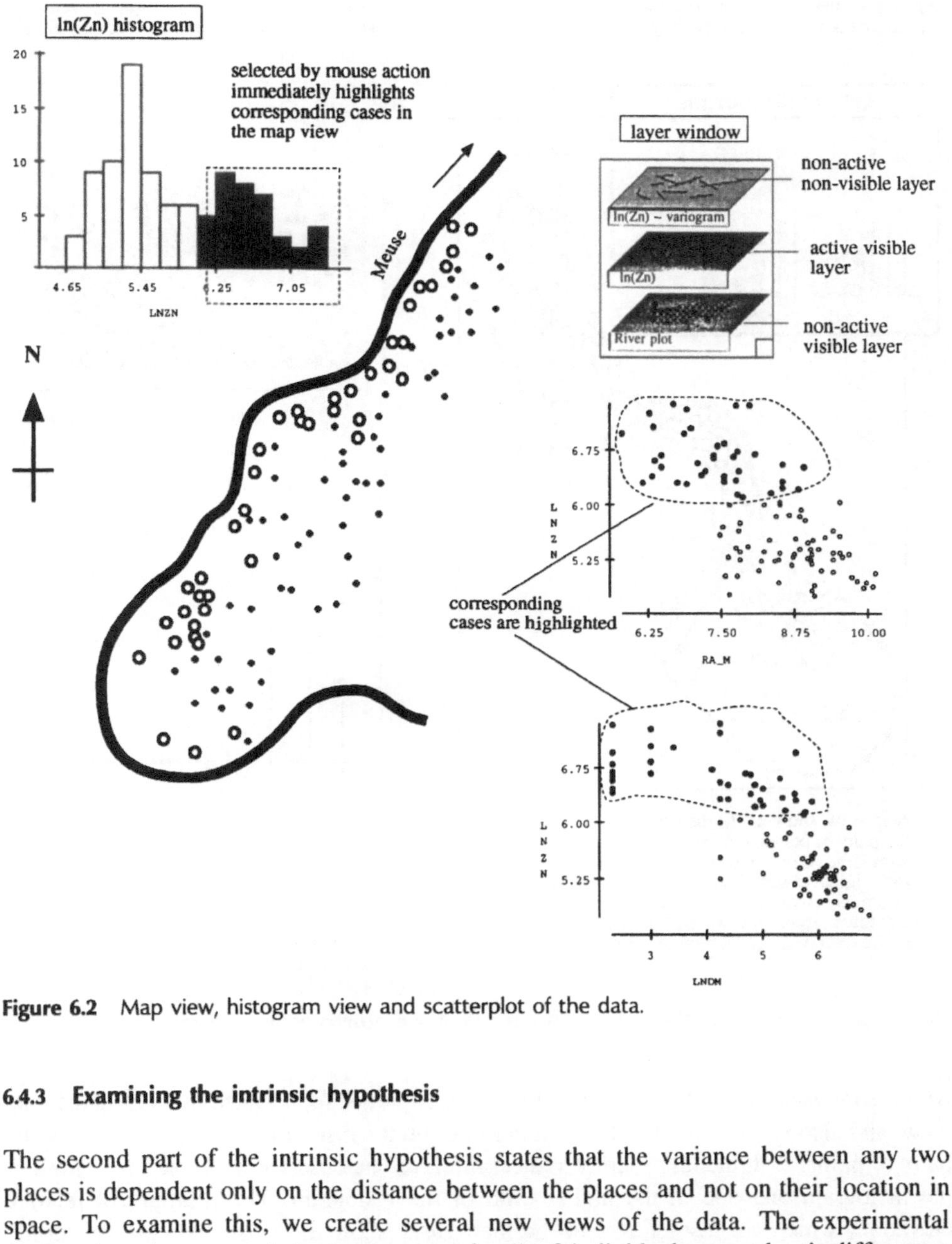

Figure 6.2 Map view, histogram view and scatterplot of the data.

6.4.3 Examining the intrinsic hypothesis

The second part of the intrinsic hypothesis states that the variance between any two places is dependent only on the distance between the places and not on their location in space. To examine this, we create several new views of the data. The experimental (average) variogram and the 'variogram cloud' of individual squared pair-differences were calculated. Lines carrying attributes of separation distance and semivariance connect each pair of points and these lines are used to visualise the spatial relationships between point pairs in the map view. Figure 6.3 shows the variogram cloud, together with the experimental variogram and the map view. Note that the experimental variogram shows the classic form of the spherical model and gives no hint of departures from theory.

We can see if the intrinsic hypothesis is valid by examining the spatial distribution of squared pair-differences over the area to see if particular classes of values are clustered in space. Selecting the largest semivariances for the first three lags in the variogram cloud

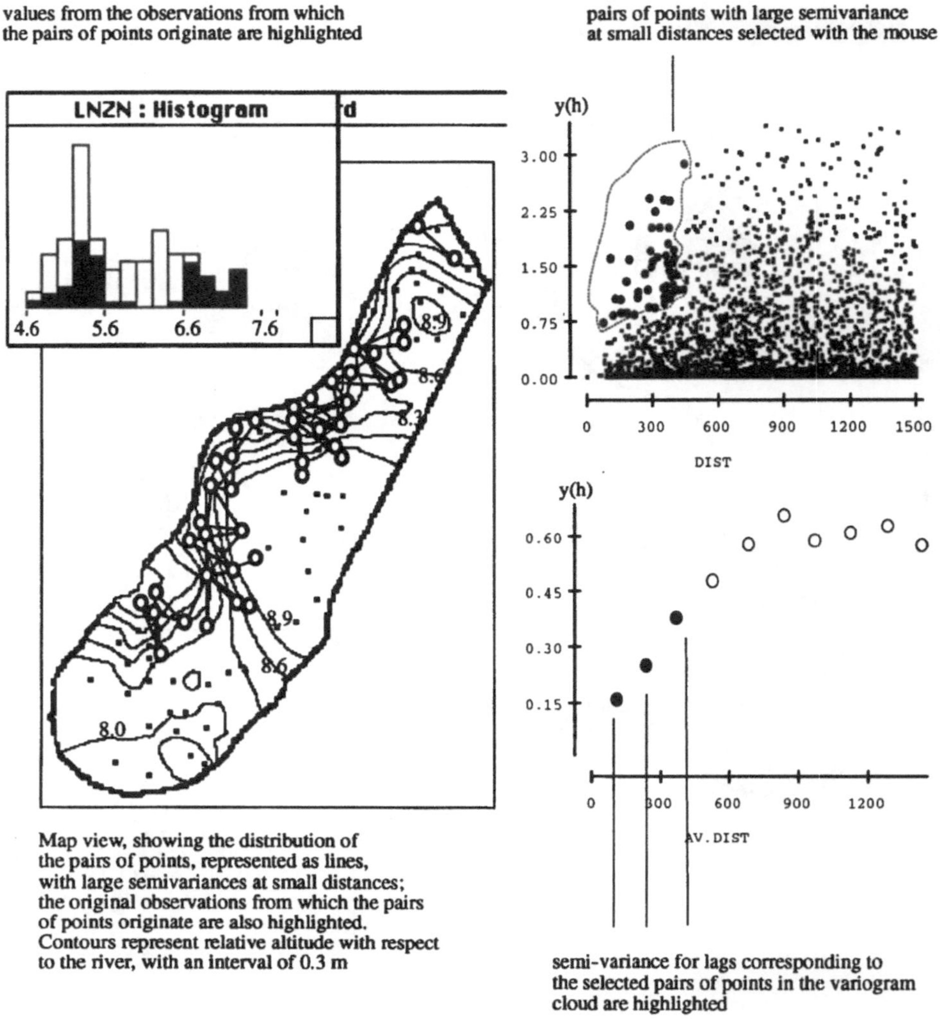

Figure 6.3 Map view together with variogram cloud and experimental variogram.

with the mouse displays the corresponding pairs of points, represented by lines in the map view, and also in the experimental variogram. From the histogram view it is clear that the pairs of points with the large semivariances all have one observation in the lower limb of the histogram and one in the higher limb of the histogram. The contours of relative elevation, which are also shown in the map view, show the reason for the clustering. There is a discontinuity in the mean and in the semivariance in the zone parallel to the river perpendicular to the steepest gradient in relative elevation.

The local nature of the anomaly can be examined further by using the method of *cross-validation*. In cross-validation, the attribute value at each original data point is predicted by ordinary kriging from the data points surrounding it, using the fitted variogram model. The goodness of fit between the original observations and the predicted values is described by the so-called Z-score for each observation:

$$Z - \mathrm{score}(x_i) = \{(z(x_i) - \hat{z}(x_i))/\sigma_i(x_i)\} \qquad (6.8)$$

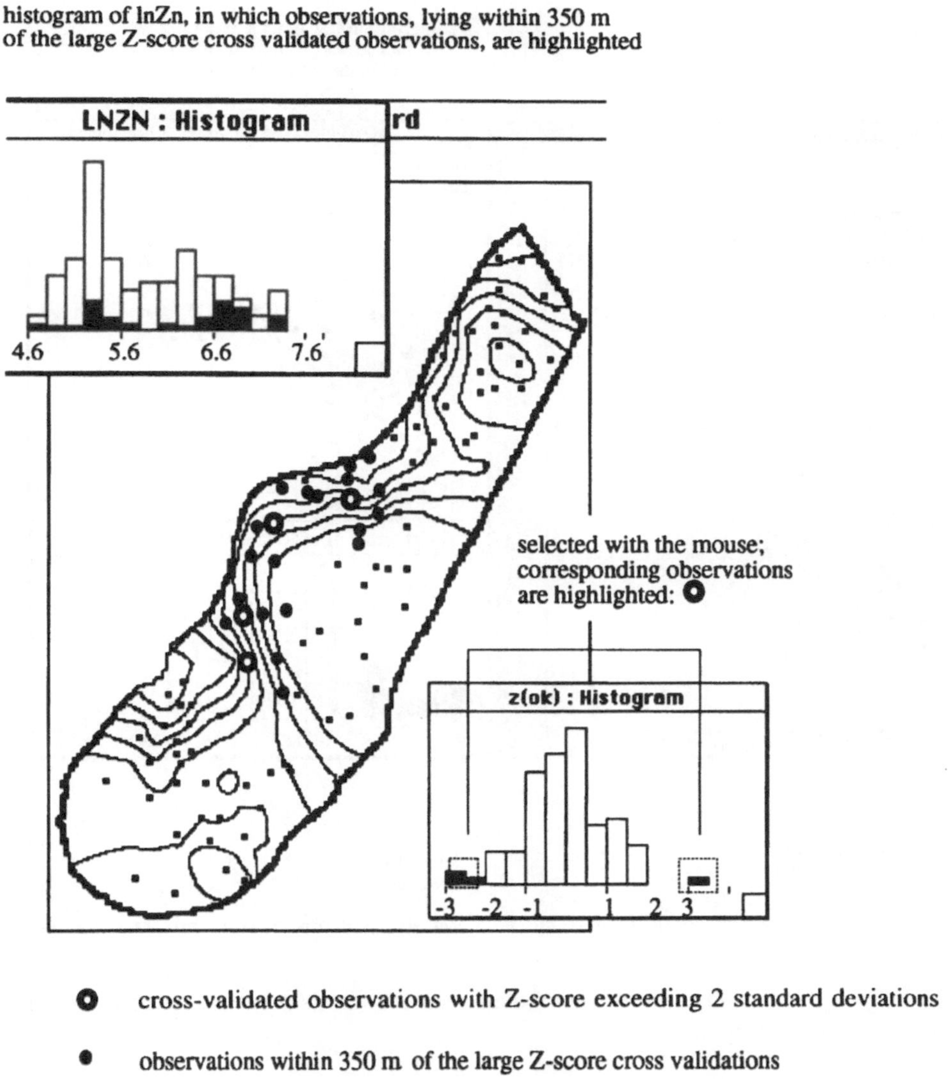

Figure 6.4 Map view and histograms of ln(Zn) and Z-scores.

where $\hat{z}(x_i)$ is the predicted value of the ith observation and $\sigma_i(x_i)$ is the standard deviation. The Z-score is a standardised value spread around a zero mean and is in units of standard deviations. The better the fit of the variogram and the closer the real pattern matches the assumptions, the smaller the distribution of Z-scores.

Figure 6.4 shows a histogram of the Z-scores: observations with scores exceeding standard deviations have been selected, highlighting the corresponding four cases in the map view. It can be seen that these most aberrant points are located in the same part of the area as before. Linking these points with all samples within 350 m highlights the observations which most influence the large deviations in the Z-scores. From both the map view and the histogram of the zinc content of the observations it is clear that the four extreme points are surrounded by observations belonging to either the higher or the lower limb of the histogram. Clearly, the main deviations from the hypothesis occur in a

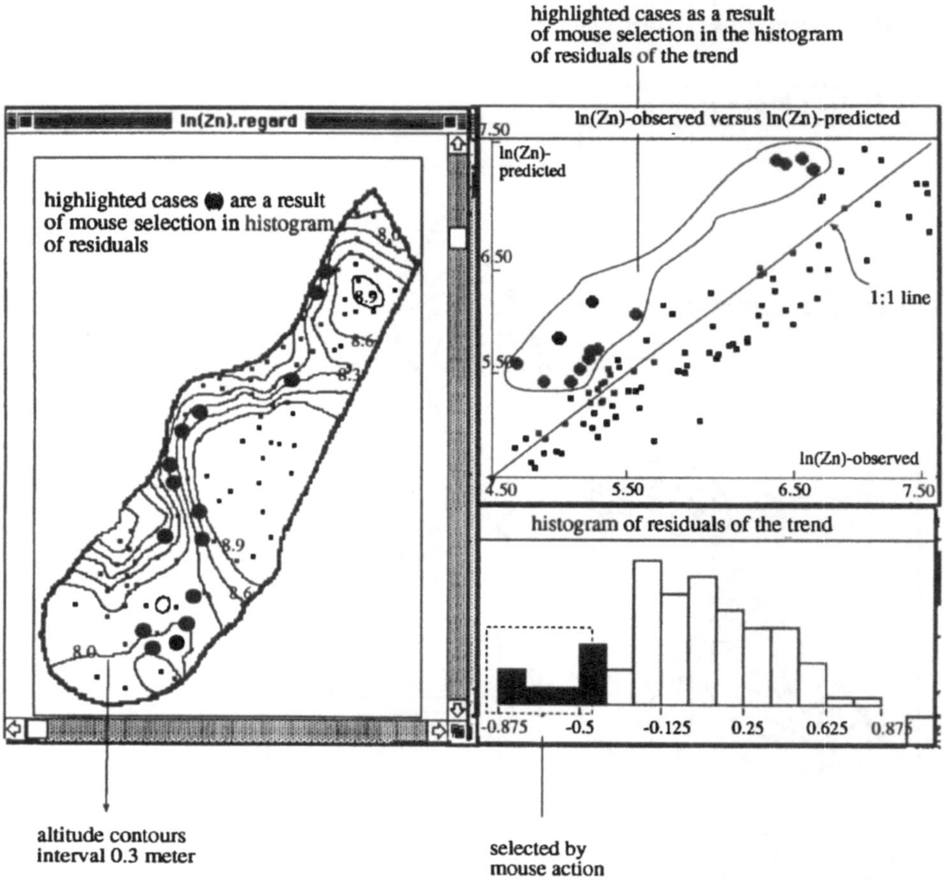

Figure 6.5 Map view, histogram of residuals and scatterplot of observed vs. predicted values of ln(Zn).

coherent part of the study area and interpolations made in this part will not be as optimal as straightforward kriging suggests.

6.4.4 Modelling spatial structure and analysing the residuals

Clearly, the assumptions of the intrinsic hypothesis are not valid for this data set. Not only is a clear trend (or *drift*) present, with large attribute values near the river and small values further away, but also the distribution in space of the variance is not homogeneous. A sensible approach is to try to model this structural variation as a trend, and then examine the residuals to see if they meet the assumptions.

The structural component in the data was described by a multiple linear regression model with ln(distance) and relative elevation as independent variables with ln(zinc) as dependent variable (Burrough *et al.*, 1994).

$$ln(zinc) = B_0 + B_1.ln(distance) + B_2.elevation + \epsilon \tag{6.9}$$

This model explained 75 per cent of the variation in the zinc content in the area. Residuals from the model were calculated for each data point and were analysed according to their spatial context. Because we are now interested in deviations from the

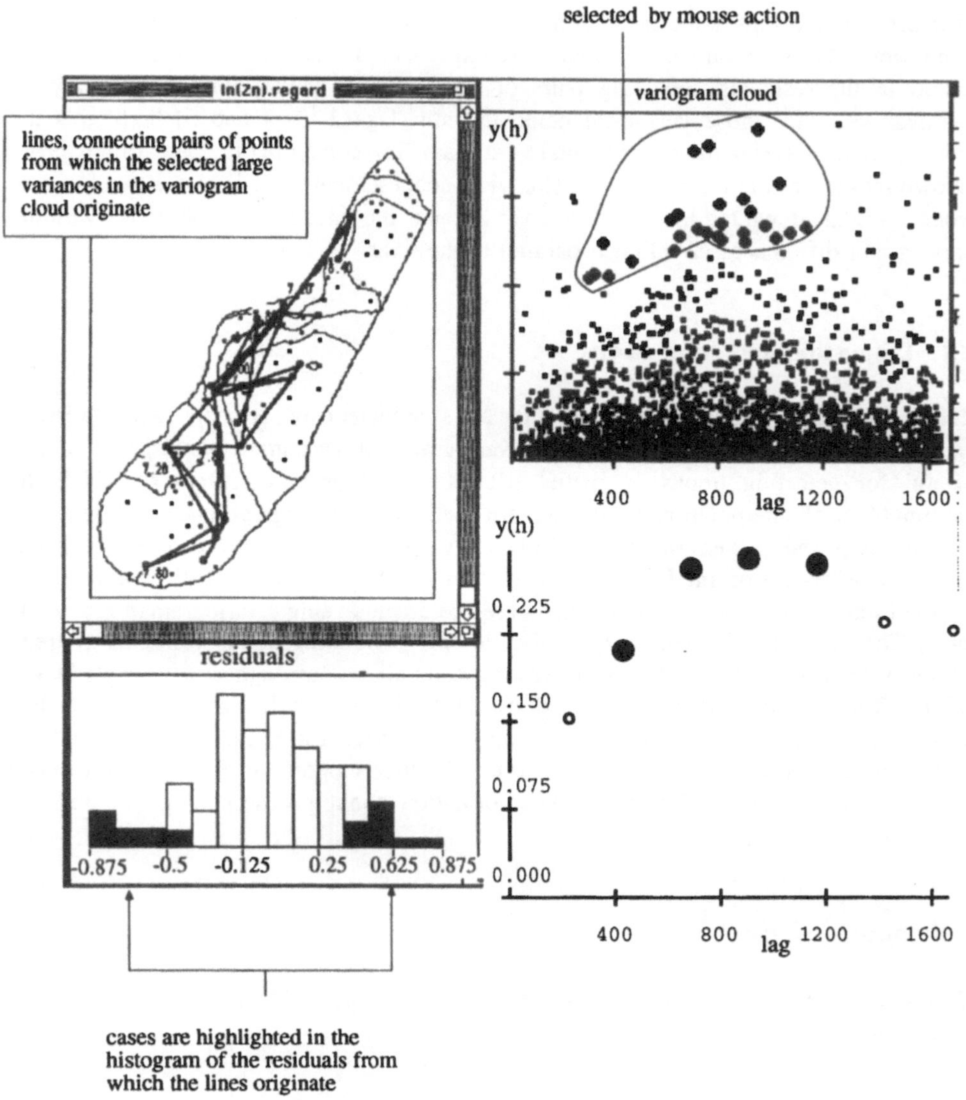

Figure 6.6 Map view, histogram of residuals from multiple regression, variogram cloud and the experimental variogram of residuals.

multiple regression model we focus on the spatial patterns of large negative and positive residuals to that model.

Figure 6.5 shows that selecting the largest negative residuals in the histogram view reveals a spatial concentration of points in zones that flank the upstream sides of locations near the river where relative elevation changes abruptly. From a geomorphological point of view the areas with large negative residuals seem to occur in an old river channel that is occasionally flooded. The highlighting of the same observations in the scattergram view clearly demonstrates that the predicted values of these points are systematically too high, which means that mapped zinc values in these areas would be overestimates.

The residuals from the multiple regression model can be analysed in the same way as the original data in Section 6.4.2 to see if the removal of the multivariate trend results in a data set that conforms to the assumptions of the intrinsic hypothesis. Figure 6.6 shows the map view, the histogram of residuals, the experimental variogram and the variogram cloud of the residuals. Selecting pairs of points in the variogram cloud with large semivariances for those lags influencing the sill (lags 2, 3, 4 and 5) highlights the corresponding cases in the experimental variogram. The corresponding pairs of points are shown by the lines in the map view. This view suggests that there is also a trend in zinc concentrations down the length of the river because the lines linking the points with the largest pair-differences are aligned parallel to the river.

6.5 Discussion and conclusions

Interactive EDA allows data to be viewed simultaneously as maps and statistical summaries like histograms, scatterplots, variograms or variogram clouds. The tool is useful for detecting trends in spatial data, for exploring the extent to which the assumptions of interpolation techniques are met, for analysing residuals and detecting irregularities and clusters of aberrant points. By repeating the analyses on the residuals of multiple regression models, information can be obtained about spatial patterns that were obscured by the averaging methods used to compute single variograms for a whole area. The procedure is essentially similar to that of looking at the residuals of trend surfaces to detect local, spatially interesting anomalies (Burrough *et al.*, 1977; Davis, 1986). These local patterns suggested by EDA should not be taken at face value, however, but should be used as indications of the need for further sampling and the setting up of hypotheses which can be tested. Further support for the patterns revealed by EDA can be obtained from 'soft' information about geomorphology or physical processes.

Acknowledgements

Thanks are due to M. Rikken and R. van Rijn for field work and laboratory analysis of the soil samples.

REFERENCES

BURROUGH, P.A. (1983). Multi-scale sources of spatial variation in soil. II. A non-Brownian Fractal model and its application to soil survey, *J. Soil Science*, **34**, 599–620.

BURROUGH, P.A. (1986). *Principles of Geographical Information Systems for Land Resources Assessment.* Oxford: Clarendon Press.

BURROUGH, P.A. (1991). Sampling designs for Quantifying Map Unit Composition, Chapter 7 in Wilding, L. and Mausbach, M. (Eds) *Spatial Variabilities of Soils and Landforms*, Soil Science Society of America Special Publication Number 28, pp. 89–125. Madison WI: SSSA.

BURROUGH, P.A. (1993). Soil variability: a late 20th century view, *Soils and Fertilizers*, **56**, 529–62.

BURROUGH, P.A., BROWN, L. and MORRIS, E.C. (1977). Variations in vegetation and soil pattern across the Hawkesbury Sandstone Plateau from Barren Grounds to Fitzroy Falls, N.S.W, *Australian J. Ecology*, **2**, 137–59.

BURROUGH, P.A., RIKKEN, M. and VAN RIJN, R. (1996). Spatial data quality and error analysis issues: GIS functions and environmental modeling, in Goodchild, M.F. *et al.* (Eds) *GIS and*

Environmental Modeling: Progress and Research Issues. Fort Collins: GIS World Books.

CHAUVET, P. (1982). The variogram cloud, *Proceedings 17th APCIM Symposium*, Denver, Colorado: Golden Colorado School of Mines.

CLEVELAND, W.S. and MCGILL, M.E. (Eds) (1988). *Dynamic Graphics for Statistics*. Pacific Grove, CA: Wadsworth & Brooks Cole.

CRESSIE, N.A.C. (1993). *Statistics for Spatial Data*, revised edition. New York: John Wiley.

DAVIS, J.C. (1986). *Statistics and Data Analysis in Geology*, 2nd edn. New York: John Wiley.

HASLETT, J., WILLS, G. and UNWIN, A. (1990). SPIDER – an interactive statistical tool for the analysis of spatially distributed data, *International Journal of Geographical Information Systems*, **4**(3), 285–96. (Note that SPIDER is now called REGARD).

ISAAKS, E.H. and SHRIVASATAVA, R.M. (1989). *An Introduction to Applied Geostatistics*. Oxford: Oxford University Press.

JOURNEL, A.G. and HUIJBREGTS, C.J. (1978). *Mining Geostatistics*. London: Academic Press.

LEENAERS, H., OKX, J.P. and BURROUGH, P.A. (1990). Employing elevation data for efficient mapping of soil pollution on floodplains, *Soil Use and Management*, **6**, 105–14.

MYERS. D.E. (1989). To be or not to be ... stationary? That is the question. *Mathematical Geology*, **21**(3), 347–62.

TUKEY, J.W. (1977). *Exploratory Data Analysis*. Reading, MA: Addison Wesley.

VELLEMAN, P.F. and VELLEMAN, A.Y. (1988). *DataDesk Professional*. Ithaca, NY: Northbrook Odesta Corporation.

WALKER, P.A. and MOORE, D.M. (1988). SIMPLE – An inductive modelling and mapping for spatially-orientated data, *International Journal of Geographical Information Systems*, **2**(4), 347–63.

CHAPTER SEVEN

Exploring spatio-temporal data

ANTONY UNWIN

7.1 Introduction

Exploratory Data Analysis (EDA) has always been carried out, but only really began to attain academic credibility with the publication of Tukey's (1977) book. Since then the lack of computing hardware that was powerful enough and the lack of computing software that was flexible enough have inhibited the development of EDA. Now we are at the stage when this excuse can no longer hold water and we have both the tools and the power to carry out effective EDA. Spatial data sets are prime candidates for exploratory analysis because they tend to have many variables, data are observational rather than experimental and neighbouring values are not independent. Several papers have shown the power of interactive graphics for spatial data (Nagel *et al.*, 1991; Unwin *et al.*, 1990, 1992).

Many data sets have a strong geographic component. The locations of mineral measurements are important to geologists, the locations of breeding sites are important to ornithologists, the locations of disease cases are important to epidemiologists. The data in those examples tend to be tied to point locations but there are many data sets in socio-economic applications where the data are summaries for regions and where it is equally important that the relative geographic locations be included in any analysis. Unemployment data, which are commonly reported by country, are usually available by large regions within the country and sometimes by much smaller areas as well. (It has been said that British politicians were particularly interested in these local data for their own constituencies just before the last UK general election.) Trade data mean little if not combined with considerable geographic and political information. Retail sales data can be difficult to interpret without the background geographic infrastructure.

Most of these data sets also have a time component. Unemployment, trade and sales are obvious examples but bird breeding patterns may change over time (possibly giving clues about environmental changes) as may disease patterns (where changes over time would be particularly useful in judging the relative importance of different causes). Up till now any time component in the data has barely been taken into account in spatial analyses and spatial information has barely been considered in analyses by time, for the statistical problems in each case are difficult enough on their own without adding the additional complexity of analysing spatio-temporal effects.

In general, analyses of spatio-temporal data should neglect neither component at the expense of the other because the interaction can be so important. Changes can take place over time in quite different ways at different locations and those locations may form an insightful geographic pattern. Data which appear to form a distinctive spatial pattern at one point in time may seem quite random at other times. Ideally, there should be ways of extracting information from the data sets easily and effectively. In practice, spatio-temporal data pose major problems as the following cases show.

7.2 Spatial data sets with a few long series

Typical data sets of this kind are unemployment data by regions, inflation rates by country, and weather or pollution data by weather or recording station. The tools needed to explore such data sets are discussed using a simple unemployment example where only one time series is used for each region. In practice it would be interesting to look at many time series for each region (unemployment by gender, by age, by occupation, other socio-economic series and so on) but in the best statistical traditions we shall just state that the approach generalises in an obvious way to multivariate analyses even though, as in the best statistical traditions, it doesn't.

German unemployment figures are published monthly by regions which in most cases are just the Bundesländer. Prior to unification there were nine regions (Hamburg and Bremen were combined with their surrounding Länder and North and South Bavaria reported separately) and data were made available for each month for each region from January 1977 to August 1990. The data are absolute numbers not rates.

There is clearly a geographic component to this data set and any analysis should look at whether there were differences between the regions (of course, there were), at what kinds of differences there were (in level, in trend, in seasonality or in special features) and at whether these differences formed a geographic pattern. As there was little spatial information and relatively more temporal information, the software Diamond Fast which provides tools for the interactive graphical analysis of time series was used to explore the data. Results were then reviewed in the light of their geographical location.

In general the series all followed the same trend over time. The large increase in unemployment in the years 1981 to 1983 was experienced by all the regions but in some there was evidence of a clear decline from that peak through the 1980s and in others not. This has been ascribed to the different industrial structures and policies of the Länder. Figure 7.1 shows two different displays resulting from overlaying the series for Berlin (which over that period was an isolated Western city in the middle of East Germany) on the series for total unemployment to assess differences. Amongst the eight Western regions there were striking differences in seasonality. The two Bavarian regions had very similar high seasonality while the other regions had much lower fluctuations. Figure 7.2 shows the two Bavarian series and one of the others.

The interactive tools employed were interrogation, editing, overlaying, aligning, zooming, rescaling, smoothing, and transformations.

Interrogation was necessary to identify the values and times of quirks in the series and to check whether common changes such as a change in trend happened simultaneously (overlaying and zooming were also applied in this connection). Interactive editing was used to correct some transcription errors (interrogation and zooming were relevant here too) and to examine sections of series in isolation (the term editing is used here to refer to detailed amendment of individual data points as well as to cutting or sectioning whole series).

Overlaying (that is, moving one series over one or more others) was used to

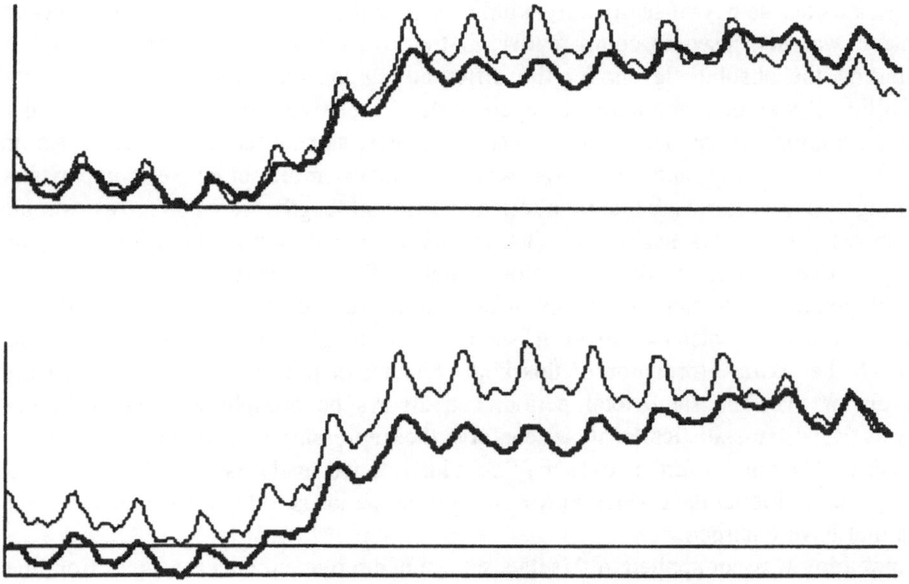

Figure 7.1 Berlin series (in bold) overlaid on Total Unemployment series (January 1977 to August 1990).

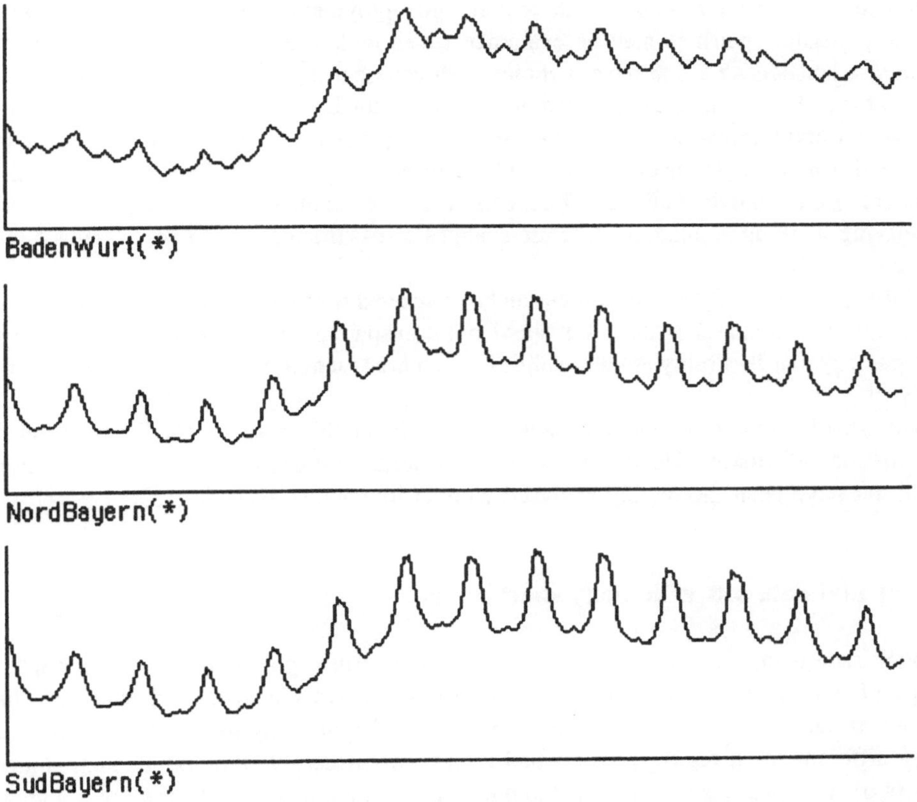

Figure 7.2 Seasonality in three regions (all series zeroed and to the same scale – January 1977 to August 1990).

investigate consistency of seasonality within series and similarity of seasonality between series. It was also used to compare overall trends. Rescaling was initially useful for comparing the absolute levels for the different regions and comparing their overall variability. It was then applied to compare series in relative terms. (Rescaling includes setting common origins and common vertical scales, standardising vertical scales and interactively growing and shrinking series to make more effective comparisons.) Aligning was used to ensure that time series of different lengths were correctly positioned (particularly after seasonality and change comparisons). Although all the regional unemployment series were the same length, figures were available for total unemployment in Western Germany which went back to the late 1940s and some financial series were also examined. It is one of the strengths of the interactive graphical approach that extra information of this kind, whether or not recorded at the same time intervals or over the same total period, can always be brought into an exploratory analysis. Reported statistical analyses can give the impression that once the analysis of a data set has begun no further data may be added. Good analyses will always consider looking for additional data which may cast light on the ideas to be investigated or on the ideas that have emerged.

Smoothing was not applied to this data set as unemployment data are, apart from their seasonality, already very smooth. Some experimentation with transformed series was carried out, looking at relative movements (one series divided by another) and shares of totals (each series divided by the sum of all). Relative movements have often proved insightful with unemployment data (there is a particularly interesting common pattern across countries in the ratio of female to male unemployment) but are not so useful when so many possible pairings may be examined (even with 9 series there are 36 pairs, 72 when you consider that you have a choice with each pairing which series to take as the numerator and which as the denominator). Shares of totals are interesting as they remove seasonality and trend which vary across all series equally (removing variable seasonality and trend from a single series is hard) but have to be interpreted carefully. Changes in the numbers of unemployed may be different from changes in the rates of unemployed as the underlying working population structure changes across the regions. In 14 years a lot can happen.

Corresponding analytic procedures can be envisaged for many of the above interactive tools, but by comparison they are slow, not transparent and very inflexible. Speed, transparency and flexibility are all qualities which are essential for successful exploratory analyses.

The small number of regions made it possible in this case to handle geographic information informally. On the other hand the large amount of temporal information could not have been thoroughly analysed without interactive analysis tools.

7.3 Spatial data sets with many short series

Annual trade data, sales data, constituency election results, pollution data by month for the past few months, annual vaccination rates for recent years and census data are all data sets where large numbers of variables may be recorded at many different points (or for many different small regions) for a small number of successive time points (sometimes only two). The methods described in the previous section are unsuitable as there is a great deal of spatial information but hardly any temporal information.

Eurostat have data on imports of all kinds into the EC countries and amongst the most important data are oil imports. Annual data were available for the years 1977 to 1990. This data set is typical in having a huge number of series (it is surprising just how many countries have exported oil to the EC at one time or another over the past 15 years) and substantial structural changes (the membership of the EC increased from 9 to 12 countries over the period with the inclusion of Greece, Portugal and Spain; there were major upheavals amongst the oil exporters, the Iran–Iraq war being the biggest of these). Not only are the time series too short for any kind of standard time series analysis but the changes that have taken place invalidate treating the data as equivalent anyway.

The interest in this data set is not so much in the absolute levels of oil flows into the EC from different countries as in the changes that have taken place in those flows over time. The shortness of the time series involved meant that tools primarily devised for the exploration of spatial data could be applied to good effect and the interactive software for exploring spatial data, REGARD, was used.

Initial editing was required to reduce the number of flows involved and any small flows were simply ignored. The decision as to what to call 'small' and identification of small flows was based on interactive exploration of the data, primarily ignoring the spatial component. For those not familiar with the EC's oil trade, most interest was probably in the differences, often historically linked, in supply partners for the various EC members. For experts, interest centred more on the changes that had taken place over the period in the sources and quantities of supplies. One notable feature was that the larger EC countries appeared to have lessened their dependence on individual suppliers by decreasing the share of any one supplier in their trade.

The data set was loaded in three ways. First, it was loaded using three layers with a regional layer having a simplified world map and the relevant countries loaded as regions, a point layer with one central point for each country and a line layer of flows between the centres in the point layer. Flow values (weight and value) were recorded for each year on each line in separate records. Second, it was loaded as in the first case but with all the flows on each line in a single record. Third, it was loaded in 14 line layers with the data for each year being given a different layer. It is surprisingly common in statistics that data sets can be looked at in different ways using different structures. The advantages of a single flexible structure are usually matched by disadvantages in directly accessing the data. At this stage, experience suggests that interactive graphics tools should have as direct access as possible. Viewing the data with each year in a separate layer had the advantage that years could be looked at singly (and the clutter of flowlines thereby substantially reduced) but the disadvantage that linked comparisons between years were not possible. That version is therefore not considered further.

The interactive tools employed were interrogation, linking (both within and between layers), subselection, zooming, animation and scatterplot 'refining'. Transformation facilities were also used extensively but are not available interactively in REGARD.

Interrogation was used to look at raw data on individual flows and to list sets of importers and/or exporters. It would have been useful to have a summarising interrogation facility to report on selected points, lines or regions as a group but although a prototype window of this kind has been designed for REGARD it has not yet been implemented. Linking was used extensively within the network layer to highlight the biggest flows, the flows with the biggest shares of trade and the biggest year on year changes. The first and last obviously only applied to the bigger EC countries while the shares gave insight into the buying patterns of the smaller countries as well.

Time was primarily included by plotting weights against year in a scatterplot. This gave a pattern of vertical parallel dot plots with most of the dots at the foot of each

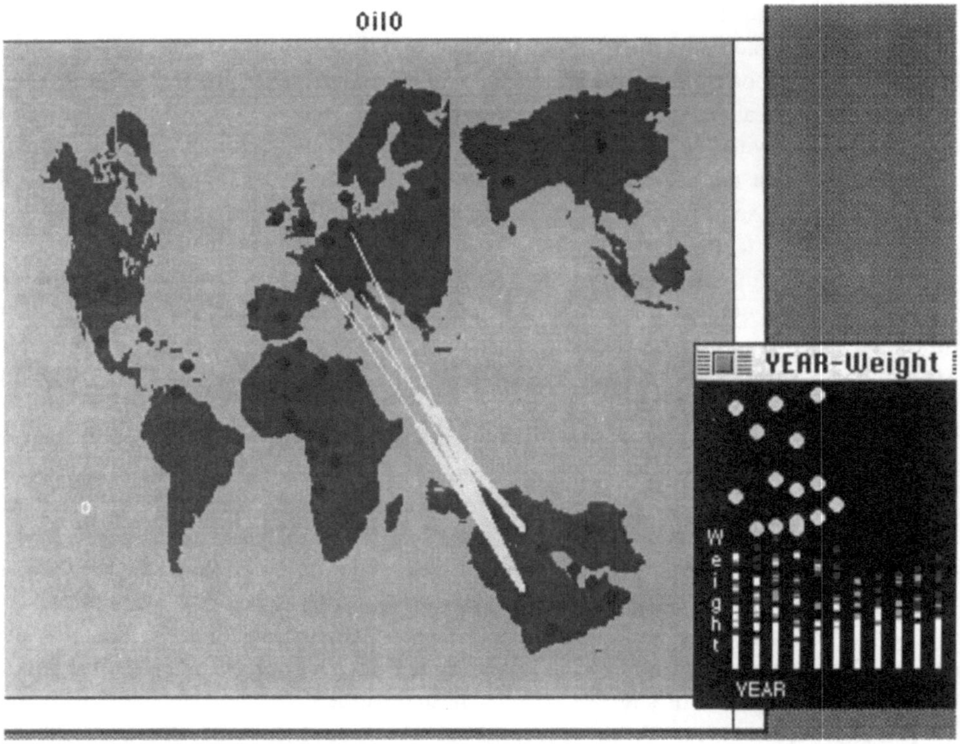

Figure 7.3 Largest EC oil imports by weight in the years 1977 to 1982.

column and the most interesting ones at the top (Figure 7.3). Linking between layers was then used to identify which countries were involved in the largest flows and, inversely, which flows were associated with which countries to identify changes in buying patterns over time. Interactive subselection was useful for selecting an importer and then restricting the resulting selection to only some of the relevant exporters.

Time was handled in an alternative way by using the second way of loading the data set. As the data were all in the same record it was possible to plot data from one year against data from another year in scatterplots and to plot data from three years together in a rotating plot. Linking box plots of the distribution in each year was also effective with this version of the data set as there was enough screen space to display the 14 box plots in parallel (a reduced version is shown in Figure 7.4). Highlighting groups of flows defined by importer or exporter or in other ways gave a sequence of highlighted box plots across the years showing the distributional changes of a group over time relative to box plots of all the data. Following a group over time is not the same as following individual changes and the parallel box plots gave another view of the data. This was noticeable in looking at all of the oil trade for a country when its trade with one other country had been unusual.

Zooming on the map was used to study areas such as the Gulf States where there are many important exporters together. It would have been useful to have been able to zoom into scatterplots as well, for the bulk of the data displayed was small values of little importance. The interactive tools that are available for scatterplots were, however, used extensively. Rather than plotting each point the same size, REGARD permits highlighted points to be drawn larger (just how much larger can be selected by the user). This is necessary in large data sets for picking out the highlighted points which can be difficult to see even with the change of colour. On top of that, REGARD allows the user interactively

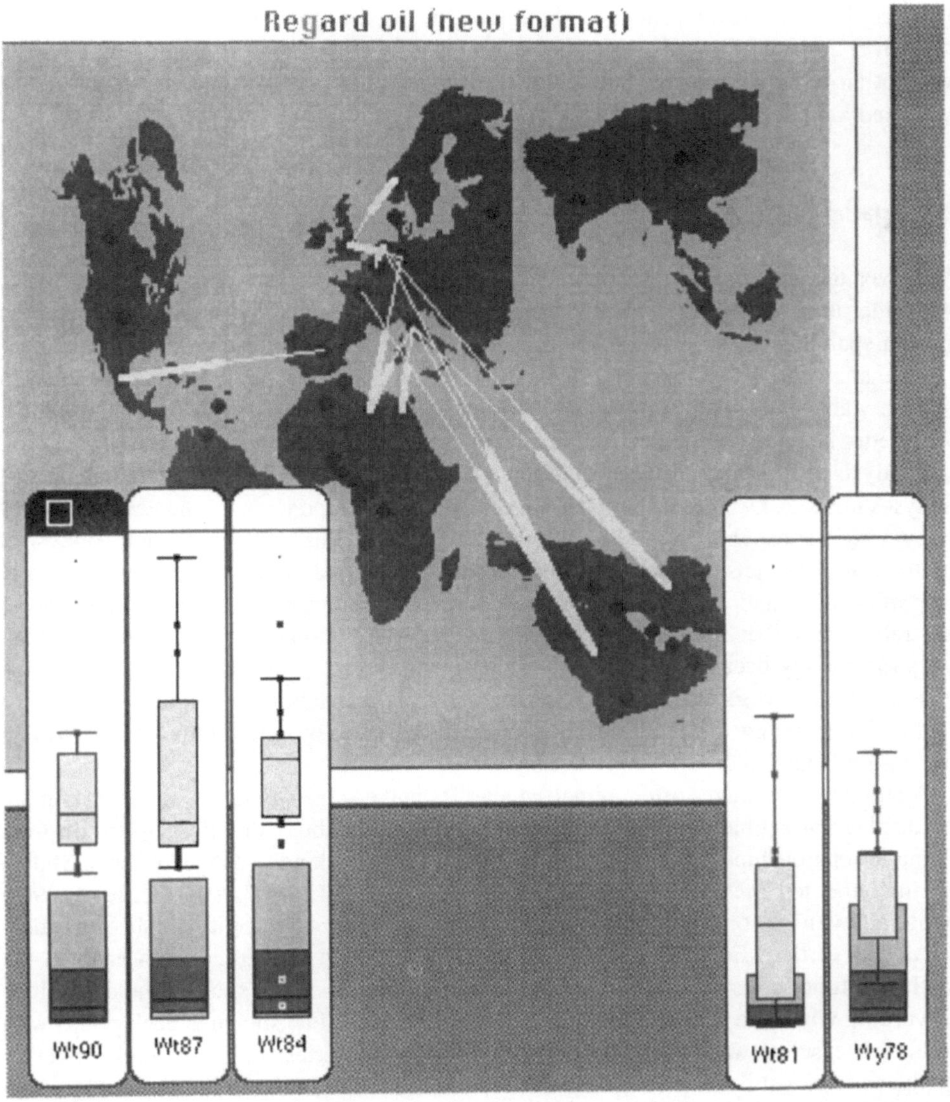

Figure 7.4 Box plots across years of oil imports by weight; flows which were big in both 1987 and 1990 have been selected.

to change the size of all points and uses a brightness addition scheme to decide how brightly to draw the resulting overlaps (the sensitivity of the scheme can be set by the user). Running through a range of different displays of the same scatterplot quickly is like trying out a range of different bivariate density estimates in which each one may give a different view of the data and convey a different idea. Scatterplots in REGARD have only two colours, one for highlighted points and one for highlighted points, though, as just outlined, the data are displayed with different brightness depending on the number of points overlapping. In some applications, most definitely with trade data, it would be helpful to have more colours so that groups could be distinguished. The difficulty is then to decide how to highlight. Data Desk uses different colours and then brightness to highlight – but how should overlapping points of different colours be combined?

Animation was used to run through the data year by year but, as a flow is shown if it exists at all and as flows from one side of the globe to the other (say, from the Phillipines to France) dominate flows between countries close by (say, from the UK to Ireland), this was not an effective tool for the data set. Network data sets are special in many other ways and animation is more interesting with point or regional data sets over time.

7.4 Spatial data sets with many long series

Whatever the success of the approaches applied in the two cases above in which there was either little spatial or little temporal information, the general case of large quantities of both spatial and temporal information will require other approaches.

Rouhani and Wackernagel (1990) used variograms and principal components to analyse water levels at a number of different sites in France. Their data set could have been better handled by interactive graphics tools as there were only some 16 sites (in fact, a reanalysis with Diamond Fast highlighted several issues which the more mathematical analyses missed). Using either variograms or principal components for the general case of many long series will never be computationally efficient and would not cope with multivariate data sets or data sets with discontinuities and series of different temporal support. Haslett and Cameron (1990) have used cross-correlations to analyse series of Australian rainfall at 30 locations. This worked well for the regularly recorded series they analysed, mainly because rainfall series can be considered to be stationary. Whether the approach would work well with far more series of less stationary character is open to doubt. Summarising the relationship between two series by one number alone will in general be inadequate.

A number of packages offer animation features and one possibility would be to animate the displays over time. Animation is often very attractive but can only help to illustrate dramatic relationships and is not good for identifying any kind of lagged effects. Sliders (as implemented in Data Desk (Velleman, 1992) and XLISP-STAT (Tierney, 1990)) would allow more control over the animation but much more flexibility would be needed.

As part of the REGARD research Graham Wills added a feature which enabled the display of time series by location. Even in its prototype form this showed the difficulties of working with time series spatially. Only one series could be shown at each point (with multivariate series we would wish to look at several), series for points close together displayed awkwardly and aligning the series spatially made comparisons by time difficult. Nevertheless some more flexible version of this seems to be one of the few possibilities with potential. Rather than displaying all series simultaneously in a single window it would be better only to show series for selected points and allow them to be grouped in different windows according to the groupings and comparisons one wants to make. Some initial exploration with cross-correlations might suggest initial groupings. This suggestion is practically equivalent to recommending that Diamond Fast and REGARD be run in parallel but with major additional organisational and tracking facilities. The power, flexibility and ease of use of those facilities would determine the value of the approach.

7.5 Conclusions

Many of the ideas presented in this chapter apply equally well to spatial data with a depth component such as occur with borehole measurements taken at different depths.

Graphical exploratory methods are even more essential in that situation as the measurements are unlikely to be taken at equal intervals (which always complicates analytic time series methods) and the borehole directions may not be vertical. If data are recorded down the same borehole at many different times we then have data with four-dimensional support and extra complications. No doubt cases with even higher dimensional support can be envisaged.

As with many areas of modern statistical research it is not the complications of dealing with well-defined special cases which cause difficulty with spatio-temporal data, it is the complexities of dealing with large amounts of data which are structured in a variety of different and interconnected ways. Providing software which goes some way to tackling these problems is not only important as a short-term solution to analysing them, it is also essential for the long-term development of better solutions. Users must be encouraged to apply the tools that are currently available to test them, to identify gaps and to suggest new tools. One of the most attractive features of interactive graphics research is the possibility of involving users directly because of the transparency of the ideas employed.

Exploratory Spatial Data Analysis is potentially extremely powerful but tools appropriate for many areas of application have still to be developed. Spatio-temporal data is a case in point. There are many new tools that could be incorporated into ESDA, probably stimulated or based on models developed out of spatial modelling.

7.6 Software

The exploratory time series software used in the first section was Diamond Fast (Unwin and Wills, 1988). The exploratory spatial analysis software used in the second section was REGARD (Wills *et al.*, 1991; Wills, 1992). Both these packages only run on Macintosh computers. Information on availability may be obtained from the author.

REFERENCES

HASLETT, J. and CAMERON, M. (1990) Geostatistical methods: recent developments and applications in surface and subsurface hydrology, in Bardossy, A. (Ed.), *Proceedings of International Workshop*, pp. 72–9, Karlsruhe, Germany.

NAGEL, M., HUBER, T.M., and HÖG, H. (1991). Hochinteraktive Graphik und ihre Anwendung auf Daten mit territorialem Bezug, in Faulbaum, F. (Ed.), *Advances in Statistical Software 3*, pp. 177–84, Stuttgart: Gustav Fischer Verlag.

ROUHANI, S., and WACKERNAGEL, H. (1990). Multivariate geostatistical approach to space–time data analysis, *Water Resources Research*, **26**(4), 585–91.

TIERNEY, L.(1990). *Lisp-Stat*, New York: John Wiley.

TUKEY, J. W. (1977). *Exploratory Data Analysis*. John Addison-Wesley: London.

UNWIN, A.R. and WILLS, G. (1988). Eyeballing time series. *Proceedings of 1988 ASA Statistical Computing Section* pp. 263–8.

UNWIN, A.R., SLOAN, B.J. and WILLS, G. (1992). Interactive graphical methods for trade flows, *New Techniques and Technologies for Statistics*, Office for Official Publications of the EC.

UNWIN, A.R., WILLS, G. and HASLETT, J. (1990). REGARD — Graphical Analysis of Regional Data, *ASA Proceedings of the Section on Statistical Graphics*, pp. 36–41.

VELLEMAN, P. F. (1992). *Data Desk*, Data Description: Ithaca, New York.

Wills, G. (1992). *Spatial Data: Exploration and Modelling via Distance-based and Interactive Graphics Methods*, Trinity College Dublin.

Wills, G., Unwin, A.R., and **Haslett, J.** (1991). Spatial interactive graphics applied to Irish socioeconomic data, in *Proceedings of the ASA Statistical Graphics Section*, pp. 37–41, ASA: Atlanta, Georgia.

CHAPTER EIGHT

The Moran scatterplot as an ESDA tool to assess local instability in spatial association

LUC ANSELIN

8.1 Introduction

As large spatial databases become increasingly available to researchers in the social and physical sciences, new tools are needed for the analysis of this information that match the sophistication in storage, retrieval and display provided by the rapidly evolving technology of geographic information systems (GIS). In many instances, the context is *data rich* but *theory poor* (Openshaw, 1991, 1993) and techniques are needed to 'let the data speak for themselves' (Gould, 1981), that is, to aid in discovering patterns, and to suggest potential relationships and hypotheses. A large battery of such methods now exists, following the pioneering ideas of Tukey (1977) on *exploratory data analysis* (EDA), which stress the interaction between the individual and the data by means of summarising displays, innovative graphics and other highly computational tools (see for example, the overview in Cleveland and McGill, 1988). EDA techniques such as box plots, Chernoff faces, Tukey star diagrams, and scatter-plot matrices are commonly used in studies that combine GIS and spatial analysis, for example, as illustrated in the applications of a so-called archaeologist's workbench in Farley *et al.* (1990) and Williams *et al.* (1990). However, such applications are *aspatial* in that they ignore the special characteristics of spatial data, such as spatial dependence and spatial heterogeneity (Anselin, 1990). As is well known, such properties will affect the validity of standard statistical techniques, and a special set of spatial statistical methods or spatial econometric methods are needed (for overviews, see Cliff and Ord, 1973, 1981; Anselin, 1988a; Cressie, 1991; Haining, 1990).

Methods of exploratory data analysis that take into account the spatial aspects of the data, that is, *exploratory spatial data analysis* (ESDA) are by no means as accepted as standard EDA tools, although they are often suggested as being an important part of the integration of spatial analysis and GIS (for example, Anselin and Getis, 1992; Bailey, 1992; Goodchild *et al.*, 1992; Fotheringham and Rogerson, 1993). An important component of such an ESDA is to measure the spatial association between observations for one or several variables. As argued in Anselin and Getis (1992) and illustrated in Anselin *et al.* (1993b), such measures can easily be incorporated in a framework that combines spatial analysis with a geographic information system. Most indices of spatial association, such as Moran's I and Geary's c spatial autocorrelation coefficients (Cliff and Ord, 1973, 1981), the variogram (Cressie, 1991), and generalised measures of spatial autocorrelation (Hubert, 1987) are global in nature. In other words, they indicate the presence or absence of a stable pattern of spatial dependence that is true for the whole data set. In practice, such a viewpoint may not be very realistic, especially when very large data sets are analysed. In these instances, the degree of spatial association between observations may show instability in the form of local non-stationarity, spatial regimes or spatial drift (for example, Anselin, 1990).

In this chapter, I suggest a simple tool to visualise and identify the degree of spatial instability in spatial association by means of Moran's I. It is based on the interpretation of this statistic as a regression coefficient in a bivariate regression of the spatially lagged variable (say, Wy) on the original variable (y). Such an interpretation readily allows for the use of a scatterplot for easy visualisation. This scatterplot may be used in isolation, in the traditional fashion, or may be integrated as an additional *view* on the data in a system of dynamic or interactive graphics, to allow for so-called scatterplot brushing (Monmonier, 1989; Haslett *et al.*, 1990, 1991; Unwin, 1993).

In the remainder of the chapter, I first briefly discuss the salient characteristics of techniques for exploratory *spatial* data analysis. Next, I review some methods that have been suggested to deal with local instability in spatial association. This is followed by an outline of the ideas behind the Moran scatter plot and a discussion of its properties and potential use. The technique is illustrated with an analysis of the spatial pattern of conflict between African countries.

8.2 Exploratory spatial data analysis

Broadly speaking, spatial data analysis can be defined as the statistical study of phenomena that manifest themselves in space. As a result, location, area, topology, spatial arrangement, distance and interaction become the focus of attention. This is well recognised in geography, for example, as expressed in Tobler's (1979) *First Law of Geography*, in which 'everything is related to everything else, but near things are more related than distant things'. In order to make this concept operational, observations must be referenced in space, that is, their locations must be specified as points, lines or areal units. The spatial referencing of observations is the salient feature of a GIS.

The important role of location for spatial data, both in terms of absolute location (co-ordinates in a space) as well as in terms of relative location (spatial arrangement, topology), has major implications for the way in which statistical analysis may be carried out. In fact, location leads to two different types of so-called spatial effects: *spatial dependence* and *spatial heterogeneity*. The former results directly from the First Law of Geography. This law will tend to result in observations that are spatially clustered, or, in other words, will yield samples of geographical data that will not be independent. From a geographical perspective, this spatial dependence is the rule rather than the exception, and it conflicts with the usual assumption of independent observations in statistics. The dependence in spatial data is often referred to as spatial autocorrelation (for a recent review from a non-geographer's perspective, see Legendre, 1993). The second, but equally important spatial effect is related to spatial (or regional) differentiation which follows from the intrinsic uniqueness of each location. Such spatial heterogeneity (or, non-stationarity) may be evidenced in spatial regimes for variables, functional forms or model coefficients (see Anselin, 1988a, Chapter 9, for a review, and, more recently, Dutilleul and Legendre, 1993).

Exploratory data analysis may be considered as *data-driven* analysis, in that it approaches the data without many preconceived ideas, theories or hypotheses. The focus is on generating insight into patterns and associations, and on describing the data by means of so-called resistant methods, that is, methods that are not (or are less) sensitive to *extreme* or *atypical* observations (for a more detailed discussion, see Tukey, 1977; Good, 1983; and, in a spatial context, Haining, 1990, Chapter 2). None of the traditional tools of EDA are especially geared to dealing with spatial data. Moreover, many EDA techniques suggested for the initial exploration of correlation between variables, such as scatter-plot matrices, or for post-model diagnostics, such as added variable plots, generate measures of fit and of significance that become invalid in the presence of spatial dependence, as pointed out in Anselin and Getis (1992).

Exploratory *spatial* data analysis (or, spatial exploratory data analysis) should focus explicitly on the spatial aspects of the data, in the sense of spatial dependence (spatial association) and spatial heterogeneity. In other words, these techniques should aim to describe spatial distributions, discover patterns of spatial association (spatial clustering), suggest different spatial regimes or other forms of spatial instability (non-stationarity), and identify atypical observations (outliers). In a general sense, all currently available indicators of spatial autocorrelation could thus be considered as part of ESDA. However, this is not very meaningful in terms of the link between ESDA and GIS. In fact, many of the *old* techniques of spatial data analysis were developed in an era of scarce computing power, small data sets and minimal computer graphics, and their current implementations take only limited advantage (if at all) of the data storage, retrieval and visualisation capabilities of a GIS. More specifically, such methods tend to summarise a complete spatial distribution into a single number, such as Moran's I coefficient of spatial autocorrelation (Moran, 1948). While this may have been useful in an analysis of small data sets, such as the classic 26 Irish counties in Cliff and Ord (1973), it is not very meaningful (or may even be misleading) in an analysis of spatial association in hundreds or thousands of spatial units. The degree of non-stationarity (spatial instability) in large spatial data sets is likely to be such that several regimes of spatial association would be present. For example, in an analysis of the Weimar elections in 1930 in Germany, O'Loughlin *et al.* (1994) found that a highly significant Moran's I at the level of 921 electoral districts in effect hides several distinct local patterns of spatial clustering and complete spatial randomness. Therefore, the sole emphasis on *global* measures of spatial association as the type of spatial statistics needed in a GIS (for example, as in Griffith, 1993) is misplaced, even though the computation of such a statistic may be implemented with currently available GIS software in a fairly straightforward manner (for example, Ding and Fotheringham, 1992). Instead, the focus of ESDA techniques used in conjunction with a GIS should be on measuring and displaying *local* patterns of spatial association, on indicating local non-stationarity, on discovering *islands* of spatial heterogeneity and so on. A few methods that have been suggested to accomplish this goal are reviewed next.

8.3 Local instability in spatial association

Measures of spatial association can be broadly classified into two groups, based on the way in which spatial interaction is conceptualised. In one approach, more commonly found in geography, the interaction is seen as a covariation between neighbouring observations. I will refer to this as the *neighbourhood view* of spatial association. Neighbours are typically defined as spatial units that have a common boundary or that are within a given critical distance of each other, although more complex definitions are possible as well (see Anselin, 1988a, for a review). The neighbourhood or contiguity structure of a data set is formalised in a spatial weights matrix W, with elements $w_{ij} = 0$ when i and j are not neighbours, and non-zero otherwise (typically, w_{ii} is assumed to be zero). In a general sense, in the neighbourhood view of spatial association, indicators are computed based on functions of the values observed at each location and the weighted average (or, spatial lag, Wy) of observations at neighbouring locations. In other words, these measures tend to deal with covariation or correlation between neighbouring values, but no interaction occurs with locations further away, that is, interaction takes the form of a step function. In the other approach, based on geostatistics, the spatial interaction is conceptualised as a continuous function of a distance metric. I will refer to this as the *distance view* of spatial association. The indicator of choice is the variogram or semi-variogram, which is based on the (squared) difference between values observed at a given distance apart (for a detailed overview, see Cressie, 1991).

The indicators of spatial association from either view that are most relevant for an exploratory approach to spatial data analysis are those that show local patterns and allow for local instabilities. Four particular strands of research are interesting in this respect. I will briefly review them next (for a more extensive review, see Anselin, 1994).

8.3.1 Indicators based on the neighbourhood view of spatial association

8.3.1.1 *G statistics*

In a recent article, Getis and Ord (1992) suggest two statistics to measure the degree of local spatial association for each observation in a data set. Their G_i and G_i^* statistics consist of the ratio of the sum of values in neighbouring locations, defined by a given distance band, to the sum over all observations (excluding the value at i for the G_i statistics, but including it for the G_i^* statistic). This statistic may be computed for many different distance bands, for example, as $G_i^*(d) = \sum_j w_{ij}(d)y_j / \sum_j y_j$, where $w_{ij}(d)$ is a binary matrix with $w_{ij} = 1$ when i and j are within a distance d from each other and zero otherwise. Getis and Ord derive the moments for the G_i and G_i^* statistics under the assumption of normality, which allows the indication of significant local spatial association for each observation. The G_i and G_i^* statistics can be easily implemented and visualised in an integrated GIS-ESDA framework, as illustrated in Ding and Fotheringham (1992) and Anselin *et al.* (1993b).

These statistics are particularly useful in the detection of potential non-stationarities, for example, when the spatial clustering of like values is concentrated in one subregion of the data. Their interpretation differs from that of other measures of spatial association (such as Moran's I) in that positive association means clustering of high values and negative association clustering of low values (and not the contiguity of opposite magnitudes). A slightly different form was recently suggested in Ord and Getis (1995), where the distributional characteristics are discussed in detail. An extension of the idea behind the G_i and G_i^* statistics to a general class of local indicators of spatial association (LISA) is presented in Anselin (1995).

8.3.1.2 *Geographical analysis machines*

In the various *geographical analysis machines* developed by Openshaw and associates (for example, as described in Openshaw, 1993; Openshaw *et al.*, 1990, 1991), the focus is on the efficient and automatic search for patterns in a spatial data base, with little interaction with the user and limited capability in terms of visualisation or statistical inference. The search for indications of association is based on computationally intensive algorithms for pattern recognition, such as neural networks, and is applied to spatial, space–time as well as multivariate association. This approach is particularly well suited to the indication of so-called *hot spots* or spatial clusters, although the extent to which such clusters are truly significant is sometimes unclear.

8.3.2 Indicators based on the distance view of spatial association

8.3.2.1 Pocket plot

The pocket plot is a device suggested in Cressie (1991) as a way to identify local *pockets* of non-stationarity in the variogram. When observations are given on a regular grid or lattice, the residual contribution of each row or column to the variogram can be computed for different lags. For each row or column, the distribution of these residuals can be described by a box plot, which indicates whether the central tendency is different from zero (which is the expected value) and also allows outliers (that is, distance lags for which the residual contribution of the row or column is extreme) to be identified. In a sense, they are the counterpart of the local indicators of spatial association in the neighbourhood view, and can be readily visualised in a linked map.

8.3.2.2 Interactive spatial graphics

Though not specifically intended to measure local spatial association, the interactive dynamic graphics tools developed by Haslett (1993) (Haslett *et al.*, 1990, 1991; Unwin, 1993), include the variogram (in the form of a semi-variogram or variogram cloud) as an additional *view* of the data, in addition to more traditional views, such as a histogram and a map. This adds a measure of spatial association to the otherwise mostly descriptive statistics and also allows the assessment of the extent to which particular locations (or, rather, pairs of locations) drive the overall measure of association. In other words, their approach, as implemented in the SPIDER-REGARD software packages allows for the combination of indicators of spatial association and spatial heterogeneity (non-stationarity) with a map view and non-spatial descriptive statistics, in a highly visual and interactive manner.

8.4 The Moran scatterplot

8.4.1 Principle

Moran's well known I statistic (Moran, 1948; Cliff and Ord, 1971, 1981) gives a formal indication of the degree of linear association between a vector of observed values y and a weighted average of the neighbouring values, or spatial lag, Wy. The linear association between y and Wy underlies the specification of spatial autoregressive processes, which are typically used to express the generating mechanism behind the spatial dependence. Formally, Moran's I can be expressed in matrix notation as:

$$I = (N/S_0)y'Wy/y'y$$

where N stands for the number of observations, S_0 is the sum of all elements in the spatial weights matrix ($S_0 = \sum_i \sum_j w_{ij}$), y are the observations in deviations from the mean, and Wy is the associated spatial lag. When the spatial weights matrix is row-standardized such that the elements in each row sum to 1, this expression simplifies to:

$$I = y'Wy/y'y$$

since in this case, $S_0 = N$.

Since the y are in deviations from their mean, I is formally equivalent to a regression coefficient in a regression of Wy on y (but not of y on Wy, which would be a more natural way to specify the spatial process). The interpretation of Moran's I as a regression coefficient provides a way to visualise the linear association between y and Wy in the form of a bivariate scatterplot of Wy against y (and not of y against Wy, which would be the usual form). I will refer to this as a *Moran scatterplot*. The Moran scatterplot can be augmented with a linear regression (as a linear smoother of the scatterplot) which has Moran's I as slope, and which can be used to indicate the degree of fit, the presence of outliers, of leverage points, and so on, in the usual fashion. It is important to note that the regression of Wy on y conforms to all the classical assumptions in regression analysis, and thus can be subjected to all the standard diagnostics for model fit (for example, Belsley *et al.*, 1980). The slope in this regression is a legitimate estimate for Moran's I, but its significance (using the standard t-test for the regression) is not appropriate. The interpretation of Moran's I in this manner clearly illustrates the way in which the statistic summarises the overall pattern of linear association, in the sense that a lack of fit would indicate important local pockets of non-stationarity.

The interpretation of Moran's I as a bivariate regression coefficient is perfectly general, and in fact applies to any statistic that can be expressed as a ratio of a quadratic form and its sum of squares. An example of this is the familiar Durbin–Watson statistic for serial correlation in time series, which takes the form $e'Ae/e'e$, that is, the coefficient in a regression of Ae on e. In spatial analysis, the same approach can be taken for Moran's I on regression residuals and the Lagrange multiplier statistics for spatial dependence in Anselin (1988b).

8.4.2 Implementation

The implementation of a Moran scatterplot is straightforward, since most statistical and many GIS software packages include a scatterplot function and an associated linear regression smoother and indication of fit. The only complicating factor is the construction of the spatial lag, Wy. In order to accomplish this, the information on the spatial arrangement of the observations, for example, as contained in a GIS, must be taken to construct a spatial weights matrix. A number of approaches to carry this out with current software are outlined in Anselin *et al.* (1993a). Once a spatial weights matrix is available, a spatially lagged variable can be computed easily (Anselin, 1992; Anselin and Hudak, 1992).

8.4.3 Interpretation

An effective interpretation of a Moran scatterplot should centre on the extent to which the linear regression line reflects the overall pattern of association between Wy and y. In other words, the indication of observations that do not follow the overall trend represents useful information on local instability or non-stationarity. Three aspects in particular merit some attention.

8.4.3.1 *Pockets of positive and negative association*

Since the variables are taken as deviations from their means, the scatter plot is centred on 0,0. The four quadrants in the scatter-plot box thus represent different types of association between the value at a given location (y_i) and its spatial lag, that is, the weighted average of the values in the surrounding locations (wy_i). The upper right and lower left quadrants represent positive spatial association, in the sense that a location is surrounded by similar valued locations. For the upper right this is association between high values (above the mean), while for the lower left quadrant this is association between low values (below the mean). Note that these two quadrants correspond to the notions of positive (high–high) and negative (low–low) spatial association of the Getis–Ord (1992) statistic. In other words, an examination of the relative densities of these two quadrants provides an indication of the extent to which the global measure of spatial association is determined by (dominated by) patterns of association between high or low values, similar to a pattern of significant positive and negative G_i^* statistics. Clearly, the substantive interpretation of such a pattern should be of interest, but it may also indicate a poor choice of the spatial weights matrix.

The upper left and lower right quadrants correspond to negative association, that is, low values are surrounded by high values (upper left) and high values are surrounded by low values (lower right). Again, the relative densities of these quadrants indicate which of these patterns of negative spatial association (in the traditional sense) dominate.

It is highly unlikely that a positive (negative) Moran's I is obtained by observations that are only in the lower left and upper right (upper left and lower right) quadrants. However, it is important to note the extent of *deviant* association and the degree to which these points influence the slope of the regression line (Moran's I). In some instances, a considerable mix of the two types of association for a given Moran's I may indicate the presence of different spatial regimes or local non-stationarity. It also indicates that the global indicator of spatial association may be a poor measure of the actual dependence in the process at hand.

8.4.3.2 *Outliers and leverage points*

Points in the scatterplot that are *extreme* with respect to the central tendency reflected by the regression slope may be outliers in the sense that they do not follow the same process of spatial dependence as the bulk of the other observations. They could thus be considered pockets of local non-stationarity, especially if they correspond to spatially contiguous locations or boundary points. The presence of outliers may also point to problems with the specification of the spatial weights matrix or with the spatial scale at which the observations are recorded. An intuitive indication of outliers can be based on the normalised residuals from the regression of Wy on y.

Similarly, observations that exert a large influence or leverage on the regression slope are of interest, again, particularly if they are spatially clustered or correspond to boundary points. The latter case provides a way to assess the influence of boundary values on the global measure of spatial association. A number of measures of leverage or influence, such as the diagonal elements of the hat matrix of Hoaglin and Welsch (1978), and Cook's (1977) measure of influence have been suggested in the literature and most statistical packages contain ways to implement them.

8.4.3.3 *Spatial regimes*

Useful insight into the extent to which the linear regression is a proper approximation to the pattern of spatial dependence in the data may be given by a robust local regression or scatterplot smoother, such as a LOWESS (locally weighted scatterplot smoother, Cleveland, 1979). A distinct non-linearity, alternating patterns of positive and negative association, or clearly different slopes in the smoother all indicate the inappropriateness of a single global measure for the spatial association in the data. When the distinct slopes may be associated with spatially clustered observations, they may indicate the presence of different spatial regimes, or spatial heterogeneity.

8.5 Illustration: spatial patterns of conflict in Africa

A geographical perspective has been increasingly applied in recent years to the analysis of international interactions in general, and international conflict in particular (for a review, see, for example, the collection of papers in Ward, 1992, and in particular Diehl, 1992). Measures of spatial association, such as Moran's I, have been applied to quantitative indices for various types of conflicts and co-operation between nation states, such as those contained in the COPDAB database (Azar, 1980). For such indices of international conflict and co-operation, both O'Loughlin (1986) and Kirby and Ward (1987) found significant patterns of spatial association indicated by Moran's I. The importance of spatial effects in the statistical analysis of conflict and co-operation was confirmed in a study of the interactions between 42 African nations, over the period 1966–78, reported in a series of papers by O'Loughlin and Anselin (1991, 1992) and Anselin and O'Loughlin (1990, 1992). For an index of total conflict in particular, there was strong evidence of both positive spatial autocorrelation (as indicated by Moran's I and the estimates in a mixed regressive, spatial autoregressive model) as well as spatial heterogeneity in the form of two distinct spatial regimes (as indicated by Getis–Ord G_i^* statistics and the results of a spatial Chow test on the stability of regression coefficients). This phenomenon is thus particularly suited for an application of the Moran scatterplot as an exploratory device.

The spatial pattern of the index for total conflict is illustrated in the quintile map in Figure 8.1 (for details on the data sources and the substantive interpretation, see O'Loughlin and Anselin, 1992, and Anselin and O'Loughlin, 1992). The suggestion of spatial clustering that follows from a visual inspection of this map is confirmed by a strong positive and significant Moran's I of 0.555, with an associated standard normal z-value of 6.99 (all computations were carried out with the *SpaceStat* software for spatial data analysis, Anselin, 1992). This statistic is computed for a spatial weights matrix based on distance contiguity, using the smallest distance cut-off such that each country has at least one neighbour (the distance cut-off is different from the one used in Anselin and O'Loughlin, 1992, hence the slightly different results; it roughly equals the distance between the centroids of Egypt and Sudan).

The countries with significant values for the Getis–Ord G_i^* statistic (using a significance level of $p = 0.01$) are depicted in Figure 8.2. The darker shade corresponds with strong positive spatial association for Egypt, Sudan, Ethiopia, and Somalia, indicating a spatial cluster of nations in Northeast Africa with high conflict indices. The lighter shade on the map corresponds with strong negative association, and again results in a spatial cluster of nations, but now with low conflict indices and in West Africa:

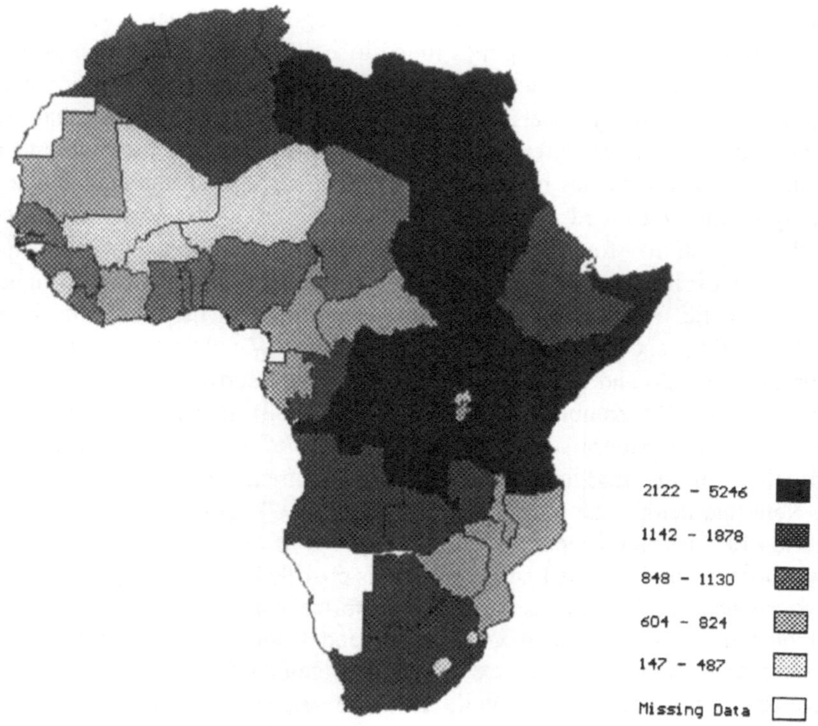

Figure 8.1 Total conflict index, Africa 1966–78.

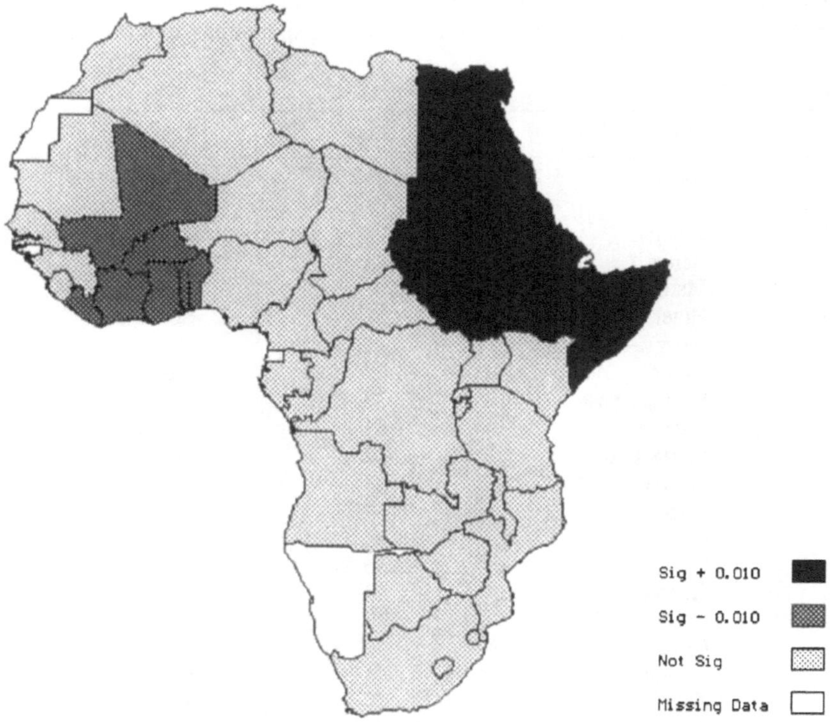

Figure 8.2 Significant G_i^* statistics for Conflict Index ($p = 0.01$).

Mali, Burkina Faso, Liberia, Ivory Coast, Benin, Togo and Ghana. This confirms the earlier suggestion of two spatial regimes found in Anselin and O'Loughlin (1992).

Figure 8.3 is the Moran scatterplot for this data, with a linear smoother superimposed. The country labels are listed in Table 8.1. More than half of the associations fall in the lower left quadrant (25 out of 42), indicating the dominance of spatial contiguity of low values for the conflict index (negative spatial association in the terminology of the Getis–Ord statistics). Of the 11 points in the upper right quadrant (high–high association), four stand out, corresponding to Egypt (43), Sudan (42), Somalia (27) and Ethiopia (28), the same four nations that obtained a highly significant positive G_i^* statistic. While the overall tendency portrayed in the scatterplot is one of positive association, six countries show the opposite: low values surrounded by high values for Burundi (25) and Rwanda (26), and high values surrounded by low values for Zaire (20), Angola (21), Zambia (29) and South Africa (33). Except for the latter, which is a special case, this pattern suggests a spatial cluster of nations with negative spatial autocorrelation (in the traditional sense) in West Africa, around and south of the equator. Note that neither the global Moran's I nor the G_i^* statistics are able to provide an indication of this phenomenon.

A closer look at the fit of the linear smoother is provided in Table 8.2, where the three most extreme observations are listed according to the normed residuals (outliers), the diagonal element in the hat matrix (leverage) and Cook's distance (influence). For comparison purposes, the three most extreme (most significant) z-values corresponding to the G_i^* statistics are listed as well. While the linear regression has an acceptable R^2 of 0.574, the results in Table 8.2 indicate the presence of outliers that strongly influence the regression slope. Specifically, Egypt (43) has the most extreme value for three of the four indicators, and the second highest for the fourth (outlier). In other words, its unusually strong pattern of spatial association with its neighbours pulls the regression line (Moran's I) upwards, providing a stronger indication of positive spatial association than warranted

Table 8.1 Country labels

Label	Country	Label	Country
1	Gambia	22	Uganda
2	Mali	23	Kenya
3	Senegal	24	Tanzania
4	Benin	25	Burundi
5	Mauritania	26	Rwanda
6	Niger	27	Somalia
7	Ivory Coast	28	Ethiopia
8	Guinea	29	Zambia
9	Burkina Faso	30	Zimbabwe
10	Liberia	31	Malawi
11	Sierra Leone	32	Mozambique
12	Ghana	33	South Africa
13	Togo	34	Lesotho
14	Cameroon	35	Botswana
15	Nigeria	36	Swaziland
16	Gabon	38	Morocco
17	CAR	39	Algeria
18	Chad	40	Tunisia
19	Congo	41	Libya
20	Zaire	42	Sudan
21	Angola	43	Egypt

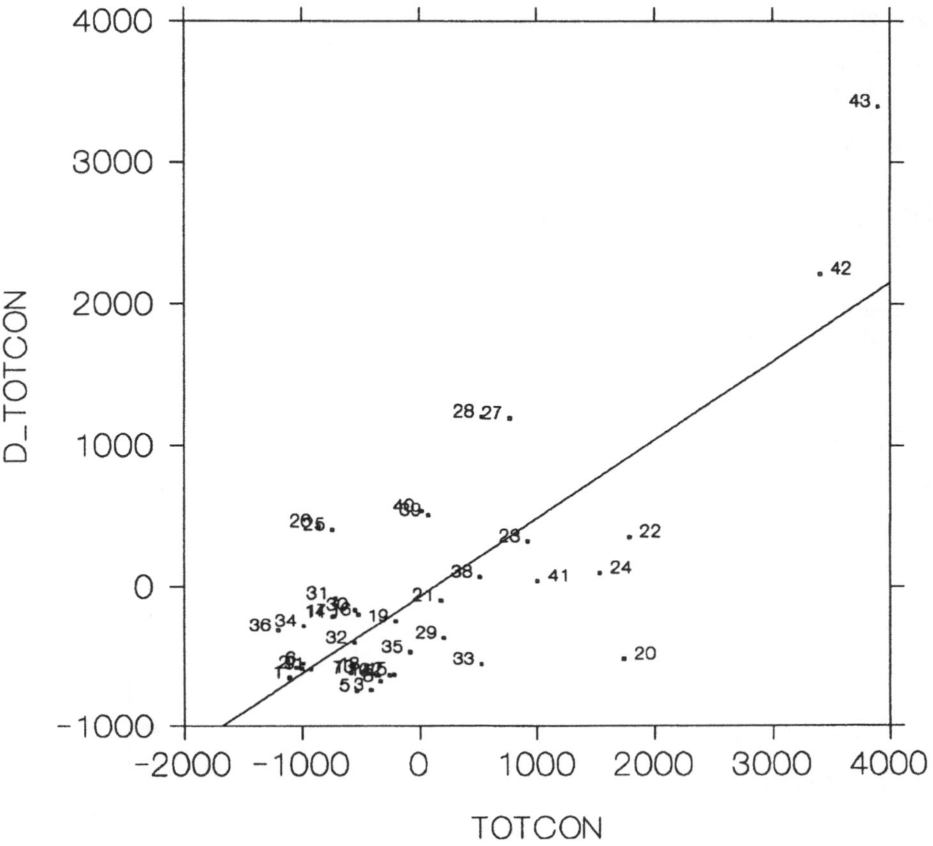

Figure 8.3 Moran scatterplot with linear smoother.

by the bulk of the other observations. A similar effect is exerted by Sudan (42). When both countries are removed from the analysis, Moran's *I* drops. It should be noted that both Egypt and Sudan are boundary observations. Moreover, due to the construction of the spatial weights matrix, Egypt only has one neighbour (Sudan), while Sudan only has two (Egypt and Ethiopia). Since these countries all have high values for the conflict index, the global measure of spatial association is unduly affected. In other words, a careful analysis of the outliers and leverage points in the Moran scatter plot may indicate problems with the specification of the spatial weights matrix, as is the case in this example.

Finally, in Figure 8.4, a LOWESS smoother is superimposed on the Moran scatterplot. Again, the strong influence of Egypt (43) and Sudan (42) on the slope of the line is made clear. Also, the dip in the curve indicates a shift from positive to negative association which points to spatial heterogeneity.

8.6 Conclusions

The interpretation of Moran's *I* as a bivariate regression coefficient and the associated Moran scatter plot suggested in this chapter turn out to be useful devices in exploratory spatial data analysis. In particular, a careful analysis of the Moran scatterplot may help to gain insight into at least six important aspects of spatial association:

Table 8.2 Indicators of extreme observations

Index	Country label	Country name	Value
Outlier	20	Zaire	0.119
(normed residual)	43	Egypt	0.110
	28	Sudan	0.082
Leverage	43	Egypt	0.412
(hat matrix)	42	Sudan	0.174
	28	Ethiopia	0.052
Influence	43	Egypt	7.023
(Cook's distance)	28	Ethiopia	0.189
	42	Sudan	0.136
Spatial association	43	Egypt	4.70
(G_i^* z-value)	42	Sudan	4.17
	28	Ethiopia	2.56

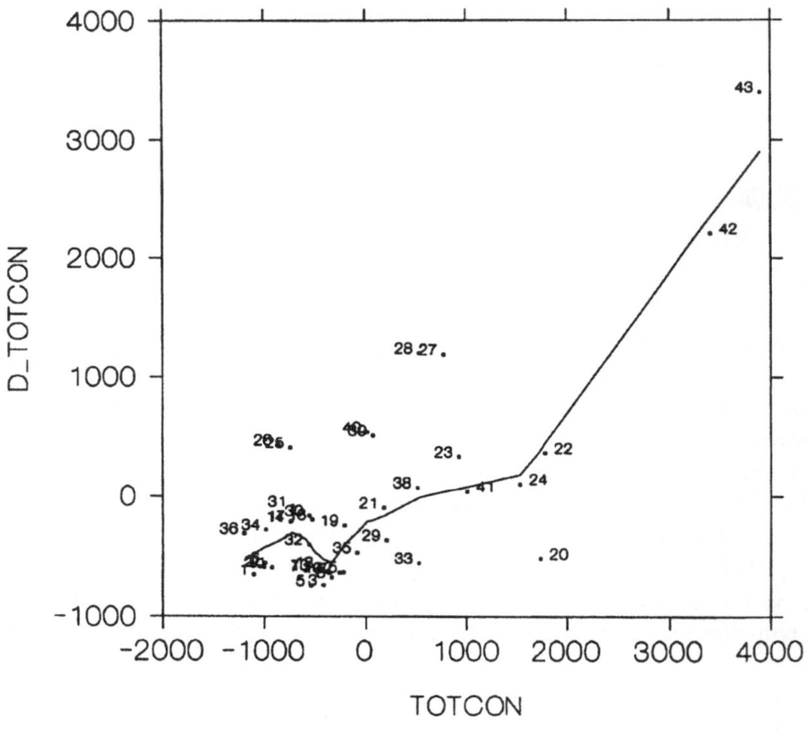

Figure 8.4 Moran scatterplot with LOWESS smoother.

- Decomposing spatial association into its four components: low–low and high–high positive association and low–high and high–low negative association,
- Identifying observations that are outliers relative to the global measure of spatial autocorrelation given by Moran's I,
- Discovering different spatial regimes in the degree (slope) of spatial association,
- Finding observations that exert a large influence (leverage) on the Moran coefficient,

- Indicating the leverage and influence of observations that suffer from boundary effects,
- Suggesting problems with the specification of the spatial weights matrix.

The indication of extreme observations in terms of spatial association is similar to that obtained from the Getis–Ord G_i^* statistics. However, the Moran scatterplot provides additional information as well and thus should be considered as a useful complement to local indicators of spatial association.

The Moran scatterplot can easily be incorporated in an integrated ESDA-GIS modelling strategy, especially one based on interactive dynamic graphics. Since a scatterplot is already part of the usual views of the data implemented in this approach, it takes little additional effort to include a special scatterplot that relates the Wy to y. Such a framework would provide a powerful tool for the exploratory analysis of spatial dependence.

Acknowledgements

The research of which this chapter is an outgrowth was supported in part by grants SES 89-21385 and SES 88-10917(from the National Center for Geographic Information and Analysis, NCGIA) from the US National Science Foundation, and by grant GA-AS 9212 from the Rockefeller Foundation. An initial version was presented at the NCGIA *Workshop on Exploratory Spatial Data Analysis and GIS*, Santa Barbara, CA, February 25–27, 1993.

REFERENCES

ANSELIN, L. (1988a). *Spatial Econometrics: Methods and Models*, Dordrecht: Kluwer Academic.

ANSELIN, L. (1988b). Lagrange Multiplier test diagnostics for spatial dependence and spatial heterogeneity, *Geographical Analysis*, **20**, 1–17.

ANSELIN, L. (1990). What is special about spatial data? Alternative perspectives on spatial data analysis, in Griffith, D.A. (Ed.), *Spatial Statistics, Past, Present and Future*, pp. 66–77, Ann Arbor, MI: Institute of Mathematical Geography.

ANSELIN, L. (1992). *SpaceStat: A Program for the Analysis of Spatial Data*, Santa Barbara, CA: National Center for Geographic Information and Analysis, University of California.

ANSELIN, L. (1994). Exploratory Spatial Data Analysis and Geographic Information Systems, in Painho, M. (Ed.) *New Tools for Spatial Analysis*, pp. 45–54, Luxembourg: Eurostat.

ANSELIN, L. (1995). Local Indicators of Spatial Association – LISA, *Geographical Analysis*, **27**, 93–115.

ANSELIN, L. and GETIS, A. (1992). Spatial statistical analysis and geographic information systems, *The Annals of Regional Science*, **26**, 19–33.

ANSELIN, L. and HUDAK, S. (1992). Spatial econometrics in practice: A review of software options, *Regional Science and Urban Economics*, **22**, 509–36.

ANSELIN, L. and O'LOUGHLIN, J. (1990). Spatial econometric models of international conflicts, in Chatterji, M. and Kuenne, R. (Eds), *Dynamics and Conflict in Regional Structural Change*, pp. 325–45, London: Macmillan.

ANSELIN, L. and O'LOUGHLIN, J. (1992). Geography of international conflict and cooperation: Spatial dependence and regional context in Africa, in Ward, M. (Ed.) *The New Geopolitics*, pp. 39–75, Philadelphia, PA: Gordon and Breach.

ANSELIN, L., HUDAK, S. and DODSON, R. (1993a). *Spatial Data Analysis and GIS: Interfacing GIS and Econometric Software*, Technical Report 93-7, Santa Barbara, CA: National Center for Geographic Information and Analysis, University of California.

ANSELIN, L., DODSON, R. and HUDAK, S. (1993b). Linking GIS and spatial data analysis in practice, *Geographical Systems* **1**, 3–23.

AZAR, E. (1980). The conflict and peace data bank (COPDAB) project, *Journal of Conflict Resolution*, **24**, 143–52.

BAILEY, T. (1992). Statistical spatial analysis and geographic information systems: A review of the potential and progress in the state of the art, *EGIS'92, Proceedings of the Third European Conference on Geographical Information Systems*, pp. 186–203, Utrecht: EGIS Foundation.

BELSLEY, D., KUH, E. and WELSCH, R. (1980). *Regression Diagnostics: Identifying Influential Data and Sources of Colinearity*, New York: John Wiley.

CLEVELAND, W. (1979). Robust locally weighted regression and smoothing scatterplots, *Journal of the American Statistical Association*, **74**, 829–36.

CLEVELAND, W.S. and MCGILL, M.E. (1988). *Dynamic Graphics for Statistics*, Pacific Grove, CA: Wadsworth.

CLIFF, A. and ORD, J.K. (1973). *Spatial Autocorrelation*, London: Pion.

CLIFF, A. and ORD, J.K. (1981). *Spatial Processes, Models and Applications*, London: Pion.

COOK, R. (1977). Detection of influential observations in linear regression, *Technometrics*, **19**, 15–18.

CRESSIE, N. (1991). *Statistics for Spatial Data*, New York: John Wiley.

DIEHL, P. (1992). Geography and war: A review and assessment of the empirical literature, in Ward, M. (Ed.) *The New Geopolitics*, pp. 121–37, Philadelphia, PA: Gordon and Breach.

DING, Y. and FOTHERINGHAM, A.S. (1992). The integration of spatial analysis and GIS, *Computers, Environment and Urban Systems*, **16**, 3–19.

DUTILLEUL, P. and LEGENDRE, P. (1993). Spatial heterogeneity against heteroscedasticity: an ecological paradigm versus a statistical concept, *Oikos*, **66**, 152–71.

FARLEY, J.A., LIMP, W.F. and LOCKHART, J. (1990). The archaeologist's workbench: Integrating GIS, remote sensing, EDA and database management, in Allen, K., Green, F. and Zubrow, E. (Eds), *Interpreting Space: GIS and Archaeology*, 141–64, London: Taylor & Francis.

FOTHERINGHAM, A.S. and ROGERSON, P. (1993). GIS and spatial analytical problems, *International Journal of Geographical Information Systems*, **7**, 3–19.

GETIS, A. and ORD, K. (1992). The analysis of spatial association by use of distance statistics, *Geographical Analysis*, **24**, 189–206.

GOOD, I.J. (1983). The philosophy of exploratory data analysis, *Philosophy of Science*, **50**, 283–95.

GOODCHILD, M.F., HAINING, R.P., WISE, S. *et al.* (1992). Integrating GIS and spatial analysis – Problems and possibilities. *International Journal of Geographical Information Systems*, **6**, 407–23.

GOULD, P. (1981). Letting the data speak for themselves, *Annals Association of American Geographers*, **71**, 166–76.

GRIFFITH, D.A. (1993). Which spatial statistics techniques should be converted to GIS functions? in Fischer, M.M. and Nijkamp, P. (Eds) *Geographic Information Systems, Spatial Modelling and Policy Evaluation*, pp. 101–114, Berlin: Springer-Verlag.

HAINING, R. (1990). *Spatial Data Analysis in the Social and Environmental Sciences*, Cambridge: University Press.

HASLETT, J. (1993). Exploratory analysis of spatial data: Networks and interactive graphics. *Position Papers, Workshop on Exploratory Spatial Data Analysis and GIS*, Santa Barbara, CA: National Center for Geographic Information and Analysis.

HASLETT, J., BRADLEY, R., CRAIG, P., UNWIN, A. and WILLS, C. (1991). Dynamic graphics for exploring spatial data with applications to locating global and local anomalies, *The American Statistician*, **45**, 234–42.

HASLETT, J., WILLS, G. and UNWIN, A. (1990). SPIDER – An interactive statistical tool for the analysis of spatially distributed data, *International Journal of Geographical Information Systems*, **4**, 285–96.

HOAGLIN, D. and WELSCH, R. (1978). The hat matrix in regression and ANOVA, *The American Statistician*, **32**, 17–22.

HUBERT, L.R. (1987). *Assignment Methods in Combinatorial Data Analysis*, New York: Marcel Dekker.

KIRBY, A. and WARD, M. (1987). The spatial analysis of peace and war, *Comparative Political Studies*, **20**, 293–313.

LEGENDRE, P. (1993). Spatial autocorrelation: Trouble or new paradigm, *Ecology*, **74**, 1659–73.

MONMONIER, M. (1989). Geographic brushing: Enhancing exploratory analysis of the scatterplot matrix, *Geographical Analysis*, **21**, 81–4.

MORAN, P.A.P. (1948). The interpretation of statistical maps, *Journal of the Royal Statistical Society B*, **10**, 243–51.

O'LOUGHLIN, J. (1986). Spatial models of international conflicts: Extending current theories of war behavior, *Annals, Association of American Geographers*, **76**, 63–80.

O'LOUGHLIN, J. and ANSELIN, L. (1991). Bringing geography back to the study of international relations: Dependence and regional context in Africa, 1966–1978, *International Interactions*, **17**, 29–61.

O'LOUGHLIN, J. and ANSELIN, L. (1992). Geography of international conflict and cooperation: Theory and methods, in Ward, M. (Ed.), *The New Geopolitics*, pp. 11–38, Philadelphia, PA: Gordon and Breach.

O'LOUGHLIN, J., FLINT, C. and ANSELIN, L. (1994). The Political Geography of the Nazi Vote: Context, Confession and Class in the 1930 Reichstag Election, *Annals, Association of American Geographers*, **84**, 351–80.

OPENSHAW, S. (1991). Developing appropriate spatial analysis methods for GIS, in Maguire, D., Goodchild, M.F. and Rhind, D. (Eds) *Geographical Information Systems: Principles and Applications, Vol. 1*, pp. 389–402, London: Longman.

OPENSHAW, S. (1993). Some suggestions concerning the development of artificial intelligence tools for spatial modelling and analysis in GIS, in Fischer, M.M. and Nijkamp, P. (Eds), *Geographic Information Systems, Spatial Modelling and Policy Evaluation*, pp. 17–33, Berlin: Springer-Verlag.

OPENSHAW, S., BRUNDSON, C. and CHARLTON, M. (1991). A spatial analysis toolkit for GIS. *EGIS'91, Proceedings of the Second European Conference on Geographical Information Systems*, pp. 788–97, Utrecht: EGIS Foundation.

OPENSHAW, S., CROSS, A. and CHARLTON, M. (1990). Building a prototype geographical correlates exploration machine, *International Journal of Geographical Information Systems*, **4**, 297–311.

ORD, J.K. and GETIS, A. (1995). Local spatial autocorrelation statistics: distributional issues and an application, *Geographical Analysis*, **27**, 286–306.

TOBLER, W. (1979). Cellular geography, in Gale, S. and Olsson, G. (Eds), *Philosophy in Geography*, pp. 379–86, Dordrecht: Reidel.

TUKEY, J.W. (1977). *Exploratory Data Analysis*, Reading MA: Addison-Wesley.

UNWIN, A. (1993). Interactive statistical graphics and GIS – Current status and future potential, *Position Papers, Workshop on Exploratory Spatial Data Analysis and GIS*, Santa Barbara, CA: National Center for Geographic Information and Analysis.

WARD, M. (1992). *The New Geopolitics*, Philadelphia, PA: Gordon and Breach.

WILLIAMS, I., LIMP, W. and BRIUER, F. (1990). Using geographic information systems and exploratory data analysis for archeological site classification and analysis, in Allen, K., Green, S. and Zubrow, E. (Eds) *Interpreting Space: GIS and Archaeology*, pp. 239–73, London: Taylor & Francis.

PART TWO

Spatial Integration Issues

CHAPTER NINE

Integration through overlay analysis

DAVID UNWIN

9.1 Map overlay

Map overlay has been used as a basic tool for spatial analysis for almost as many years as thematic maps have been drawn. Although we do not usually think of it in GIS terms, the famous detection by John Snow of a source of infected water in the mid-nineteenth Soho cholera outbreak was an early example of the overlay of two point maps (for cholera cases and for water pumps) followed by what would nowadays be called a 'visualisation' of the newly created 'proximity' relationships between cases and sources of contaminated water. Had Snow lived in the 1960s and had a degree qualification in academic geography, it is almost certain that he would have used some form of spatial statistical analysis to confirm or deny his hypothesis. Had he been alive today, he would probably have used the overlay function in a proprietary GIS. Overlay is nowadays frequently used as a basic form of cartographic modelling and is undertaken for a number of reasons. It is used, for example, to establish the patterns of 'unique conditions' when multinomial fields are overlain in a sieve map analysis (McHarg, 1969; Bonham-Carter *et al.*, 1988) and it also forms the basis of methods by which an entirely new map is predicted as some function of the distributions of other observed attributes.

In this chapter, I develop a typology of GIS operations in which the concept of overlay is absolutely central, examine the usual implementation of the method, summarise the case that almost always this should be regarded as an operation in statistical modelling, and, finally, give an overview of some of the methods that have been adopted. The individual points are illustrated by reference to case studies involving the use of GIS to predict areas at risk from landsliding.

9.2 Map overlay as a basis for classification of analytical functions in GIS

In the literature, several authors (see, for example, Dangermond, 1990) have presented lists of functions which presumably differentiate geographical information systems from other, generic information systems. More often than not, these appear as lists of facilities that users might want from their system or even the so-called 'functionality' provided by specific systems. It is clear that, given the wide range of application of GIS, such lists can be very long indeed.

More rarely, attempts have been made by some authors to classify functions, notably by Burrough (1992) and Tomlin (1990). A summary of Burrough's scheme is given in Table 9.1, from which it can be inferred that overlay can be located in several of his classes. Usually, it is implemented as a class 1 operation involving prediction of new attributes from the values of existing attributes at the same place without reference to considerations of probability or of values in neighbouring areas. Much of the substantive content of this chapter consists of an argument that in most cases overlay is a class 2 operation, that is, where the transformation allows for non-exact values. However, overlay also underpins the determination of topological relationships in class 3, the surface operations of class 4, the buffer and point-in-polygon operations of class 5, the creation of new spatial objects in class 6, and the derivation of geometrical attributes in class 7. With the exception of operations on the co-ordinates of the spatial data themselves to change scale or projection, classes 8 and 9 of Burrough's typology are general database operations which are not unique to GIS.

Table 9.1 Nine classes of operation in Geographical Information Systems

Class 1. Operations by which a new attribute and its value are derived from existing attributes by using exact-valued non-geographical attributes of exact objects (points, lines, areas and fields). Symbolically, this operation can be presented as $U_X = f(A, B, C, D\ldots)$. The linking function can be Boolean, arithmetical or numerical taxonomic.

Class 2. Operations as class 1 but where the transformation allows non-exact values of non-geographical attributes of exact objects. This can be represented as $P(U_X) = f(A, B, C, D\ldots)$ but now the linking function is probablistic, for example a fuzzy logical operation, Bayesian prior, predicted value from a regression, or most probable class from a likelihood-based classification.

Class 3. Operations by which the value of an attribute at a point or in a given grid cell is derived from values of the same or other attributes within a given neighbourhood, N, of that point, that is $U_X = f(A_N, B_N, C_N, D_N\ldots)$. Examples are as adjacency, connection and proximity.

Class 4. Operations by which the value is derived from values of the same or other attributes within a given neighbourhood surrounding the point as in class 3 but the attributes are those of a continuous surface. Examples include convolution operators such as geometric filters, local statistics, interpolation, estimates of derivatives of the surface at the point, and other surface properties such as intervisibility.

Class 5. Operations by which an area or neighbourhood N surrounding a point, line or area receives new attributes according to the value of the spatial entity at location x. Examples of this type of function are buffering and point-in-polygon operations.

Class 6. Operations by which new spatial objects are created from existing objects or the objects are modified. This class includes polygon overlay with logical operations, buffering in vector maps, centroid computation, tessellations, contouring and various smoothing operations.

Class 7. Operations in which new attributes are derived from the geometrical attributes G_n of (vector) objects or sets of objects as $U_X = f(G_1, G_2, \ldots G_n)$. Examples include shape measures, area and the determination of topological relationships.

Class 8. Operations in the database which do not create new attributes such as counts, histograms, windowing and so on.

Class 9. Data management operations such as rectification, scale change and map join.

Source: Modified after Burrough (1992).

Table 9.2 GIS operations considered as variations on an 'overlay' theme

	POINTS	LINES	AREAS	FIELD
POINT	2D Regression Nearest neighbour	2D Interpolation Adjacency Connection	Proximal mapping Thiessen and other polygons	Buffer interpolation Kriging Density and potential estimation
LINE	Intersections	Smoothing and generalisation	Vector to raster conversion	Line buffer
AREA	Centroids	Skeletons Raster to vector conversion	Areal interpolation Polygon overlay Adjacency	'Raster' modelling Density estimation
FIELD	VIP Predicted height Surface Specific points	Surface network Drainage network	Watersheds determination Viewshed	Vector fields and derivatives

Tomlin's (1990) map algebra provides a language for defining operations in cartographic modelling and has been implemented as part of the command set of several systems but it can also be considered as a form of classification of GIS functions using the spatial scope of the operation as its basis. Although much of map algebra turns out to be a renaming of functions that have long been used in the world of image processing and is most suited to the raster data model, the same principles can be extended to network models (Yarwood, 1993) and hence, by implication, to patterns of adjacency between 'resels' in area-valued data with irregular zone shapes and sizes. It thus has the potential to describe most, if not all, GIS operations.

Table 9.2 presents a different view of GIS functions which makes the concept of map overlay absolutely central and develops from earlier ideas of 'map comparison' (Unwin, 1981). In this, GIS operations are seen to result from transformations which occur when maps displaying one type of spatial object are overlain by maps displaying another, or the same. This transformational view cross tabulates the differing types of spatial object, considered geometrically (point, line, area and surface/field) to recognise 16 general classes of operation. Each class represents a particular sort of overlay and the table provides some examples of these; readers may be able to amuse themselves by inserting other analysis functions into it. For example, the transformation from line objects (rows of Table 9.2) to area objects (columns) is exemplified by the familiar vector to raster conversion, whereas the inverse operation, from areas to lines gives both raster to vector conversion and skeletonisation. Similarly, the transformation from points to fields is exemplified by point buffering (choosing a specific distance contour on a continuous surface of distance away from the point), conventional interpolation from a valued point sample, and density estimation. The diagonal cells of Table 9.2 represent classes of transformation that use overlay to produce the same type of spatial object, and in this cell lie the familiar area to area transformations of areal interpolation and standard polygon overlay. This way of classifying GIS analyses is not ideal, nor is it totally comprehensive, but it has the merit of providing a unified framework which makes the integration of information by overlay a totally central analytical operation in GIS.

9.3 A statistical perspective

The paper by MacDougall (1975) represents an early attempt to isolate some of the problems that occur when maps are overlaid, and a comprehensive reassessment relating MacDougall's ideas to a digital environment is provided by Chrisman (1987). The ideas they present remain as relevant today as when they were formulated. In this section I suggest that, for at least three basic reasons, 'cartographic modelling' by map overlay should be treated as a statistical operation.

9.3.1 The data

Most GIS provide an overlay operation that clearly reflects their data model which assumes that the real world can be adequately described by discrete point, line, area or field entities that have exact attributes. When these entities are combined in an overlay operation it is also assumed that the result must have similar characteristics (Heuvelink and Burrough, 1993). Indeed, given the data models adopted, it would be difficult for things to be otherwise. Any overlay is driven by a logical model in which each polygon or pixel which represents a land area is scored 'true' or 'false' on its attribute values according to whether or not it is in or out of a defined set.

A simple example is the use of map overlay to identify areas at risk from landslides where there are steep (greater than 30°), north-facing slopes developed on loess bedrock (Wang and Unwin, 1992). A sustained example in the social sciences is provided by Openshaw *et al.* (1989) who use a sieve operation to identify areas suitable for the disposal of hazardous nuclear waste where there are few people (less than 490 km^{-2}) with good rail access (less than 3 km from a line) that are not already designated as conservation areas. Similar observations relate to the other classes of overlay outlined in Table 9.2, for example, the point-in-polygon operation is regarded as a totally deterministic operation that does not allow for uncertain polygon boundaries, or, for that matter, uncertain locations of the point objects.

As is well known, this essentially Boolean procedure makes a number of assumptions about the data used and the implied relationships between the attributes used (Veregin, 1989). Most have been studied individually at some time.

1. The relationships really are Boolean. In almost all the examples I have looked at, this is not only scientifically absurd, it frequently also throws away a great deal of metric information. Although geomorphologists talk of 'threshold' slope values for some processes, there is nothing particularly important about, for example, using a 30° value to represent 'steep' above which landslides are deemed to be common. Clearly, the relationship between propensity to slide and slope should be modelled by a continuous function and the slope attribute represented by a ratio-scaled attribute.
2. Interval or ratio-scaled attributes are known without significant measurement error (Burrough and Heuvelink, 1992). This is, of course, frequently assumed in statistical

modelling, but in the spatial world of the GIS many of the individual input attributes are themselves the results of estimation from sample data that may or may not also provide an estimate of the magnitude of the error involved. Goodchild (1989) deals with the obvious case where the data are a continuous geographic field established by interpolation from control point sample data. Kriging provides an estimate of the associated error and simulation can be used to get a feel for its effects (Fisher, 1990), but in most cases it will be unknown and almost certainly greater than the authors of the problem either know about or care to admit. The problem is further compounded when the scalar field is used to estimate first, even second, derivatives of altitude by a variety of often poorly understood and documented methods. The most careful analysis of this problem remains that of Evans (1980).

In Wang and Unwin (1992), and quite typical of many such studies, the 'slope' attribute was derived by computing the gradient vector field from a digital elevation matrix using whatever algorithm the system used happened to implement. Moreover, and again equally typically, the input field had itself been derived from sample data in the form of digitised contours and spot heights. It is patently absurd to think that the slope values used are without error and, therefore, that a specific 30° isoline can be threaded through them to establish discrete regions of steep slopes. For a comprehensive treatment of contour interpolation accuracy in a GIS context see Wood (1994). A useful exercise to illustrate the magnitude of this problem is to generate a gradient field from a digital elevation matrix of an area to which you have access and then to take a surveyor's level to a selection of identifiable points and attempt to measure the real gradient through them (Walsh *et al.*, 1987). Comparison with the GIS results on which you might have relied will be salutary and will illustrate in a practical way an obvious, but in a GIS context little considered, problem to do with the concept of a limit.

3. Any categorical attribute data are also known exactly (Goodchild *et al.*, 1992). However, a very large number of 'categorical coverages' or 'multinomial fields' are the products of either a classification (for example satellite derived land use; see Mead and Szajgin, 1982) or an interpretative mapping (as in soil or land-use survey). In the first case it is clear that any categorical assignment should have associated with it a vector of probabilities of that pixel or polygon's membership of the recognised classes (Goodchild *et al.*, 1994). In the second, it is certain that, because of the need to generalise both in the field and on the resulting maps, there will be 'inclusions' of land within mapped units that do not have the mapped property (Fisher, 1991). The level of generalisation involved in a conventional soils map is much higher than those not involved seem to appreciate. Compare, for example Smee's map of the soils of the parish of Haselbech in Northamptonshire, England, based on the objective recording of what was actually found into over seventy distinct soil types at 75 000 auger holes over a 1500 acre area (Smee, 1975), with the extreme generalisation and interpretive interpolation that goes into the production of a standard soil map. Although generically this is often thought of as the 'soil map' problem, similar difficulties occur in many other contexts and make it almost certain that map overlay products that involve them as inputs will be most uncertain. In Wang and Unwin (1992), the land cover type was obtained by classification of a LANDSAT MSS image by standard unsupervised maximum likelihood into one of four types and is clearly subject to considerable error.
4. Finally, it is clear that the boundaries of the discrete objects recognised in the data may be uncertain, either as a result of real uncertainties in locating gradual transitions

in interpretive mapping (Mark and Csillag, 1989) or due to errors introduced in digitising a caricature of these boundaries off paper maps (Bolstad *et al.*, 1990).

A consequence of all these sources of error is that the 'results of overlay ... unravel more questions about data quality and boundary mismatch than they solve' (Leung and Leung, 1993, p.189). These problem are intrinsically statistical, and, as has been indicated, several are relatively well understood, even if GIS practitioners continue to ignore them. What is less well understood is how individual errors interact with each other in frequently long and complex overlay analysis to affect the quality of the final solutions. Clearly, however, if the attribute data are subject to error, then any products of an overlay must themselves also have errors, but a consideration of the very many sources of error in a typical study will indicate that they are unlikely to be capable of being 'washed away' in a single, global error statistic.

One possible general framework for error modelling that has shown promise is the use of Zadeh's concept of fuzzy sets (Zadeh, 1965) which has been explored, amongst others, by Burrough (1989) and his co-workers (Heuvelink and Burrough, 1993). In this the sharp class boundaries implied by the Boolean approach are made 'fuzzy' by replacing the Boolean sets by continuous class sets and assigning objects to them by means of a probabilistic membership function. In Burrough's typology of GIS operations (Burrough, 1992) this change to the conceptual data model leads to overlay using fuzzy operations, Bayesian prior probabilities, regression models and the whole gamut of multivariate statistical operations for classification and discrimination, in fact, almost anything that might be thought of as probabilistic. The fundamental step is to regard the process as inherently probabilistic in nature.

9.3.2 Incorporating prior knowledge

A second reason for regarding almost all map overlay operations as fundamentally statistical is that some additional prior knowledge of the phenomenon being investigated almost always exists. A variety of methods have been adopted which make use of this prior knowledge in several ways. In the conventional Boolean approach, it is incorporated in the attributes selected and in the choice of the individual thresholds used in the overlaid maps. In effect, the overlay evaluates a function formed by the product of a series of binary weights set 1 if the area is 'suitable', 0 if it is not (Burrough, 1992).

In the basic form of analysis it is assumed that the individual attributes have equal weight, yet this is exceedingly unlikely. For example, in almost every study of landslide potential, attention is drawn to the extreme importance of slope relative to, say, slope aspect which, although it is an influence, has far less importance (Wadge, 1988). An obvious improvement is to weight the inputs in some way which accords with this prior knowledge. A simple scheme, implemented by Gupta and Joshi (1990) in a study of landslides in a part of the Himalaya is to score each attribute on an ordinal scale (low=0; medium=1; high=2) and sum to give the overall index.

This gives an additive linear model in which all the inputs are transformed onto an assumed single scale of landslide risk and it has the advantage of being implementable in almost all proprietary GIS by use of reclassification operations. A simple and often used scheme is to use the z score transformation. Note that if the length of the measurement scale used is the same for all attributes, this retains equal weighting and in no way does it use prior knowledge to guide the choice of inputs. In Wang and Unwin (1992) we refer to this as the 'weighting factors' approach.

It is a relatively small step from this to argue that in general any such underlying scale will be one in the probability of union (U), P(U) thus measuring the r_i on a continuous scale from 0 to 1 and to use Bayes' theorem to estimate the required probability given some observed evidence. Agterberg (1989) discusses this method and it has been formalised as the 'weights of evidence' approach (Bonham-Carter, 1991). Recently, a version of this method has been implemented in extensions to the IDRISI GIS (Eastman *et al.*, 1993). Chapter 11 by Fabbri and Chung deals with this approach in more detail.

In the weights of evidence approach the probability that a pixel contains a point event, such as a landslide, is evaluated starting with the prior P(H) assumed to be constant over the entire region. Bayes' theorem is then used to calculate $P_i(H_i'E)$, the posterior probability that a pixel will contain an event given the presence of the *i*th binary pattern. Importantly, in the Appendix to Bonham-Carter (1991, p. 183) this formulation is shown to be related to the logarithm of the odds of the event and its complement. In Wang and Unwin (1992) we adopted a modification of this approach which made use of an extensive database describing 210 past landslides containing an engineering assessment of each slope into one of three stability classes. The probability of each assignment was calculated for each category of the three environmental variables analysed (slope, lithology and aspect). Normalising each by the average for all categories of that variable led to an index which expressed the relative strength of that category in determining the propensity to landslide and then using simple addition led, eventually, to a map of landslide risk.

A variety of other methods taken from decision science which attempt to provide optimum weights for the attributes incorporated into an overlay have also been experimented with (Carver, 1991; Banai, 1993; Pereira and Duckstein, 1993), including some that involve participatory decision making. Again, Eastman *et al.* (1993) provide appropriate software to implement a number of these procedures. Similarly, expert knowledge can be incorporated as a set of rules in an IKBS using Bayes' theorem as its basic engine (Skidmore *et al.*, 1991) or making use of fuzzy logic (Leung and Leung, 1993).

9.3.3 Linear models?

The use of the methods outlined in (Section 9.3.2) enables expert knowledge to be incorporated into the overlay process by use of a common scale using probability and by providing sensible weights for each input attribute. However, in their basic form they do not guide the investigator as to which attributes should be used from an available set, nor do they explicitly allow for possibly important interactions between the attributes used. In particular, most applications will have available, or be able to collect, sample data on the phenomenon whose distribution is being modelled. An obvious stratagem is to use general linear models to screen the possible attributes and to estimate appropriate weights for the overlay.

In the case of landslides, Jibson and Keefer (1989) used conventional OLS in a stepwise model to estimate the relative risk in an area of Tennessee and Kentucky where landslides are common. In this, the best OLS solution was used to suggest which attributes to include and to provide weights for a simple map overlay. Recently, geologists have experimented with the use of canonical correlation analysis to develop 'favourability models' for cases where a set of maps are to be predicted by the overlay, for example, a series of rock properties, and where extensive sample data exist to allow their calibration (Pan, 1993). Recognising that the dependent variable in this type of work will often be interpreted as a probability, Vincent (1990) used a logit model to predict the chance of finding the Corn Bunting in each 10 km square of the UK using a series of environmental variables as predictors. It should be noted that the interpretation of 'probability' in this spatial sense may not be straightforward and depends greatly on the context.

Similar methods have been used by ecologists mapping the distribution of various flora and fauna. An extensive review is given by Pereira and Itami (1991, compare Pereira and Duckstein, 1993) in the introduction to a study in which they model the distribution of the Red Squirrel on Mt Graham, Arizona. In fact, they calibrated two kinds of model. First, a logistic multiple regression in which the probability of a squirrel being observed in a grid cell was estimated as a function of some 14 environmental variables. Second, in a rare application, they used Wrigley's (1976) method of trend surface probability mapping to model the same thing solely as a function of spatial location. As an interesting technical extension, both models were integrated into a Bayesian analysis using the trend surface estimates as prior and the environmental model estimates as evidence.

The logistic model is appropriate when the predictor attributes are measured on interval or ratio scales, but in a great many GIS applications this will not be the case and the predictors are essential multinomial fields of, for example, geological formation, or land use type. The appropriate model is thus the log-linear (Wrigley, 1985), but with attention concentrating on assessing the effects of a set of explanatory attributes on the response. In Wang and Unwin (1992), our predictors were slope aspect (coded 'north' or 'south' facing), lithology (coded 'loess', 'bedrock' and 'mixed') and slope angle (coded 'shallow', 'medium' and 'steep') and the results, calibrated using GLIM, were a set of landslide probabilities for each of the 18 ($2 \times 3 \times 3$) possible unique conditions combinations of these attributes. As might be expected, the results were largely in accord with geomorphological experience, but of particular utility was the ability to examine the predictive effect of interactions between the predictor attributes, especially the varying effect of slope aspect with different lithologies. We regard this use of the log-linear model as a means by which a replicatable approach to map overlay can be achieved.

9.4 Conclusion

Almost all the alternatives discussed above are preferable to the standard Boolean map overlay and all involve conceptualising the operation as a problem in statistical estimation. In essence, what they do is to use traditional statistical modelling techniques to inform the cartographic modelling methods implemented in most GIS in two ways. First, they allow sample data to be used to decide which attributes should be used in an overlay and, second, they provide weights which are in some sense optimal. In themselves, they do not provide a solution to the problems of data quality discussed above, but, in recognising the outcome as a probability of the area having the predicted attribute, they go some way towards this. Almost all involve research decisions of a type that makes it extremely unlikely that they could ever be incorporated directly into a proprietary GIS. Indeed, it is doubtful whether such a direct incorporation is either necessary or desirable.

REFERENCES

AGTERBERG, F.P. (1989). Computer programs for mineral exploration, *Science*, **245**, 76–81.

BANAI, R. (1993). Fuzziness in Geographical Information Systems: contributions from the analytic hierarchy process, *International Journal of Geographical Information Systems*, **7**, 315–29.

BOLSTAD, P.V., GESSLER, P. and LILLESAND, T.M. (1990). Positional uncertainty in manually digitised map data, *International Journal of Geographical Information Systems*, **6**(6), 469–78.

BONHAM-CARTER, G.F. (1991). Integration of geoscientific data using GIS, in Goodchild, M.F., Rhind, D.W. and Maguire, D.J. (Eds), *Geographical Information Systems: Principles and Applications*, pp. 171–84, London: Longman.

BONHAM-CARTER, G.F., AGTERBERG, F.P. and WRIGHT, D.F. (1988). Interpretation of geological data sets for gold exploration in Nova Scotia, *Photogrammetric Engineering and Remote Sensing*, **54**, 1585–92.

BURROUGH, P.A. (1989). Fuzzy mathematical methods for soil survey and land evaluation, *Journal of Soil Science*, **40**, 477–92.

BURROUGH, P.A. (1992). Development of intelligent geographical information systems, *International Journal of Geographical Information Systems*, **6**(1), 1–11.

BURROUGH, P.A. and HEUVELINK, G.B.M. (1992). The sensitivity of Boolean and continuous (fuzzy) logical modelling to uncertain data, *Proceedings, EGIS 92, Munich* (Vol. 2), pp. 1032–41.

CARVER, S. (1991). Integrating multi-criteria evaluation with geographic information systems, *International Journal of Geographical Information Systems*, **5**, 321–39.

CHRISMAN, N.R. (1987). The accuracy of map overlays: a reassessment, *Landscape and Urban Planning*, **14**, 427–39.

DANGERMOND, J. (1990). A classification of the software components currently used in geographic information systems. Reprint of 1983 article in Peuquet D.J. and Marble D.F. (Eds), *Introductory Readings in Geographic Information Systems*, pp. 30–51, London: Taylor & Francis.

EASTMAN, J.R, KYEM, P.A.K., TOLEDANO, J. and WEIGEN, J. (1993). *GIS and Decision Making. Explorations in Geographic Information Systems Technology, Vol.4*, Geneva: UN Institute for Training and Research.

EVANS, I.S. (1980). An integrated system of terrain analysis and slope mapping, *Zeitschrift für Geomorphologie*, Suppl-Bd, **36**, 274–95.

FISHER, P. (1990). Simulation of error in digital elevation models, *Papers and Proceedings of Applied Geography Conference*, **13**, 37–43.

FISHER, P.F. (1991). Modelling soil map-unit inclusions by Monte Carlo simulation, *International Journal of Geographical Information Systems*, **5**(2), 193–208.

GOODCHILD, M.F. (1989). Modelling error in objects and fields in Goodchild, M.F. and Gopal, S. (Eds), *Accuracy of Spatial Databases*, pp. 107–44, London: Taylor & Francis.

GOODCHILD, M.F., CHI-CHANG, L. and LEUNG, Y. (1994). Visualizing fuzzy maps. *Visualization in Geographical Information Systems*, in Hearnshaw, H. and Unwin, D. (Eds), pp. 158–167, London: Belhaven.

GOODCHILD, M.F., SUN, G. and YANG S. (1992). Development and test of an error model for categorical data, *International Journal of Geographical Information Systems*, **6**(2), 87–104.

GUPTA, R.P. and JOSHI, B.C. (1990). Landslide hazard zoning using the GIS approach – a case study from the Ramganga catchment, Himalaya, *Engineering Geology*, **28**, 119–45.

HEUVELINK, B.M. and BURROUGH, P.A. (1993). Error propagation in cartographic modelling using Boolean logic and continuous classification, *International Journal of Geographical Information Systems*, **7**(3), 231–46.

JIBSON, R.W. and KEEFER, D.K. (1989). Statistical analysis of factors affecting landslide distribution in the new Madrid seismic zone, Tennessee and Kentucky, *Engineering Geology*, **27**, 509–42.

LEUNG, Y. and LEUNG, K.S. (1993). An intelligent expert system shell for knowledge based geographical information systems. 1. The tools, *International Journal of Geographical Information Systems*, **7**(3), 189–99.

MACDOUGALL, E.B. (1975). The accuracy of map overlays, *Landscape Planning*, **2**, 23–30.

MCHARG, I. (1969). *Design with Nature*, New York: Natural History Press.

MARK, D.M. and CSILLAG, F. (1989). The nature of boundaries on 'area class' maps, *Cartographica*, **26**, 65–78.

MEAD, R.A. and SZAJGIN, J. (1982). Landsat classification accuracy assessment procedures, *Photogrammetric Engineering and Remote Sensing*, **48**, 139–41.

OPENSHAW, S., CARVER, S.J. and FERNIE, F. (1989). *Britain's Nuclear Waste, Safety and Siting*, London: Belhaven.

PAN, G. (1993). Canonical favorability model for data integration and mineral potential mapping, *Computers & Geosciences*, **19**, 1077–1100.

PEREIRA J.M.C. and ITAMI, R.M. (1991). GIS-based habitat modelling using logistic regression: a study of the Mt. Graham Red Squirrel, *Photogrammetric Engineering and Remote Sensing*, **57**, 1475–86.

PEREIRA, J.M.C. and DUCKSTEIN, L. (1993). A multiple criteria decision-making approach to GIS-based land suitability evaluation. *International Journal of Geographical Information Systems*, **7**, 407–24.

SKIDMORE, A.K., RYAN, P.J., DAWES, W., SHORT, D and O'LOUGHLIN, E. (1991). Use of an expert system to map forest soils from a geographical information system, *International Journal of Geographical Information Systems*, **5**, 431–45.

SMEE, D. (1975). *Soil Mapping. A Case Study at Haselbech, Northamptonshire 1954–1978. Part 1: Memoir, Part 2 : Soil Illustrations, Part 3 : Soil Map*, Salisbury, UK: Compton Russell.

TOMLIN, C.D. (1990). *Geographic Information Systems and Cartographic Modeling*, Englewood Cliffs, NJ: Prentice Hall.

UNWIN, D.J. (1981). *Introductory Spatial Analysis*, pp. 187–206, London: Methuen.

VEREGIN, H. (1989). Error modelling for the map overlay operation. Goodchild, M.F. and Gopal, S. (Eds), *Accuracy of Spatial Databases*, pp. 3–18, London: Taylor & Francis.

VINCENT, P. (1990). Modelling binary maps using ARC/INFO and GLIM, Harts, J., Ottens, H.F.L. and Scholten, H.J. (Eds), *Proceedings of EGIS '90*, (Vol. 2), pp. 1108–16, Ultrecht: EGIS Foundation.

WADGE, G. (1988). The potential of GIS modelling of gravity flows and slope instabilities, *International Journal of Geographical Information Systems*, **2**, 143–52.

WALSH, S.J., LIGHTFOOT, D.R. and BUTLER, D.R. (1987). Recognition and assessment of error in geographic information systems, *Photogrammetric Engineering and Remote Sensing*, **53**, 323–34.

WANG, S. Q. and UNWIN, D.J. (1992). Modelling landslide distribution on loess soils in China: an investigation, *International Journal of Geographical Information Systems*, **6**(5), 391–405.

WOOD, J. (1994). Visualizing contour interpolation accuracy in digital elevation models, in Hearnshaw, H. and Unwin, D. (Eds), *Visualization in Geographical Information Systems*, pp. 168–80, London: Belhaven.

WRIGLEY, N. (1976). *An Introduction to the use of logit models in geography, Concepts and Techniques in Modern Geography*, **10**, Norwich: GeoAbstracts.

WRIGLEY, N. (1985). *Categorical Data Analysis for Geographers and Environmental Scientists*. London: Longman.

YARWOOD, D.S. (1993). Cartographic Modelling in network space? An investigation through the design and implementation of a network analysis prototype. Unpublished M.Sc. Thesis, Department of Geography, University of Edinburgh.

ZADEH, L.A. (1965). Fuzzy sets, *Information and Control*, **8**, 338–53.

CHAPTER TEN

Descriptive modelling and declarative modelling for spatial data

BIANCA FALCIDIENO, CATERINA PIENOVI and MICHELA SPAGNUOLO

10.1 Introduction

In spatial data handling research, a great deal of effort has gone into the improvement of computational techniques and the solution of several problems, whereas not enough attention has been paid to the modelling aspects and to architectural considerations for the development of more general and comprehensive Spatial Information Systems (SIS). Owing to this lack of communication between modelling experts and users, many of the commercial systems seem to be mere databases customised for the particular context and spatial data are often considered as attributes of application-specific information. In contrast to this, the peculiar nature of spatial data suggests a more careful choice of the system architecture, based on the opposite approach, where the model of the spatial data takes the leading role and outlines the framework for the organisation of the application-specific knowledge. Obviously, the spatial data model should not be considered as a static structure which simply codes the measures of the natural phenomena, but rather as a set of interrelated models each corresponding to a different level of interaction with the data (Laurini and Thompson, 1992). Traditional spatial data models are important for the development of specific algorithms but must be further explored to provide higher levels of abstraction that can relate directly to certain properties of the data and can establish an efficient link between geometry and application-specific information (Weibel and Heller, 1989).

The aim of this chapter is to give an overview of our research on the modelling aspects of spatial data handling. The basic idea is that a general system architecture for spatial data should highlight the main aspects of modelling, which could be summarised by the concepts of representation and simulation of reality.

The representation of reality is performed by providing suitable structures to store the data and powerful tools to analyse and synthesise the information. In this chapter emphasis is put on the need for abstraction mechanisms which should generate high-level data descriptions based on prominent features. For example, the analytical extraction of land-forms from a digital terrain model is a complex task which has the aim of generating a rich symbolic description which is one step closer to representing the morphological structure of the surface than the row elevation data and requires less storage. This approach is here considered as a *descriptive modelling*, which corresponds to a bottom-up approach to the information coding. In dealing with spatial data and spatial problems we need some sort of conceptual modelling as developed in computer science, to provide valuable intellectual aids for solving different spatial problems.

In devising a general SIS, however, not only tools for analysing the data should be provided, but a *declarative modelling* should also be supported to provide means to simulate reality by interacting with high-level concepts. Let us give an example. Suppose that we need to simulate a terrain with certain morphological characteristics in order to study the effect of some natural phenomena. A concise description of a landscape could be easily given in terms of high-level concepts, for example, two chains of mountains, a narrow valley between them, a river at the bottom of the valley and a broad meadow at the end of the valley. A SIS supporting this kind of declarative definition of the scene should be able to generate a surface model by decomposing the described terrain into geometric primitives and by automatically building the corresponding geometric model, thus allowing the user to ignore low-level operations. In this view, we propose an integrated architecture for SIS, which is based on both geometric models and conceptual models. In this architecture, descriptive and declarative modelling gives the possibility to interact with high-level abstractions which can be linked to application concepts and can provide a support to predictive tasks, a very important aspect of GIS.

10.2 Descriptive modelling

Descriptive modelling can be seen as the bottom-up approach to spatial data which has the main aim of upgrading knowledge of the observed natural phenomena by extracting information which is implicitly contained in the initial set of data. A typical flow of operations involved in a descriptive interaction with data is depicted in Figure 10.1.

In the first phase, which is common and necessary in any SIS, sets of data are acquired from the real world. For example, in the context of digital terrain modelling, data correspond to measures of terrain elevation, while for air pollution studies sets of statistical data will be collected. More precisely, this first level of information consists of original observations of the natural phenomenon, which are generally characterised by *geometric attributes* related to the position (typically triplets of co-ordinates) and form (dimension, shape and so forth) and by *non-geometric attributes* which affect only the geographical meaning (geological data, historical notations, land use, for example). These basic data reflect the best level of knowledge available: if only elevations are collected, then the SIS can handle only this limited information. However, this initial model is not yet a structure suitable for surface modelling and it should be structured and upgraded with the introduction of further information.

Geometric models correspond to a second level of spatial data modelling and are derived from the basic model through the generation of relationships between individual observations (point, line, polygon topology). The geometric model reflects and highlights the particular nature of spatial data, which can be synthesised by two of its properties, continuity and coherence. In the real geographical world a topographic surface can be visited at any location and thus the geometric model should reflect an underlying view of space as continuous and the need to determine elevation at any of arbitrary precise position. In mathematical terms, the discretised surface should be represented by a continuous approximation of the surface. Spatial coherence is here meant as the tendency for nearby locations to influence each other and to possess similar attributes. Consequently, the geometric model should make the adjacency relationships between

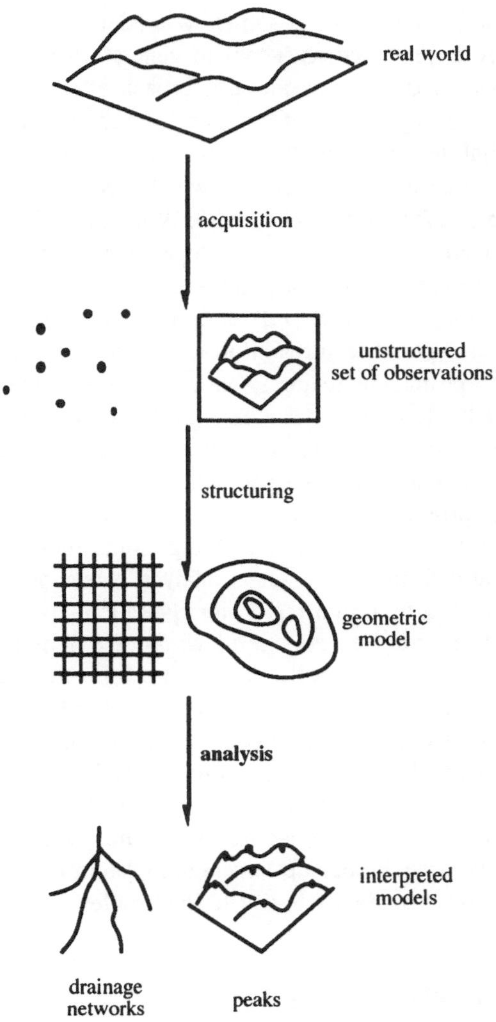

Figure 10.1 An example of descriptive interaction with spatial data.

its primitive entities (for example, points, vertices and polygons) explicit, because the adjacency could be used as a key to access data.

Geometric models permit users to answer specific spatial questions, such as: How much terrain can be seen from a particular watch tower? What is the area covered by a given type of vegetation? In general, each question falls into one of these two categories (Laurini and Thompson, 1992):

- find what is in a particular place
- find the position of an object

The determination of both categories of information involves spatial measurements and checking spatial relationships which can be directly deduced from the geometric model. Several geometric models exist, such as regular grids, triangulations or contours, which act as interfaces to subsequent modelling procedures but the resulting model is still at a low conceptual level because it does not support an overall comprehension of the represented geographical phenomena. For example, the qualitative nature of a terrain given by its morphological features is somehow hidden in the geometric model. Thus, *descriptive modelling* is proposed as a tool to generate a further level of spatial data model, and it is considered as a bottom-up approach to spatial data modelling, which allows the user to extract high-level information from the low-level geometric model. In the context of digital terrain modelling, for example, the automated extraction of topographic features is a rather complex task which has been studied by several authors (Peucker and Douglas, 1975; Laffey *et al.*, 1983; Wolf, 1989; Tribe, 1990; Falcidieno and Spagnuolo, 1991). Methods have been developed mainly for regular grids and with particular attention to the extraction of point and line features.

Descriptive modelling uses powerful interpretation algorithms which perform the analytical extraction of several features in order to generate a richer and more synthetic description, based on prominent features, which codes information in an efficient and effective way. Symbolic descriptions can be regarded as frameworks through which it is possible to point out some properties of interest the surface may have for a specific analysis. The same approach has been adopted in the picture processing context in relation to the similar problem of segmentation. There, the goal is to subdivide an image into maximal disjoint regions each satisfying some uniformity predicate. Image segmentation is usually followed by scene analysis when more global information is used to merge regions and to assign region interpretation.

Symbolic descriptions can be very useful within a SIS, not only for the modelling aspects but also because they behave as a directory associated with a computer representation of a natural phenomenon that can be used to facilitate selective access and retrieval of information from the representation. Suppose one wishes to analyse a terrain around the highest peak. Instead of searching the entire digital model to locate this feature, the same information could easily be obtained by interrogating a high-level terrain description based on its morphological features. Indeed, when the number of observed points is very large, it becomes essential to determine a procedure to access and retrieve efficiently only those data relevant to the problem at hand.

10.3 Declarative modelling

Descriptive modelling is a basic issue in spatial data systems, but alone it is not sufficient to devise comprehensive and efficient systems. Indeed, in each phase of a descriptive interaction, users have a rather limited role, meaning that they can handle only those data measured from the reality and no tools are given to simulate or predict any natural phenomenon. For this reason, we would like to introduce a new approach to spatial data models that is called *declarative modelling*. The aim is to represent facts, to simulate processes, to express judgements or to provide effective descriptions of geographic phenomena, through sets of properties or constraints. The computer has to generate the potential answers to these descriptions and to present them to the users.

To move from description to prescription, however, implies the development of tools allowing users to describe, generate, explore, visualise and understand classes of geographical phenomena. These techniques broaden the role of modelling from a relatively passive inquiry to a much more active intent. Descriptive models answer questions whereas declarative models solve problems (Tomlin, 1990).

The problems addressed by declarative models generally involve two different uses of them for exploration and for generation (see Figure 10.2).

The first use requires a selective exploration of the spatial data model using geometric,

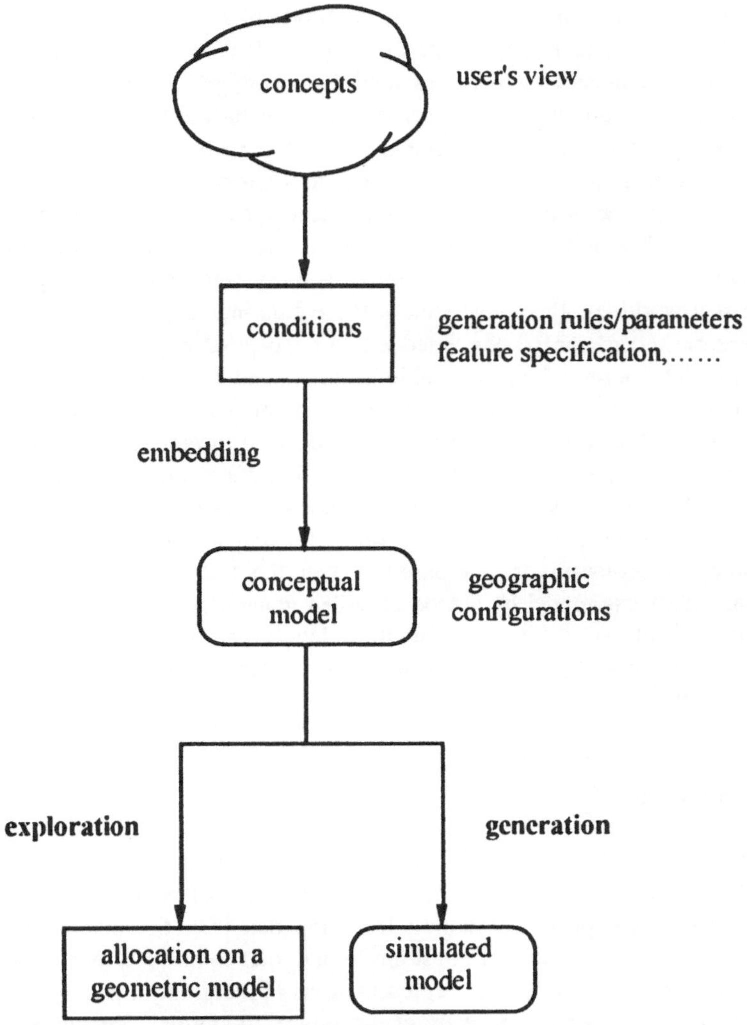

Figure 10.2 An example of declarative interaction with spatial data.

topological, geographical properties in order to satisfy stated objectives. This process, also called cartographic *allocation* (Tomlin, 1990), may range from the siting of a proposed land use to the positioning of fire towers. The initial statement of an allocation problem is a descriptive task which consists of an explicit specification of some geographic conditions necessary to achieve the stated objective. Different problems occur when stating the set of conditions, properties, or constraints, which should be satisfied by the area of interest.

- Define local or global properties. The former are associated with atomic parts of geographic space, while the latter permit links to be established between groups of locations as integrated wholes.
- Provide mechanisms to check the consistency of a set of constraints or the derivation of new properties when controlling given ones.

The set of conditions expressed by the user defines the *conceptual model* of the spatial phenomena in a suitable formalism. While it is possible to give a well-defined characterisation of a geometric model, the definition of a conceptual one is harder to give precisely because a detailed definition of the latter depends on the user's needs and the nature of the phenomenon to be represented. However, geometric models play a leading role in the proposed architecture as they represent the background knowledge used by all the different applications. The meaning and the structure of a conceptual model has been roughly described in the previous section, where it has been explained that it is a tool to link sets of geometric primitives to an application-specific interpretation.

The user can handle these two levels of representation using either a descriptive or a declarative interaction. A typical descriptive activity starts with the acquisition of data from the real world and the conversion of these data into any suitable geometric model. The geometric model can be converted into a conceptual model through understanding processes which interpret the low-level information and extract high-level descriptions of the phenonomen represented by the geometric model. Also, geometric models of different levels of accuracy or fairness can be derived through refinement processes.

The declarative approach corresponds to a different interaction with data. The first step is the definition of a set of conditions (constraints, needs and so forth) which will be embedded into a conceptual model, described using a consistent formalism (entity-relationship, object-based or rule-based). Then the system automatically builds the underlying geometric model by a process called generation. As for the geometric level, conceptual models of different abstraction levels can be derived through analysis processes. End-users or programs can interrogate both kinds of model according to the application needs.

10.4 Conclusions

Up to now the emphasis in spatial data systems has been on descriptive models. On the one hand, this approach is essential for the development of intelligent SIS as it supports tools to retrieve complex information from the data. On the other hand, it has the limitation that it can only answer questions which refer to observations of the real world. Also, descriptive models are static representations of a current or past situation and little help is given to represent some view of the future, to simulate known processes, or to predict the behaviour of geomorphological phenomena. We believe that future applications of SIS will necessitate more flexible models, which better formalise the users' different needs and views. To do this, we propose a different approach to designing spatial information systems based on the declarative modelling. Developing such systems requires us to provide description, generation, and understanding tools, together with abstraction and refinement mechanisms. The major point of this approach is that the user has only to specify the set of conditions the overall model has to obey through a list of properties and/or constraints. Understanding this formal description and translating into a classical geometric model are left to the computer. To be more effective and efficient, declarative modelling is integrated with the descriptive way, here considered as a bottom-up approach to spatial data modelling which permits us to generate symbolic descriptions from low-level geometric models.

These concepts can be summarised as depicted in Figure 10.3, where the system architecture is sketched. In this modelling scheme, a typical descriptive activity can start with the acquisition of data from the real world and the conversion of these data into a suitable geometric model. The geometric model can be converted into a conceptual model through *understanding* processes which interpret the low-level information and

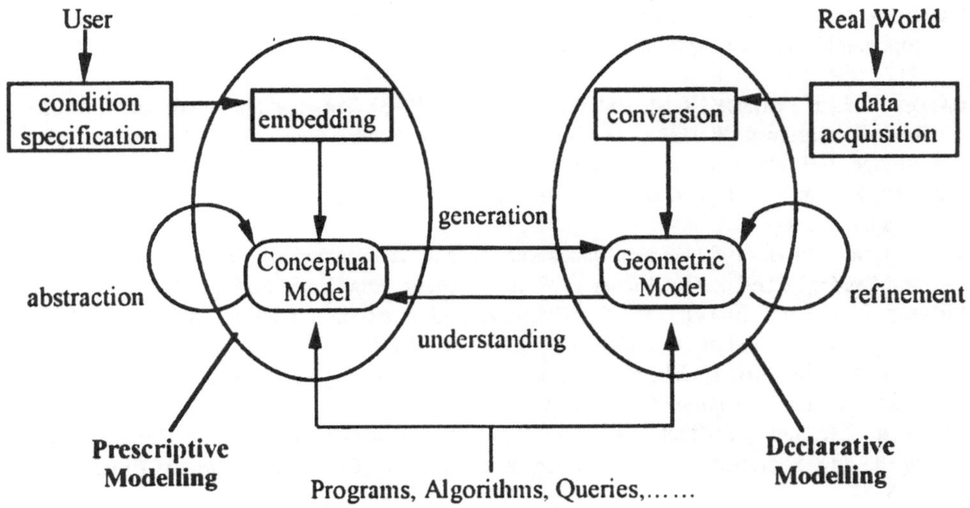

Figure 10.3 The proposed system architecture.

extract high-level descriptions of the phenomenon represented by the geometric model. Also, geometric models of different levels of accuracy or fairness can be derived through *refinement* processes. The declarative approach corresponds to a different interaction with the data. The first step is the definition of a set of conditions (for example, constraints and needs) which will be embedded into a conceptual model, described using a consistent formalism (entity-relationship, object-oriented or rule-based, for example). Then, the system automatically builds the underlying geometric model by a process called generation. As for the geometric level, conceptual models of different abstraction levels can be derived through analysis processes. End-users or programs can interrogate both kinds of model according to the application needs.

The proposed system architecture has been partially implemented in the context of digital terrain modelling. In particular, we have focused our attention on topographic terrain, i.e. surfaces in which altitude is a function of position in a geographic domain (Falcidieno *et al.*, 1992). Spatial analysis for the extraction of morphological features has been studied for different kinds of geometric model: triangulations, grids and contour maps. Based on the results of the recognition step, conceptual models have been defined mainly in the form of surface networks and characteristic region configuration graphs. Moreover, the process of refinement and generation has been implemented using constrained triangulation as a tool to keep in the geometric model the features declared by the users (Pienovi and Spagnuolo, 1994).

REFERENCES

De Floriani, L., Falcidieno, B. and Pienovi, C. (1985). Delaunay-based representation of surfaces defined over arbitrarily shaped domains, *Computer Vision, Graphics and Image Processing*, **32**, 127–40.

Falcidieno, B. and Pienovi, C. (1990). Natural surface approximation by constrained stochastic interpolation, *Computer-Aided Design*, **22**(3), 167–72.

Falcidieno, B., Pienovi, C. and Spagnuolo, M. (1992). Discrete surface models: constraint-based generation and understanding, in Falcidieno, B., Herman, I., and Pienovi, C. (Eds), *Computer Graphics and Mathematics*, pp. 245–61, Germany: Springer Verlag.

FALCIDIENO, B. and SPAGNUOLO, M. (1991). A new method for the characterization of topographic surfaces, *International Journal of Geographical Information Systems*, **5**(4), 397–412.

LAFFEY, T.J., HARALICK, R.M. and WATSON, L.T. (1983). Topographic classification of image intensity surfaces, *Workshop on Computer Vision: Representation and Control*, pp. 171–7, Rindge, NH: Computer Society Press.

LAURINI, R. and THOMPSON, D. (1992). *Fundamentals of Spatial Information Systems*, London: Academic Press.

MANDELBROT, B.B. (1989). Fractal landscapes without creases and with rivers, in Peitgen, H.O. and Saupe, D. (Eds), *The Science of Fractal Images*, New York: Springer Verlag.

MORRIS, D.G. and FLAVIN, R.W. (1989). A digital terrain model for hydrology, *4th Int. Symposium on Spatial Data Handling, Zürich*, **1**, 250–62.

MUSGRAVE, F.K., KOLB, C.E. and MACE, R.S. (1989). The synthesis and rendering of eroded fractal terrains, *Computer Graphics*, **23**(3), 41–50.

PEUCKER, T.K. and DOUGLAS, D.H. (1975). Detection of surface specific points by local parallel processing of discrete terrain elevation data, *Computer Graphics and Image Processing*, **4**, 375–87.

PIENOVI, C. and SPAGNUOLO, M. (1994). Handling discrete surfaces by analysis and simulation, *Computer and Graphics*, **18**(6), 785–93.

TOMLIN, C.D. (1990). *Geographic Information Systems and Cartographic Modelling*, Englewood Cliffs, Prentice Hall.

TRIBE, A. (1990). Automated recognition of valley heads from digital terrain models, *Earth, Surface Processes and Landform*, **15**(8).

WEIBEL, R. and HELLER, M. (1989). A framework for digital terrain modelling, *4th Int. Symposium on Spatial Data Handling, Zürich*, **1**, 219–29.

WOLF, G.W. (1989). Metric surface networks, *4th Int. Symposium on Spatial Data Handling, Zürich*, **2**, 884–56.

CHAPTER ELEVEN

Predictive spatial data analysis in the geosciences

ANDREA G. FABBRI and CHANG-JO F. CHUNG

11.1 Introduction

In the earth sciences during the last 28 years, much research work has been focused on the capture, management and computer analysis of spatially distributed data. Such data, ranging from topographic data and field observations, to geophysical and geochemical survey recordings and to satellite and airborne imagery, has reached earth scientists in many diverse formats, encodings and resolutions which, to be of predictive utility, require decoding, correction, enhancement and integration. Efforts in these processing tasks have sometimes obscured the fact that the final target of most spatial databases is to support decisions on resources, hazards and environmental impacts.

A decision process for spatial data can have one or more of the following purposes: (a) assessment, (b) search for patterns, (c) understanding of natural or social phenomena, (d) management of information, and (e) prediction of location and intensity of the phenomena under study. This chapter discusses the conceptualisation of predictive modelling in spatial data analysis (SDA). SDA is generally performed on GIS. A broad view of a GIS as a management tool by Aronoff (1989) contains the definition: 'A GIS is designed for the collection, storage and analysis of objects and phenomena where geographic location is an important characteristic or critical to the analysis' (p.1). Clearly, the reference to analysis and phenomena is indicative of modelling and suggestive of predictive representation in the spatial domain.

An analysis of how SDA can support a spatial perspective on data for explanation or exploration, understanding and insight is provided by Goodchild *et al.* (1992) who stated that:

1. space provides a convenient indexing scheme;
2. a spatial perspective allows easy access to information;
3. it permits different types of event to be associated in a GIS as overlays, and
4. it enables spatial distribution of objects and phenomena to be used to represent spatial interaction.

Spatial distribution, however, can be misleading if the apparent spatial pattern is used to make inferences without a realistic conceptual model of the multidimensional relationships between spatial objects. Techniques for SDA were developed in the

1960s and 1970s when the main concern in GIS applications was in developing infrastructures and facilities to store and manage data on resources, and less in the analysis of the data. While this is indeed a fact, it can also be argued that in the earth sciences some of the main survey institutions such as the US Geological Survey, the Geological Survey of Kansas, and the Geological Survey of Canada, have maintained support for research in SDA since the 1960s, and they are still continuing to do so (Sampson 1975a, b; Agterberg and Chung 1975; Duda *et al.*, 1978). Such work has recently come to the attention of the GIS community (Burrough, 1986).

The greatest difficulty in SDA is represented by the complexity of formulating a conceptual model of a process into a computational form. A discussion of such difficulties by Green and Craig (1984) highlights the need to develop a rationale to associate different types and layers of information within a multiple dataset. Such a rationale for integration signifies that what is stored within a modelling spatial database is not data but knowledge. It is the end result of analytical work where relationships between layers of spatial information are established, for instance, by geophysical models for integration. According to those authors, spatial data models, in order of increasing complexity can be classified as:

- those dealing with objects (for example, rocks) which produce uniform sensor response;
- those dealing with a mixture of objects influencing a sensor;
- those in which the sensor is influenced more by adjacent objects than by distant ones;
- those in which additional data sets have to be integrated to explain the spatial patterns on a plane (for example, integrating vegetation distribution to explain the contribution of pixel brightness in Landsat data).

It is this latter type of model which represents the greatest challenge in data integration to develop a physical rationale in support the spatial relationships between the different data types (continuous and non-continuous), their transformations and their normalisations.

To counter the often made remark that GIS modelling tools are not always adequate for many applications, it can be argued that the programming of a satisfactory set of analytical functions within a GIS is probably of lesser difficulty than the construction of a modelling rationale adequate to predict natural complexity. A brief overview of representative approaches and applications, and of modelling tools and systems is provided as a context for discussing a unified approach to modelling structures and strategies.

11.2 SDA approaches in the earth sciences

Traditionally in the earth sciences the spatial representation and interpretation of physical or chemical events are fundamental. Quantitative map analysis is older than the computer itself, but it was in the 1960s that more quantitatively oriented geologists started to use computers to capture data such as geochemical or sedimentary sample sets, structural orientation values (Fox, 1967), thickness or depth of stratigraphic units related to oil reservoirs (Merriam and Harbaugh, 1964), and assay values in mine plans to detect and measure trends and anomalies by means of conceptual models (Agterberg, 1968). More and more powerful computers and methods of data processing had to be developed to capture and store the data, to process raw data for corrections and integration, and to perform the analysis for information extraction and automated display. Methods such as trend surface analysis (Agterberg and Chung, 1975), graphic analysis and interpolation systems (Sampson, 1975a, b; Harbaugh *et al.*, 1977), regression (Agterberg *et al.*, 1972; Chung, 1983; Bonham-Carter and Chung, 1983; McCammon and Kork, 1992), characteristic analysis (McCammon *et al.*, 1983, 1984), and artificial intelligence (Duda, 1980; Duda *et al.*, 1978) were used for mineral resource assessment and exploration. Digital image processing, image analysis by mathematical morphology and interactive techniques for data capture and integration became more widely used in geology (Fabbri, 1984). More recently, new techniques in geometric probability, statistical pattern integration and spatial reasoning are being used in mineral exploration (Bonham-Carter *et al.*, 1988). Table 11.1 summarises some of the main spatial models used in the earth sciences.

As we have seen, SDA and spatial modelling have a long tradition in the earth sciences. One of the main reasons is the need to represent information spatially in order to discover and characterise spatial relationships and use them for prediction. For this reason, tools for spatial data analysis exist in large number and most GIS have now a collection of tools for spatial manipulation and analysis. A typical example of interactive commands for spatial analysis or map algebra is shown in Table 11.2.

Such a modelling algebra, typical in many raster-based GIS, when supported by database management tools and spreadsheet capabilities to handle attribute tables, enables a rather extensive modelling within a GIS without the need of external software. As will be discussed later, it seems that the major difficulty in performing SDA may not be the limitations of processing algebra and tools in a GIS, but the complexity in

Table 11.1 Types of spatial model developed in the earth sciences

Analysis	Data structures	References
Trend surface analysis	Points, cells	Agterberg and Chung (1975); Sampson (1975a, b).
Multivariate regression	Cells	Agterberg *et al.* (1972); Chung and Agterberg (1980); Chung (1978, 1983); Bonham-Carter and Chung (1983).
Characteristic analysis	Cells	McCammon *et al.* (1983, 1984); Botbol *et al.* (1978); Suslick and Figueredo (1990).
Geostatistics	Points, cells and blocks	Davis (1986); Ma and Royer (1988) Hohn (1988); Isaaks and Srivastava (1989); Deutsch and Journel (1993).
Image analysis	Rasters	Agterberg and Fabbri (1978); Fabbri (1980, 1984); Kasvand (1983).
Min–max auto-correlation	Rasters	Switzer and Green (1984); Chung and Fabbri (1990); Chung *et al.* (1992).
Expert systems	Cells, rasters	Duda (1980); McCammon (1989); Usery *et al.* (1988, 1989); Campbell *et al.* (1982); Katz (1991).
Fractal analysis	Vectors, rasters	Turcotte (1992).
Bayesian probability models	Rasters	Bonham-Carter *et al.* (1988); Agterberg *et al.* (1990, 1993); Chung *et al.* (1993); Sing *et al.* (1993); Bonham-Carter (1994).
Favourability function	Vector, raster, cells polygons	Chung and Fabbri (1993); Chung and Moon (1991).
Weighted multivariate regression	Vector, raster, cells	Chung *et al.* (1995).

Table 11.2 Interactive expressions from the modelling language of ILWIS, a GIS with extensive analytical tools (ITC, 1993).

1 MAP2:=TABLE.COLUMN (Map 1)

Reclassification of a raster map, MAP1, according to attribute values in column COLUMN of table TABLE, associated to MAP1.

2 MAP_C:= IF (MAP_A >20, MAP_B+3, MAP_B-2)

If the values in MAP_A are larger than 20 THEN add 3 to MAP_B, ELSE subtract 2 from MAP_B.

3 SLOPE()=SQRT(@1*@1+@2*@2)*100

Create function SLOPE. @1 and @2 indicate parameters to be substituted by image names during interaction. Here they represent the height difference (gradient) per pixel in the X resp. Y direction.

SL_MAP:=SLOPE(MAP1, MAP2)

MAP1 and MAP2 are the two variables for @1 and @2.

4 MAP2:=NBMIN(MAP1#)

NBMIN is a neighbourhood aggregation function which finds the minimum value. #[] is a select neighbourhood operator for a 3×3 window where the neighbour positions are:

1 2 3

4 5 6

7 8 9

alone indicates the entire 3×3 window.

5 ITER:MAP=NBMIN(MAP1#+NBDIS)

Calculate a distance map from an initial map, MAP1. Stops only if map MAP does not change any more. ITER is the iteration function and NBDIS is a distance filter which uses the following constant values:

7 5 7

5 0 5

7 5 7

representing real problems in a computational form. For instance, given a multiple data set for the assessment of geologic hazard, can we assume 'conditional independence' between the different layers of information to construct a Bayesian model using conditional probabilities? The next section introduces several methods of representation for predictive data integration.

11.3 A unified approach: favourability functions

11.3.1 Introduction

A study region **A** and m layers, $\mathbf{L}_1, \ldots, \mathbf{L}_m$ of spatial data in **A** can be subdivided into small cells or pixels or polygons. The m layers of map data each contain one set of map units in **A** when a search target for a specific type of object **D** is sought (in mineral exploration **D** can be the occurrence of mineral deposit, in geologic hazard **D** can be the occurrence of a specific type of landslide, and in environmental impact analysis **D** can be a specific type of impact). As proposed by Chung and Fabbri (1993), for a pixel p in **A** we can consider a proposition:

$\mathbf{T}_p$: "p contains an object of type $\mathbf{D}$"

or

$\mathbf{T}_p$: "p is contained in an object of type $\mathbf{D}$" (11.1)

In general, some layers will consist of continuous measurements and some other layers will consist of non-continuous measurements. For each layer $\mathbf{L}_k$ containing continuous measurements, the quantitative value at p is a finite interval [$\min_k$, $\max_k$]. For each layer $\mathbf{L}_k$ containing non-continuous measures we may assume that the quantitative value at p takes one integer value among $\{1, 2, \ldots, n_k\}$ where n_k is the maximum number of map units in $\mathbf{L}_k$. The m layers of map data at every p in $\mathbf{A}$ can be represented, in a quantised form, by:

$$\{(\nu_k(p), k = 1, \ldots m), p \in \mathbf{A}\}, \tag{11.2}$$

where $\nu_k(p)$ is the quantised value of L_k at p. The m quantised values ($\nu_k(p)$, k=1, . . . , m) in (11.2) are regarded as m pieces of evidence for the proposition in Eq. (11.1).

We may regard the quantisation, ν_k, as a function of $\mathbf{A}$ into a finite interval for the kth layer:

$$\nu_k: \begin{cases} \mathbf{A} \to [\min_k, \max_k], & \text{if } \mathbf{L}_k \text{ is continuous pattern} \\ \mathbf{A} \to \{1, 2, \ldots, n_k\}, & \text{if } \mathbf{L}_k \text{ is non-continuous pattern.} \end{cases} \tag{11.3}$$

In quantitative modelling for data integration in environmental studies we propose a relative favourability function, r_k, for each kth layer $\mathbf{L}_k$, defined as:

$$r_k: \begin{cases} [\min_k, \max_k], \to [\alpha\beta] & \text{if } \mathbf{L}_k \text{ is continuous pattern} \\ \{1, 2, \ldots, n_k\} \to [\alpha\beta], & \text{if } \mathbf{L}_k \text{ is non-continuous pattern.} \end{cases} \tag{11.4}$$

where α and β are known constants, and $r_k(\delta)$ ($\in [\alpha, \beta]$) represents a measurement related to the 'sureness' that the proposition in Eq. (11.1) is true given the evidence of δ at each p in the kth layer $\mathbf{L}_k$. We define the compound function f_k of r_k and ν_k at each p in $\mathbf{A}$, $f_k(p) = r_k(\nu_k(p))$ for all $p \in \mathbf{A}$:

$$f_k: \begin{cases} \mathbf{A} \xrightarrow{\nu_k} [\min_k, \max_k] \xrightarrow{r_k} [\alpha, \beta] & \text{if } \mathbf{L}_k \text{ is continuous pattern} \\ \mathbf{A} \to \{1, 2, \ldots, n_k\} \to [\alpha, \beta] & \text{if } \mathbf{L}_k \text{ is non-continuous pattern.} \end{cases} \tag{11.5}$$

where f_k is termed a favourability function for the kth layer $\mathbf{L}_k$. If $f_k(p)$ has a value near α, the sureness that the proposition in Eq. (11.1) is true is very low. If $f_k(p)$ is near β, the sureness that the proposition in Eq. (11.1) is true is very high. This favourability function can have many different interpretations, such as a probability, a certainty, a belief, plausibility or possibility.

If we have defined m favourability functions, $f_k(p)$ (k=1, 2, . . . , m), $f_k(p)$ contains a significant meaning with respect to the proposition (differently from the quantised values $\nu_k(p)$ of $\mathbf{L}_k$ at p). For instance, we can have relationships like $f_k(p) < f_k(q)$ or $f_k(p) < f_l(q)$.

At each pixel p we have:

$$(f_1(p), f_2(p), \ldots, f_m(p)), \tag{11.6}$$

which indicate how each of the m pieces of evidence ($\nu_k(p)$, k=1, 2, . . . , m) in Eq. (11.2) support the sureness that the proposition in Eq. (11.1) is true at p. The m layers of map data are represented by,

$$\{(f_k(p), k = 1, 2, \ldots, m),\ p \in \mathbf{A}\}, \tag{11.7}$$

instead of Eq. (11.2).

11.3.2 Probabilistic interpretation

Given the proposition in Eq. (11.1), $f_k(p)$ is interpreted as the conditional probability, $\text{Prob}_k\{\mathbf{T}_p \text{ is true} \mid \nu_k(p)\}$ that p contains at least one object $\mathbf{D}$ given the evidence at p in $\mathbf{L}_k$; that is, for k=1, 2, ..., m,

$$f_k(p) = \text{Prob}_k\{\mathbf{T}_p \text{ is true} \mid \text{evidence } \nu_k(p) \text{ at } p\}. \tag{11.8}$$

However, we could have defined it as,

$$f_k(p) = \text{Prob}_k\{\text{evidence } \nu_k(p) \text{ is observed at } p \mid \mathbf{T}_p \text{ is true}\}, \text{instead.} \tag{11.9}$$

Using Bayes' theorem, we establish the relationship between the two interpretations in Eqs (11.8) and (11.9):

$$\text{Prob}_k\{\mathbf{T}_p \mid \nu_k(p)\}\text{Prob}_k\{\nu_k(p)\} = \text{Prob}_k\{\nu_k(p) \mid \mathbf{T}_p\}\text{Prob}_k\{\mathbf{T}_p\}, \tag{11.10}$$

where $\mathbf{T}_p$ denotes the proposition that p contains an object of type $\mathbf{D}$. $\text{Prob}_k\{\mathbf{T}_p\}$ is the prior probability that a pixel p contains an object $\mathbf{D}$ before we have any evidence, and $\text{Prob}_k\{\ \nu_k(p)\}$ is the probability that p has evidence $\nu_k(p)$.

11.3.3 Certainty factor interpretation

A certainty factor (**CF**) at p for the kth layer, $\mathbf{CF}_k(p)$ is defined as the change in certainty that the proposition in Eq. (11.1) is true, from a position without the evidence $\nu_k(p)$ at p to one given the evidence $\nu_k(p)$ at p in the kth layer $\mathbf{L}_k$. Certainty factors were originally proposed by Shortliffe and Buchanan (1975) for the medical expert system MYCIN. The **CF** ranges in value between -1 and $+1$. Heckerman (1986) discussed the propagation of uncertainties through inference networks and proposed the following certainty factor in terms of probabilities:

$$\mathbf{CF}_k(p) = \begin{cases} \dfrac{\text{Prob}_k\{\mathbf{T}_p \mid \nu_k(p)\} - \text{Prob}_k\{\mathbf{T}_p\}}{\text{Prob}_k\{\mathbf{T}_p \mid \nu_k(p)\}\ (1 - \text{Prob}_k\{\mathbf{T}_p\})} & \text{if } \text{Prob}_k\{\mathbf{T}_p\}|\{\nu_k(p)\} > \text{Prob}_k\{\text{T}_p\} \\[2ex] \dfrac{\text{Prob}_k\{\mathbf{T}_p \mid \nu_k(p)\}\ - \text{Prob}_k\{\mathbf{T}_p\}}{\text{Prob}_k\{\mathbf{T}_p\}(1 - \text{Prob}_k\{\mathbf{T}_p\})} & \text{if } \text{Prob}_k\{\mathbf{T}_p|\nu_k(p)\} > \text{Prob}_k\{\text{T}_p\} \end{cases} \tag{11.11}$$

11.3.4 Belief function interpretation

The interpretation of the favourability function according to the Dempster–Shafer belief function (Shafer, 1976) requires that two functions be defined for each layer,

$\mathrm{Bel}_k\{\mathbf{T}_p \mid \nu_k(p)\}$ and $\mathrm{Pls}_k\{\mathbf{T}_p \mid \nu_k(p)\}$, which represent the minimum and maximum degree of belief that the evidence $\nu_k(p)$ of $\mathbf{L}_k$ supports the proposition that a pixel p contains objects of type $\mathbf{D}$. They range in value from 0 to 1, and are termed the belief function and the plausible function. The difference $\mathrm{Pls}_k\{\mathbf{T}_p \mid \nu_k(p)\} - \mathrm{Bel}_k\{\mathbf{T}_p \mid \nu_k(p)\}$ represents ignorance of one's belief that p contains objects of type $\mathbf{D}$ given evidence $\nu_k(p)$ in $\mathbf{L}_k$.

11.3.5 Fuzzy set interpretation

We can consider a fuzzy set S consisting of all the pixels ($p \in \mathbf{A}$) where the proposition in Eq. (11.1) is 'likely' to be true (Zadeh, 1965). The fuzzy set S is defined by a membership function μ_S:

$$\mu_s : \mathbf{A} \rightarrow [0, 1], \tag{11.12}$$

which represents grade of membership or degree of compatibility. We may also interpret the membership function as a possibility function (Zadeh, 1978) of p, where the proposition is likely to be true and is denoted by a set of ordered pairs,

$$S = \{(p, \mu_s(p) : p \in \mathbf{A}\}. \tag{11.13}$$

At each kth layer $\mathbf{L}_k$, we define a fuzzy set S_k by a membership function μ_k based on the evidence in $\mathbf{L}_k$,

$$\mu_k : \mathbf{A} \rightarrow [0, 1], \tag{11.14}$$

that is, at each layer we define a fuzzy set S_k,

$$S_k = \{(p, \mu_s(p) : p \in \mathbf{A}\}. \tag{11.15}$$

11.3.6 Estimation of favourability function

A crucial task in predictive modelling in a study area is to construct and estimate the favourability function for data integration. For this a first step is to define a proposition, $\mathbf{T}_p$, for which supporting data are available, for example,

"A pixel p contains a mineral occurrence of a given type"
or
"A pixel p will be part of a landslide of a given type"
or
"A pixel p will be affected by the environment."

A precise definition of the proposition will facilitate the construction of the favourability function which has to be defined for each layer in the study area. The situation is complex because no unique procedure exists. For example, according to Chung and Fabbri (1993) who provided ample discussion of such estimates, general situations can be as follows.

1. Few or no recorded occurrences (landslides or environmental impacts) are available for the proposition in environments similar to the one in the study area. The construction depends on the evidence in the study area and on the opinion of experts. For this reason it is largely subjective.

2. There are known occurrences within the study area. If we know N instances in which the proposition is true in the study area, among N we have n known events (or observations where the proposition is true) and $N - n$ unknown events (or unobserved situations in which the proposition is true). Using the n known events, in combination with expert knowledge, the task is to identify the areas where $N - n$ unknown events are likely to occur. Traditionally, multivariate statistical analysis such as regression analysis, discriminant analysis and canonical correlation analysis is used to study such a situation (Agterberg *et al.*, 1972; Chung 1978, 1983).
3. There are known occurrences in similar environments only outside the study area. Similarly to the situation in (2) above, the established multivariate relationships between the known events in other areas where environments are similar to the study area, and the combination of all input layers, are computed and applied to the study area to identify the areas where unknown events are likely to occur.

11.3.7 Combination rules and assumptions

The following expressions exemplify some of the combination rules that can be developed for the different interpretations of the favourability function.

Probability,

$$P\{D \mid A \cap B\} = P\{D \mid A\}P\{D \mid B\}\frac{P\{A\}P\{B\}}{P\{A \cap B\}}\,\frac{1}{P\{D\}}. \tag{11.16}$$

Certainty factor,

$$\mathbf{CF}_{\mathrm{AB}}(p) = \begin{cases} \mathbf{CF}_{\mathrm{A}}(p) + \mathbf{CF}_{\mathrm{B}}(p) - \mathbf{CF}_{\mathrm{A}}(p).\mathbf{CF}_{\mathrm{B}}(p) & \text{if } \mathbf{CF}_{\mathrm{A}}(p), \mathbf{CF}_{\mathrm{B}}(p) > 0; \\ \dfrac{\mathbf{CF}_{\mathrm{A}}(p) + \mathbf{CF}_{\mathrm{B}}(p)}{1 - \mathrm{minimum}(|\mathbf{CF}_{\mathrm{A}}(p)|, |\mathbf{CF}_{\mathrm{B}}(p)|)} & \text{if } \mathbf{CF}_{\mathrm{A}}(p) \text{ and } \mathbf{CF}_{\mathrm{B}}(p) \text{ have opposite sign} \\ \mathbf{CF}_{\mathrm{A}}(p) + \mathbf{CF}_{\mathrm{B}}(p) + \mathbf{CF}_{\mathrm{A}}(p).\mathbf{CF}_{\mathrm{B}}(p) & \text{if } \mathbf{CF}_{\mathrm{A}}(p), \mathbf{CF}_{\mathrm{B}}(p) < 0. \end{cases} \tag{11.17}$$

Dempster's rule to combine two belief functions $\mathrm{Bel}_{\mathrm{A}}$ and $\mathrm{Bel}_{\mathrm{B}}$ (Chung and Moon, 1991):

$$\mathrm{Bel}_{\mathrm{AB}}\{\mathbf{T}_p\} = \frac{\mathrm{ab} + \mathrm{a}(1 - \mathrm{b} - \mathrm{b}') + b(1 - \mathrm{a} - \mathrm{a}')}{1 - \mathrm{ab}' - \mathrm{a}'\mathrm{b}},$$

and

$$\mathrm{Bel}_{\mathrm{AB}}\{\overline{\mathbf{T}}_p\} = \frac{\mathrm{a}'\mathrm{b}' + \mathrm{a}\prime(1 - \mathrm{b} - \mathrm{b}') + \mathrm{b}\prime(1 - \mathrm{a} - \mathrm{a}')}{1 - \mathrm{ab}' - \mathrm{a}'\mathrm{b}}, \tag{11.18}$$

where $\mathrm{Bel}_{\mathrm{A}}\{\mathbf{T}_p\} = \mathrm{a}$, $\mathrm{Bel}_{\mathrm{A}}\{\overline{\mathbf{T}}_p\} = \mathrm{a}'$, $\mathrm{Bel}_{\mathrm{B}}\{\mathbf{T}_p\} = \mathrm{b}$, $\mathrm{Bel}_{\mathrm{B}}\{\overline{\mathbf{T}}_p\} = \mathrm{b}'$, and $1 - \mathrm{ab}' - \mathrm{a}'\mathrm{b}'$ is an indicator of conflict between $\mathrm{Bel}_{\mathrm{A}}$ and $\mathrm{Bel}_{\mathrm{B}}$ (Shafer, 1976).

Some of the basic fuzzy set operations applicable to spatial information are (Zadeh, 1965; Moon *et al.*, 1991):

1. Min-operator,
2. Max-operator,

3. Algebraic sum and product operator, and
4. γ-operator (Zimmermann and Zysno, 1980).

Each interpretation of the favourability function is not unique and it requires different assumptions, estimations and combination rules. For example, the probabilistic and the certainty factor interpretations require conditional independence, that given the presence of the object of type **D** in proposition (Eq. 11.1), $\mathbf{T}_p$, the attributes A and B are independent (not related). In many real-life situations dealing with physical phenomena within a study area, such assumption cannot be made and it may, therefore, be objectionable to test for such independence using the spatial database. It is likely that such a test, as proposed by Agterberg *et al.* (1990), while it satisfies the computational requirements, does not satisfy the logical model to allow the strengthening of the evidence.

Real applications are also bound to include a mixture of conceptual modelling, expert knowledge and statistical analysis of the available data. Predictive spatial data analysis, however, has to be based on the knowledge obtained from existing situations for which sufficient data are available and which form a physical or genetic environment similar to the one in the area where the prediction is sought.

11.4 Recent trends and new challenges

Much research activity in SDA is now directed towards:

1. data modelling, information integration and decision-support tools,
2. spatial statistics and inference,
3. new training methods and material for technology transfer.

Under (1), object-oriented programming and GIS (van Oosterom, 1993), knowledge-based GIS approaches for engineering geology mapping (Usery *et al.*, 1988, 1989; Cress and Deister, 1990), decision-support tools and the formalisation of data representations for integration modelling (Chung and Fabbri, 1993; Chung *et al.*, 1993, 1995) will enable the study of more complex spatial relationships. In (2), the study of spatial statistics and inference will lead to a better understanding of spatial processes and to the formulation of spatial predictive models (Cressie, 1991, 1993; Deutsch and Journel, 1993). Finally, in (3), advanced training methods are now using case studies in the fields of renewable and non-renewable resources, natural and man-induced hazard, global and local environments (IAEA, 1994; Bonham-Carter, 1994; van Westen 1993; van Westen *et al.*, 1993, Patrono *et al.*, 1994).

While, in the earth sciences, applications to mineral potential mapping, to data integration and to environmental management will lead to much more comprehensive approaches in the use of applied spatial statistics and decision systems, new multi-disciplinary modelling approaches (Goodchild *et al.*, 1993) are now proposed in the environmental field and in the design of expert system strategies for GIS (Armstrong, 1993; Maidment and Evans, 1993; Navinchandra, 1993; Wright, 1993). These future challenges, however, will still remain in the area of image understanding and in our ability to express problems in computational form in the light of imperfections in knowledge acquisition, representation and decision processes. At present, one of the limitations of decision-support systems is the lack of spatial characterisation, while GIS have only limited reasoning tools. Spatial decision support systems will allow the inclusion of spatial reasoning and expert procedures within a single interactive environment. It is to be expected that more progress in data representation, modelling of data fusion, and extensive applied research towards feasible processing strategies, will make spatial decision support systems such as advanced GIS of even greater relevance and utility in the future.

REFERENCES

AGTERBERG, F.P. (1968). Application of trend analysis in the evaluation of the Whalesback Mine, Newfoundland, *Can. Inst. Min. Metall., Spec.*, **9**, 77–88.

AGTERBERG, F.P., BONHAM-CARTER, G.F., CHENG, Q. and WRIGHT, D.F. (1993). Weight of evidence modelling and weighted logistic regression for mineral potential mapping, in Davis, J.C. and Herzfeld, U.C. (Eds), *Computers and Geology – 25 Years of Progress*, pp. 13–32, New York: Oxford University Press.

AGTERBERG, F.P., BONHAM-CARTER, G.F. and WRIGHT, D.F. (1990). Statistical pattern integration for mineral exploration, in Gaal, G. and Merriam, D.F. (Eds), *Computer Applications in Resource Estimation: Prediction and Assessment for Metals and Petroleum*, pp. 1–21, New York: Pergamon Press.

AGTERBERG F.P. and CHUNG, C.F. (1975). *A Computer Program for Polynomial Trend-Surface Analysis*. Geol. Surv. Canada, Paper 75-21.

AGTERBERG F.P., CHUNG, C.F., FABBRI, A.G., KELLY, A.M. and SPRINGER, J.S. (1972). *Geomathematical Evaluation of Copper and Zinc Potential of the Abitibi Area, Ontario and Quebec*. Geol. Surv. Canada Paper 71-41.

AGTERBERG F.P. and FABBRI, A.G. (1978). Spatial correlation of stratigraphic units quantified from geological maps, *Computers and Geosciences*, **4**(2), 285–94.

ARMSTRONG, M.P. (1993). Database integration for knowledge-based ground water quality assessment, in Wright J.R, Wiggins, L.L., Jain, R.K. and Kim, T.J. (Eds), *Expert Systems in Environmental Planning*, pp. 145–62, Berlin: Springer-Verlag.

ARONOFF, S. (1989). *Geographic Information Systems: A Management Perspective*. Ottawa, Canada: WDL Publications.

BONHAM-CARTER, G.F. (1994). *Geographic Information Systems for Geoscientists: Modeling with GIS*. New York: Pergamon Press.

BONHAM-CARTER, G.F., AGTERBERG. F.P. and WRIGHT, D.F. (1988). Integration of geological datasets for gold exploration in Nova Scotia, *Photogrammetric Engineering and Remote Sensing*, **54**(11), 1565–92.

BONHAM-CARTER, G.F. and CHUNG, C.F. (1983). Integration of mineral resource data for Kasmere Lake area, Nonthwest Manitoba, with emphasis on Uranium, *J. Math. Geol.*, **15**(1), 15–45.

BOTBOL, J.M., SINDING-LARSEN, R., MCCAMMON, R.B. and GOTT, G.B. (1978). A regionalized multivariate approach to target selection in geochemical exploration, *Economic Geology*, **73**, 534–46.

BURROUGH, P.A. (1986). *Principles of Geographic Information Systems for Land Resources Assessment*. Oxford: Clarendon Press.

CAMPBELL, A.N., HOLLISTER, V.F. and DUDA, R.O. (1982). Recognition of a hidden mineral deposit by an artificial intelligence program, *Science*, **217**(3), 927–9.

CHUNG, C.F. (1978). *Computer Program for the Logistic Model to Estimate the Probability of Occurrence of Discrete Events*. Geological Survey of Canada, Paper 78–11.

CHUNG, C.F. (1983). SIMSAG: integrated system for use in evaluation of mineral and energy resources, *Mathematical Geology*, **15**(1), 47–58.

CHUNG, C.F. and AGTERBERG, F.P. (1980). Regression models for estimating mineral resources from geological map data, *Mathematical Geology*, **12**(5), 473–88.

CHUNG, C.F. and FABBRI, A.G. (1990). Two approaches to parsimony of multi-channel remotely sensed data, in Hill, G.W. (Ed.), *Advanced Data Integration in Mineral and Energy Resource Studies*. US Geological Survey Bulletin, in press.

CHUNG, C.F. and FABBRI, A.G. (1993). The representation of geoscience information for data integration, *Nonrenewable Resources*, **2**(2), 122–39.

CHUNG, C.F., FABBRI, A.G. and VAN WESTEN, C.J. (1993). Bayesian approaches and multivariate statistical models in GIS based landslide hazard zonation. Paper presented at the 1993 IAMG Silver Anniversary Meeting, Prague, Czech Republic, October 10-15, 1993 (unpublished proceedings).

CHUNG, C.F., FABBRI, A.G. and VAN WESTEN, C.J. (1995). Multivariate analysis in landslide hazard zonation, in Carrara, A. and Guzzetti, F. (Eds), *Geographical Information Systems in Assessing Natural Hazards*, pp. 107–33, 346–47, Dordrecht: Kluwer Academic Publishers.

CHUNG, C.F. and MOON, W.M. (1991). Combination rules of spatial geoscience data for mineral exploration, *Geoinformatics*, **2**(2), 159–69.

CHUNG, C.F., RENCZ, A.N. and WATSON, G.P. (1992). Identifying geological structure in remotely sensed imagery, *Sciences de la Terre, Serie Informatique Geologique*, **31**, 439–52.

CRESS, J.J and DEISTER, R.R.P. (1990). Development and implementation of a knowledge-based geological engineering map production system, *Photogrammetric Engineering and Remote Sensing*, **56**(11), 1559–59.

CRESSIE, N.A.C. (1991). *Statistics for Spatial Data*, New York: John Wiley and Sons.

CRESSIE, N.A.C. (1993). Geostatistical analysis of spatial data, in National Research Council, *Spatial Statistics and Digital Image Analysis*, pp. 87–108, Washington DC: National Academic Press.

DAVIS, J.C. (1986). *Statistics and Data Analysis in Geology*, pp. 287–467, New York: John Wiley and Sons.

DEUTSCH, C.V. and JOURNEL, A.G. (1993). *GSLIB: Geostatistical Software Library and User's Guide*, Oxford: Oxford University Press.

DUDA, R.O. (1980). *The Prospector System for Mineral Exploration*. Stanford Research Institute International, SRI, Final Report.

DUDA, R.O., HART, P.E., BARRETT, P., GASHNIG, J.G., KONOLIGE, K., REBOH, R. and SLOCUM, J. (1978). *Development of the Prospector Consultation System for Mineral Exploration*. Stanford Research Institute International, SRI, Final Report.

FABBRI, A.G. (1980). GIAPP: geological image analysis program package for estimating geometrical probabilities, *Computers and Geosciences*, **6**(1), 153–61.

FABBRI, A.G. (1984). *Image Processing of Geological Data*. New York: Van Nostrand Reinhold.

FOX, W.T. (1967). *FORTRAN IV Program for Vector Trend Analyses of Directional Data*. Kansas Geological Survey Computer Contribution. 11: 36 p.

GOODCHILD, M., HAINING, R. and WISE, S. (1992). Integrating GIS and spatial data analysis: problems and possibilities, *International Journal of Geographical Information Systems*, **6**(5), 407–23.

GOODCHILD, M.F., PARKS, B.O. and STEYAERT, L.T. (Eds), (1993). *Environmental Modelling with GIS*, Oxford, Oxford University Press.

GREEN, A.A. and CRAIG, M. (1984). Integrated analysis of image data for mineral exploration. *Proc. Intl. Symp. on Remote Sensing of the Environment, Third Thematic Conf., Remote Sensing for Exploration Geology*, pp. 131–7.

HARBAUGH, J.W., DOVETON, J.H. and DAVIS, J.C. (1977). *Probability Methods in Oil Exploration*, pp. 90–118, New York: John Wiley and Sons.

HECKERMAN, D. (1986). Probabilistic interpretation of MYCIN's certainty factors, in Kanal, L.N. and Lemmer, J.F. (Eds), *Uncertainty in Artificial Intelligence*, pp. 167–96, New York: Elsevier.

HOHN, M.E. (1988). *Geostatistics and Petroleum Geology*, New York: Van Nostrand Reinhold.

IAEA (1994). *Spatial Data Integration for Mineral Exploration, Resource Assessment and Environmental Studies: A Guidebook*, International Atomic Energy Agency, Vienna: IAEA-TECDOC-782, 192.

ISAAKS, E.H. and SRIVASTAVA, R.M. (1989). *An Introduction to Applied Geostatistics*. New York: Oxford University Press.

ITC (1993). *ILWIS 1.4: User's Manual*, Enschede, The Netherlands: International Institute for Aerospace Surveys and Earth Sciences (ITC).

KASVAND, T. (1983). Computerized vision for the geologist, *Mathematical Geology*, **15**(1), 3–23.

KATZ, S.S. (1991). Emulating the Prospector expert system with a raster GIS, *Computers and Geosciences*, **17**(7), 1033–50.

MA, Y.Z. and ROYER, J.J. (1988). Local geostatistical filtering application to remote sensing, *Sciences de la Terre, Serie Informatique Geologique*, Nancy, France, 17–36.

MAIDMENT, D.R. and EVANS, T.A. (1993). Regulating the municipal environment using an expert geographic information system, Wright J.R., Wiggins, L.L., Jain, R.K. and Kim, T.J. (Eds), *Expert Systems in Environmental Planning*, pp. 163–86, Berlin: Springer-Verlag.

MCCAMMON, R.B. (1989). Prospector II – The redesign of Prospector, *AI Systems in Government*, pp. 88–92, Washington, DC.

MCCAMMON, R.B., BOTBOL, J.M., SINDING-LARSEN, R. and BOWEN, R.W. (1983). Characteristic analysis-1981: final program and possible discovery, *Mathematical Geology*, **15**(1), 59–84.

MCCAMMON, R.B., BOTBOL, J.M., SINDING-LARSEN, R. and BOWEN, R.W. (1984). *The New Characteristic Analysis (NCHARAN) Program*, US Geological Survey, Bulletin 1621.

MCCAMMON, R.B. and KORK, J.O. (1992). One-level prediction – A numerical method for estimating undiscovered metal endowment, *Nonrenewable Resources*, **1**(2), 139–47.

MERRIAM, D.F. and HARBAUGH, J.W. (1964). *Trend Surface Analysis of Regional and Residual Components of Geological Structure of Kansas*. Kansas Geological Survey Special Distribution Publication 11.

MOON, W.M., CHUNG, C.F. and AN, P. (1991). Representation and integration of geological, geophysical and remote sensing data, *Geoinformatics*, **2**(2), 177–82.

NAVINCHANDRA, D. (1993). Observations on the role of artificial intelligence techniques in geographic information processing, in Wright J.R, Wiggins, L.L., Jain, R.K. and Kim, T.J. (Eds), *Expert Systems in Environmental Planning*, pp. 85–118, Berlin: Springer-Verlag.

VAN OOSTEROM, P. (1993). *Reactive Data Structures for Geographic Information Systems*, Oxford: Oxford University Press.

PATRONO, A., VELKAMP, H. and FABBRI, A.G. (Eds). (1994). *A Study in Environmental Impact Assessment (EIA)*. ITC, Enschede, The Netherlands, Contribution prepared for European Community's Human Capital and Mobility Program, Contract n. ERBCHRXCT930311, Project 'Geomorphology and Environmental Impact Assessment – a network of researchers in the European Community', unpublished manuscript.

SAMPSON, R.J. (1975a). The SURFACE II graphic system, in Davis, J.C. and McCullagh, M.J. (Eds), *Display and Analysis of Spatial Data*, pp. 244–66, London: John Wiley and Sons.

SAMPSON, R.J. (1975b). *SURFACE II Graphic System*, Kansas Geological Survey, Series on Spatial Analysis 1.

SHAFER, G. (1976). *A Mathematical Theory of Evidence*, Princeton NJ: Princeton University Press.

SHORTLIFFE, E.H. and BUCHANAN, G.G. (1975). A model of inexact reasoning in medicine, *Mathematical Biosciences*, **23**, 351–79.

SINGH, P.K., GRUNSKY, E.C., VANDER-FLIER KELLER, E. and KELLER, C.P. (1993). Porphyry copper potential mapping using probabilistic methods and geographical information systems in British Columbia, *Proc. GIS'93 Symposium*, pp. 381–94, Vancouver, British Columbia.

SUSLICK, S.B. and FIGUEREDO, B.R. (1990). Use of characteristic analysis coupled with other quantitative techniques in mineral-resources appraisal of Precambrian areas in Sao Paulo, Brazil, in Gaal, G. and Merriam, D.F. (Eds), *Computer Applications in Resource Estimation: Prediction and Assessment for Metals and Petroleum*, pp. 155–83, New York: Pergamon Press.

SWITZER, P. and GREEN, A.A. (1984). *Min-max Autocorrelation Factors for Multivariate Spatial Imagery*, Technical Report No. 6, Dept. of Statistics, Stanford University, 14 p.

TURCOTTE, D.L. (1992). *Fractals and Chaos in Geology and Geophysics*. New York: Cambridge University Press.

USERY, E.L., ALTEIDE, P., ROBIN, R.R.P. and BARR, D.J. (1989). Knowledge-based techniques applied to geological engineering, *Photogrammetric Engineering and Remote Sensing*, **54**(11), 1623–28.

USERY, E.L., DEISTER, R.R. and BARR, D.J. (1988). A geological engineering application of a knowledge-based geographic information system, *Proc. American Congress on Surveying and Mapping/American Society for Photogrammetry and Remote Sensing Annual Convention 'The World in Space.' Volume 2: Cartography*, pp. 176–85, St. Louis, Missouri.

VAN WESTEN, C.J. (1993). *GISSIZ: Training Package for Geographic Information Systems in Slope Instability Zonation. Volume 1 – Theory*. International Institute for Aerospace Survey and Earth Sciences (ITC), Publication no. 15, Enschede, The Netherlands.

VAN WESTEN, C.J., VAN DUREN, I., KRUSE, H.M.G. and TERLIEN, M.T.J. (1993). *GISSIZ Training Package for Geographic Information Systems in Slope Instability Zonation. Volume 2 – Exercises*. International Institute for Aerospace Survey and Earth Sciences (ITC) Publication no. 15, Enschede, The Netherlands.

WRIGHT, J.R. (1993). Probabilistic inferencing and spatial decision support systems, in Wright, J.R., Wiggins, L.L., Jain, R.K. and Kim, T.J. (Eds), *Expert Systems in Environmental Planning*, pp. 119–44, Berlin: Springer-Verlag.

ZADEH, L.A. (1965). Fuzzy sets, *IEEE Information and Control*, **8**, 338–53.

ZADEH, L.A. (1978). Fuzzy sets as a basis for a theory of possibility, *Fuzzy Sets and Systems*, **1**, 3–28.

ZIMMERMANN, H. and ZYSNO, P. (1980). Latent connectives, human decision making, *Fuzzy Sets and Systems*, **4**, 37–51.

CHAPTER TWELVE

Problems of integrating GIS and hydrological models

U. STREIT and K. WIESMANN

12.1 GIS for hydrological modelling

The application of GIS for the modelling of hydrological processes has started in hydrology (as opposed to soil science or geomorphology) rather late. An essential cause of this delay was the concentration of research activities on the development of predominantly deterministic, physically based process models for slope segments or small river catchments with a high temporal, but low spatial, resolution. These deterministic models for the simulation and prediction of hydrological processes (for example, evapotranspiration, run-off formation, infiltration) have achieved a high standard of development. Indeed, these hydrologically very complex, but theoretically sound, models require voluminous amounts of input data of high quality. This complicates their transfer from small, predominantly homogeneous, research basins to the large river basins that are very important for water resources management and planning purposes. For macro-scale applications such as global climate models or global change models (GCM) they are practically unusable.

Thus the research interest of hydrologists must increasingly be directed towards the simplification of these physically based models to make them applicable for large-area, inhomogeneous river basins with spatially sparse monitoring networks and with minor quantity or quality in the basic input data. Doing so, it will be necessary to approximate physical laws by statistical relationships. Clearly definable hydraulic parameters must be replaced by so-called effective parameters that can meet the simplified model requirements by numerical optimisation of the input–output relations. Time-dependent parameter functions and spatially variable parameter fields must be reduced to distribution functions for a simplified representation of the spatial and temporal variability, or if necessary, to simple averaged values. The unavoidable loss of hydrological explanatory power of such 'soft' models is compensated by their practical benefit for the solution of current environmental problems to ensure sustainable development. Because of this, although the frequently cited statement 'data rich and theory poor' may be correct for socio-economic and many ecological models, for hydrological models the reverse statement is more correct.

The shortage of suitable basic data for spatially distributed hydrological modelling in large geographically inhomogeneous catchments is an important reason for the late, but now increasing, turn of hydrologists towards GIS. In this process, the application of digital terrain models (DTM) for the automatic calculation of watersheds, flow paths, river networks and hydraulically effective terrain parameters is one of the main points of interest. Moreover, the use of remote sensing data, especially for the survey of current land use, for the analysis of the spatial distribution of snow cover, and for the rough estimation of water contents in atmosphere, vegetation cover and soil, plays an important part. Digital soil maps, geological and ecological maps will quickly gain significance with increasing availability.

A second reason, which is very important for the rapidly growing interest of hydrologists in GIS, are the excellent possibilities for visualising spatial data, particularly the quick creation of hydrological maps for input and output data, as well as for model parameters. Explorative data analysis, visual evaluation of the data quality, and map-based interpretation of intermediate or final results of the model calculations are the most important application fields of GIS visualisation tools.

In contrast to the traditional domains of GIS application (data acquisition, management and cartographic representation), GIS-based methods for analysing spatial data and parameter distributions have played a minor part in hydrology up to now, although they constitute the real core of the GIS methodology. On the one hand, this is due to deficiencies in suitable numerical and statistical methods, which are criticised by many experienced GIS users. Some examples for absolutely necessary but, as a rule, not available GIS functions, are: methods for spatial interpolation, averaging, aggregation or disaggregation; methods for correlation and regression based on spatial units taking into account effects of spatial persistence (spatial autocorrelation); procedures for the numerical or knowledge-based classification (by neural nets) of small partial areas to hydrologically homogeneous response units; methods for generating topological relationships for partial basins according to their run-off contribution in the total river catchment. For example, it is not possible to estimate anisotropic semivariograms with the GRID module of Arc/INFO in a simple way, but anisotropic characteristics of spatial persistence are essential for a spatial interpolation that is dependent on distance as well as on direction. On the other hand, it is a fact that a profound knowledge of applying numerical, statistical and stochastic methods for analysing and modelling spatial data within GIS is not yet widespread in hydrology. The same is true for the ability of hydrologists effectively to use the various other GIS functions already available.

As the first International HydroGIS conference held in 1993 in Vienna has shown, with more than 80 contributions to GIS applications in hydrology and water-resources management, understanding of the necessity and usefulness of GIS techniques is increasing rapidly. The progressive development from spatially lumped models to distributed hydrological models seems to be reasonable only if based on GIS technology. In addition to the predominating type of loose coupling between hydrological models and GIS by means of data interfaces, more than 20 papers have discussed varied opportunities for a tighter integration of models and GIS.

12.2 Approaches to the integration of GIS and hydrological models

According to the objectives and complexity of hydrological modelling, the need for basic data and GIS functions, the availability of interfaces and compatibility of data models, the hardware environment, the system architecture of GIS and modelling software and so on, the method and intensity of integration should be different. The possibilities for integration stretch from very loose coupling by exchanging ASCII files up to a complete embedding of the hydrological model into the GIS.

Nyerges (1992) has developed a conceptual framework for the coupling of spatial analytical models with GIS. He distinguishes between four categories with increasing intensity of coupling:

1. *Isolated applications*. The GIS and the hydrological model may run in a different hardware environments; the data transfer between the possibly different data models is performed manually (for example, by ASCII files) and usually off-line. The user plays the role of the interface; the additional programming expenditure is very low, but the effectiveness of the coupling is limited.
2. *Loose coupling*. Here, too, the coupling is carried out by means of special files (for example, binary files); the user has to care for a suitable structuring and formatting of these files and a cross-index with references between both data models can be useful. Contrary to isolated applications, this kind of coupling is carried out on-line on the same workstation or, if necessary, on different computers in the local network. With a relatively small additional programming effort, the effectiveness of this type of coupling is a little bit greater than with isolated applications, because access to the transferred data can be in a joint computer environment.
3. *Tightly coupled applications*. In this case the data models can still be different, and an automated mutual access to the data must be possible without user interventions. However, a standardised interface can considerably increase the effectiveness of data exchange between both systems, but it requires more programming effort (for example, macro language programming). The user is still responsible for the integrity of the data.
4. *Integrated application*. This type of fully integrated linkage operates like a homogeneous system with a joint design from the user's point of view. The data accesses are based on a coherent data model and a common database management system; interactions between both subsystems are very simple and effective in this case; however, the expenditure of development may be huge; using a joint programming interface (as much as possible with a standard computer language like C) the integrated system can be extended by further GIS modules or additional model functions.

Chou and Ding (1992) present another classification of integration methods for GIS and spatial models. They give emphasis to three aspects of software design for a taxonomy of system integration.

1. *Data sharing*. The simplest, and nowadays most frequently used, way of a shared access to GIS and model data is data transferring by temporary files. The advantages are those of flexible handling and simple design. The disadvantages are slow data access and the indirect use of the GIS data model. Direct access of the GIS and the model to a shared database avoids these disadvantages, but requires an essentially higher expenditure for data modelling and implementation.
2. *Modelling technique*. An external modelling offers the advantages of an independent and flexible development and testing of the model, but requires a more or less loose coupling by some execute-function or macro-functions of the GIS afterwards. At present, this low-level integration is the dominant mode of linking hydrological models with GIS. In contrast, an internal modelling would have the important advantages of the full usability of all functions and data resources within the GIS, but only a few of the present GIS offer sophisticated numerical and statistical functions

and, especially, direct interfaces for standard computer languages (like C, PASCAL, FORTRAN) which are necessary for internal modelling.
3. *User interface*. Shifting between different user interfaces, which is frequently still necessary today, should be outdated. Standardised graphical user interfaces (GUI) like Windows or X-Windows can simplify the user's work. At least the developers of different systems should try to design the same 'look and feel' in their user interfaces. However, for the long term integrated user interfaces, from which the user can call up all GIS tools and modelling functions, will prevail. In Section 12.4 the development of such an integrated user interface is described.

Chou and Ding (1992) recommend a five-step procedure to couple GIS and spatial models.

1. Identify the integration purpose and performance requirements.
2. Design a conceptual model.
3. Map the conceptual model on to software data models.
4. Evaluate different integration alternatives.
5. Implementation.

This step-by-step procedure agrees with common sense and corresponds to basic principles of software engineering. Other papers dealing with classification problems of GIS integration (for example, Goodchild *et al*., 1992; Brilly *et al*., 1993) will not be discussed here in detail, because they do not differ substantially from the approaches mentioned above.

12.3 Problems of integration

The HydroGIS conference of 1993 showed that most of the currently used ideas of coupling GIS and hydrological models can be assigned to the types of isolated applications or loose coupling (Nyerges, 1992). At present many conceptual, methodological, technical and practical problems complicate the tighter coupling which is demanded by GIS users in applied hydrology and water resources management (for example, Fedra, 1993).

1. The greatest conceptual obstacle to a tighter linkage of GIS and hydrological models is the missing temporal dimension of GIS (Kemp, 1993). Time has special importance for modelling hydrological processes, because many relevant partial processes of the hydrological cycle (particularly precipitation, infiltration, evapotranspiration, run-off generation, and run-off concentration) progress with a high dynamic. In a technical sense the time loop has priority over spatial variability in almost all hydrological models. Only in rare cases is the spatial variability of processes really the dominant feature of the model architecture. In today's GIS temporal variation can be modelled only as a sequence of singular spatial conditions. To do this would need ideas of four-dimensional GIS within which three-dimensional spatial structures (for example, for soil and groundwater simulation) could be used, as well as the time domain. GIS running on massive parallel-

processing computers would be an optimal solution for hydrological simulation. In this, every subprocess in every hydrological response unit would have a processor of its own, a well co-ordinated cluster of several processors would be responsible for the entire hydrological process (for example, transformation of rainfall to run-off) in every response unit, and a set of clusters, structured according to hydrological–topological aspects, would simulate the total process within the spatio-temporal GIS. Hopefully this is not science fiction.

2. The greatest practical obstacle to integrating GIS and hydrological models is the almost closed-system architecture of commercial GIS. Even in widespread relational database management systems (RDBMS), import or export interfaces are not completely available or are usable only with difficulties. Direct access to such RDBMS from GIS is feasible at best for attribute data. For geometry data, a standardised conceptual and logical data model and a standardised data definition and acquisition language similar to the well-established SQL are in a phase of research rather than practical testing. The power of GIS-specific macro languages (for example, AML in Arc/INFO) is usually not sufficient for purposes of advanced hydrological modelling (for example, solving differential equations). Programming interfaces for standard computer languages with access to internal GIS functions are an exception rather than common or standard. Numerical and statistical procedures in GIS are frequently not documented with adequate precision to enable the user to judge the conditions and restrictions of their application.
3. The personnel and financial expenses of a tight coupling of hydrological models, or model components, with GIS can be very large. University groups can at best develop small prototypes because of frequent personnel fluctuations and low financial support. For GIS companies, the turnover expectations in hydrology and water-resources management to date obviously have been too low to give a push to such developments. On the other hand, this integration of hydrological methods and models, which is still insufficient, makes a broad breakthrough of GIS application more difficult. A solution to this problem seems to be possible only if university working groups, GIS companies and potential GIS users co-operate more intensively than during the past years, and if the market-dominating GIS companies are willing to promote an open GIS design giving innovative working groups a better chance for experimental and specialised solutions.

In addition to these conceptual, methodological, technical and practical problems, user-related obstacles must be considered as well. The level of knowledge concerning the chances of problem solving, and especially practical experience with GIS, have been very limited in hydrology. For methodically rather conservative officials of a water resources administration or a hydraulic engineer in a planning institution, a multimedia GIS may mean a show effect rather than a practical benefit. Because of the obviously unsurmountable misunderstanding of 'intelligent' computers, knowledge-based components seem to be the more successful the more skilfully they are camouflaged as normal software. The 'promotion' of GIS to spatial decision-support system (SDSS) or Geoscientific Information Systems (GSIS as three-dimensional scientifically oriented expansion of the traditional GIS) have a counterproductive effect on pragmatically thinking and operating decision makers in the water authorities. Moreover, some hydrologists worry about the risks of having sophisticated hydrological models with very sensitive parameters without any control at the disposal of GIS users unacquainted with the pitfalls of hydrologic modelling (Grayson *et al.*, 1993). Indeed, integrated systems with user-friendly interfaces and excellent possibilities of visualisation may lead hydrologically inexperienced users to a careless handling of very complex models.

12.4 T_4HIM: tools for hydrological information and modelling

The authors of this chapter have seen the need for a greater integration of GIS and hydrological models, but have also seen the conceptual and practical problems involved through their participation at a special research initiative of the German National Science Foundation (DFG) concerning regionalisation in hydrology. For four years, 16 university research groups in Germany have worked together to study problems and methods of transferring hydrological process models from small research catchments to large river basins. Our study group from the University of Münster has taken over the task of defining a common GIS platform and turning GIS to practical use. For the hydrologists, hydraulic engineers, ecologists and geoscientists involved in the special research initiative, advanced GIS training courses are carried out. However, the main focus of the Münster study group is the development of special GIS tools for the modelling of hydrological processes as well as for supporting regionalisation. The ultimate goal is the construction of a standardised GIS-supported system that is useful not only for the initiative, but also for the solution of practical environmental problems.

12.4.1 Definition of a common GIS platform

The first important step consisted of the selection of a suitable GIS for the special problems of regionalisation in hydrology. A thorough analysis of the GIS market and a detailed test of performance of the various GIS are basic requirements for this step. Essential boundary conditions were the executability on UNIX workstations, the existence of interfaces to relational database systems (particularly to ORACLE as the common DBMS), and a reasonable relationship of costs and performance. However, a thorough analysis of the demands of the different study groups on the GIS functionality was equally important. The decision process was performed in the three steps involving pre-selection, rough selection and concluding selection of the GIS. A comprehensive description of the complete analysis is found in Büscher *et al.* (1992).

The best solution for the purpose of the special research initiative has proved to be the construction of a jointly used GIS platform that is composed of the two GIS GRASS and Arc/INFO. The public-domain system GRASS is available in its source code and can be extended easily by self-written modules in C. Arc/INFO is a powerful and widespread commercial GIS. It offers a wide range of vector- and raster-based GIS functions and provides with ArcView an user-friendly tool for interactive data queries and visualisations. The same combination of GIS is used in some other research projects (for example, Tuttle, 1991).

At the beginning of the research initiative, the practical experiences of the hydrological study groups with GIS varied considerably. To ensure an effective application of the GIS platform, several advanced GIS training courses were carried out by the Münster study group. In the meantime, the knowledge in all study groups has been strengthened so much that the development of special GIS tools for modelling hydrological processes can be discussed between modellers and GIS developers.

12.4.2 Coupling of the GIS-platform with hydrological models

The motivations of the hydrological study groups for an application of GIS are very different. They extend from a pure spatial-data management and cartographic representation to numerical and statistical analysis using GIS standard functions (for example, polygon overlay and spatial interpolation). With increasing experience in GIS application, an extension of the usual GIS functions by more specific hydrological GIS tools is requested. A complete integration of hydrological models into the GIS platform, in the sense of the classification of Nyerges (1992), is not really wanted by the hydrological study groups at this moment in time. The concentration of the study groups on the hydrological problems of the spatial transfer and simplification of micro-scale models to large river basins is the most important response. Moreover, a completely integrated application is complicated by the closed data model of Arc/INFO for geometry data and by the ORACLE interface of GRASS, which is still missing.

Therefore, the aim of the current work is the realisation of a tight coupling of hydrological models and GIS to an extent as large as possible. The conceptual framework of this coupling is presented in the Figure 12.1.

The hydrological models are bound into the 'modular hydrological modelling system' MHMS of the US Geological Survey. MHMS is conceived as a tool to assist the programmer of sequential time-step hydrological models by providing essential features like data handling, user interface, sensitivity analysis and optimisation routines (Leavesly *et al.*, 1992). A special 'model builder' is used to assist the hydrologist in interconnecting

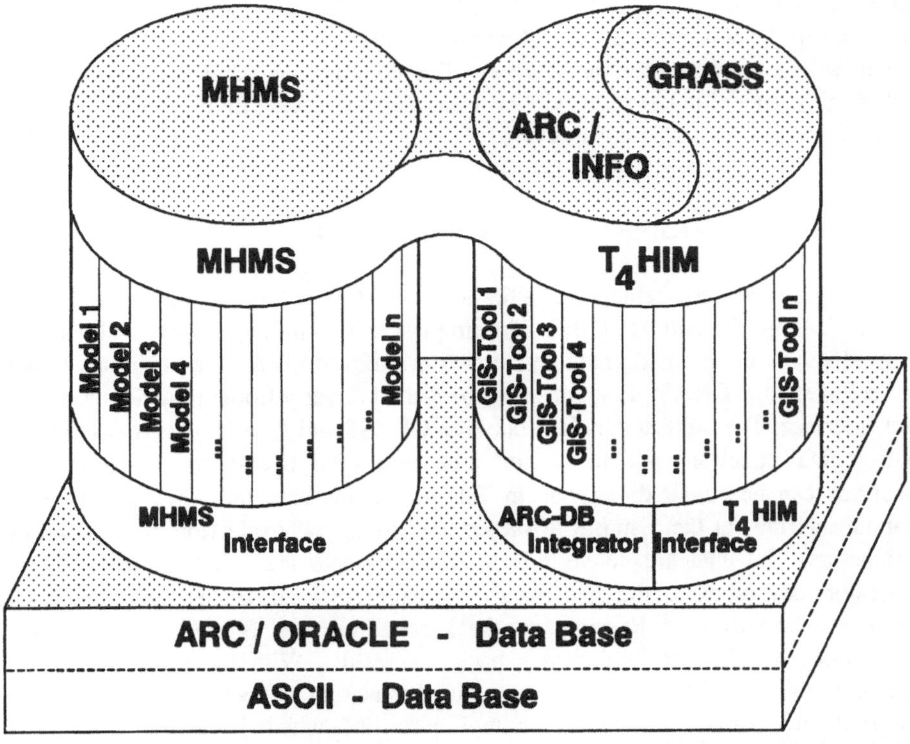

Figure 12.1 Coupling of the modular hydrological modelling system MHMS (US Geological Survey) with GIS-toolbox T_4HIM.

modules from a module library (precipitation run-off modelling system PRMS of the US Geological Survey), thus producing an executable MHMS model. Additional hydrological models, as for instance TOPMODEL (Beven and Moore, 1992), are embedded into MHMS.

The Münster study group is developing a GIS toolbox called T_4HIM (Tools for Hydrological Information and Modelling) for the integration of spatial hydrological GIS tools with the GIS platform. The GIS workshops have shown the necessity of developing such a flexible toolbox with an ergonomic graphical user interface.

T_4HIM is realised using the Arc macro language (AML) of Arc/INFO and adjusted completely to the current standard of development defined by Arc/INFO's developing company ESRI. The advantages of such a standardisation are the guarantee of a high standard of software quality and the compatibility with new versions of Arc/INFO. Moreover, already available commercial and public domain GIS tools (for example, Arctools, Alacarte) can be integrated easily. Unnecessary work with the development of GIS modules can be avoided in this way. T_4HIM runs on different UNIX systems with X-Windows based user interfaces (OSF/Motif and Open Look are standard).

As far as possible, T_4HIM and MHMS use the same data model and a common database. Spatially referenced data are managed in the GIS-specific databases in Arc/INFO and GRASS. MHMS models do not use any spatial geometry, thus an updating of the spatial entities in the process of hydrological modelling is the user's task. The management of the hydrological attribute and time series data is carried out by the DBMS ORACLE and an additional file system respectively. Since an ORACLE interface in GRASS is still missing and the Arc/INFO–ORACLE interface is difficult to handle (especially in the case of data trees with many branches), automated data transfer takes place partly by direct access to the RDBMS ORACLE and partly by ASCII exchange formats. For the new version 7.0 of Arc/INFO some improvements have been announced which will simplify direct data access. The coupling of the GIS platform and the hydrological modelling system will be improved in co-operation with the developers' group of the US Geological Survey.

12.4.3 Coupling of Arc/INFO and GRASS within T_4HIM

According to the requirements of GIS tools, T_4HIM offers different options for the coupling of GRASS and Arc/INFO. The simplest way of linking is a parallel operation of both GIS with an automatic data conversion. Moreover, GRASS applications consisting of a set of special GRASS functions may be started directly from the AML-based T_4HIM user interface. The tightest coupling between GRASS and Arc/INFO is realised for those hydrological functions which execute pure numerical operations in GRASS. These functions can be called directly from T_4HIM. In this case the AML interface puts a graphic interface at the disposal of the user, enabling easy selection of data and model parameters. The data are converted automatically into the suitable format. A script-generator, developed in the Arc macro language AML, produces the necessary shell scripts for the GIS tools. Running as a background process, GRASS subsequently takes over, processes the data and writes back the results. With the standardised T_4HIM interface, the user has access to the voluminous tool collection of two completely different GIS. An example is given in Figure 12.2 which shows the GRASS-based calculation of a so-called 'time–area map' for the modelling of the process of run-off concentration.

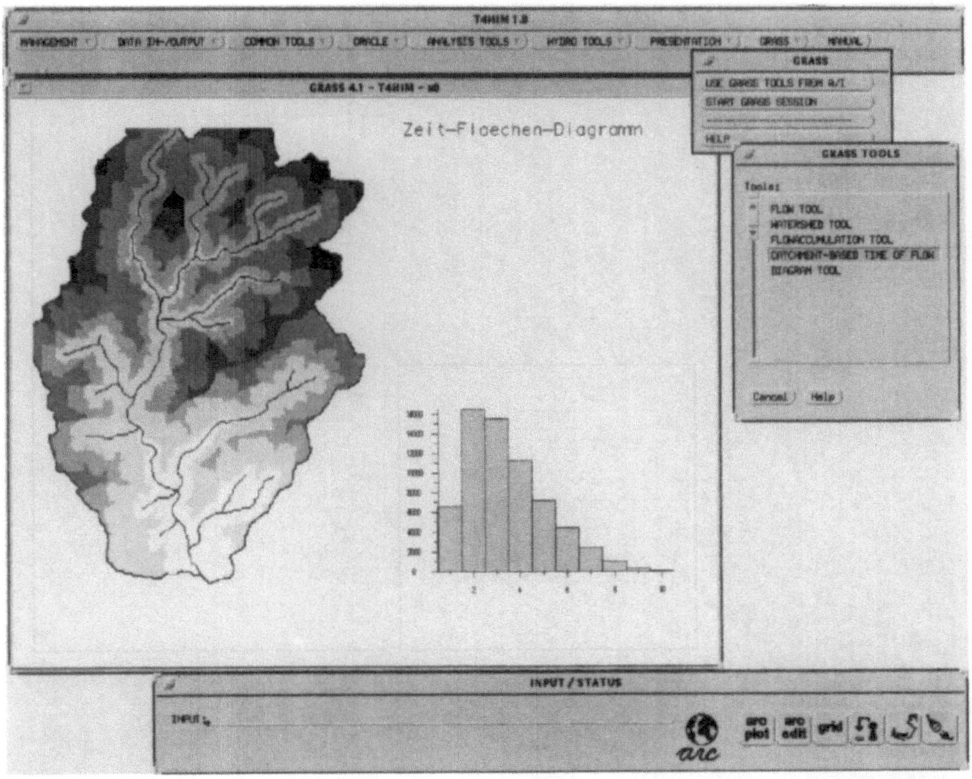

Figure 12.2 Calculation of time–area maps for modelling the process of run-off concentration using the GIS GRASS under T_4HIM.

12.4.4 State of the development of T_4HIM

The development of the GIS toolbox T_4HIM started in July 1992. As quickly as possible, a first prototype was needed to be put at the disposal of the other study groups of the special research initiative. This prototype T_4HIM 1.0 has been distributed since September 1993. It offers a wide range of standard tools for the management and processing of geocoded data sets and for the cartographic representation of basic data and hydrological results. Examples are (Figure 12.3):

- Automated import and export of data to/from different database systems and GIS
- Interactive manipulation of attribute and geometry data
- Logical and algebraic overlays
- Standard layouts for hydrological maps.

The finished hydrological tools focus mainly on the DTM-based calculation of hydromorphometric parameters, the automatic delimiting of sub-basins and methods of spatial interpolation. Hydrological standard procedures for the calculation of flow directions and indices of run-off accumulation have been implemented for different GIS and data models (Arc/INFO-raster, GRASS-raster, Arc/INFO-TIN CASCA-DING). Self-defined expansions (for example, divided flow accumulation in the case of ambiguous flow directions) have also been realised. For the T_4HIM prototype most of the GIS tools have been developed by pure macro-language programming based on Arc/INFO's GIS functions. Already completed are GIS tools for the automated detection of river networks

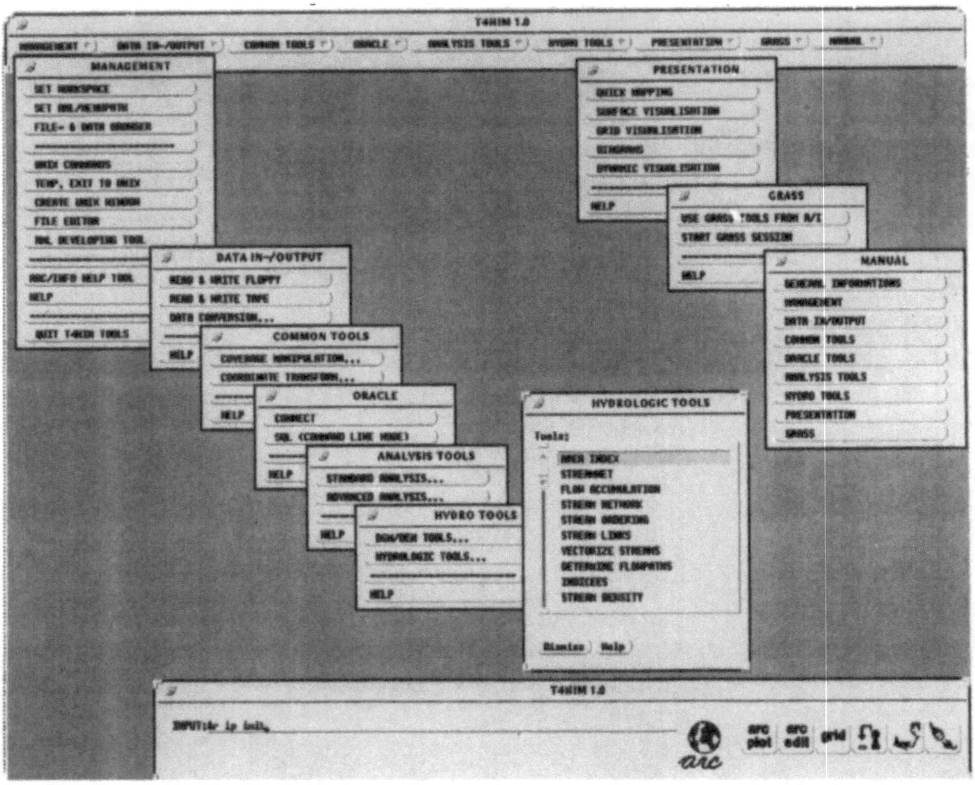

Figure 12.3 User interface of T_4HIM with various tools for data manipulation, GIS functions, hydrological modelling and visualisation.

and for the areal partitioning of river basins according to different user-defined criteria (predefined or interactively marked reference points, given thresholds for the area of the sub-basins). The topology of a set of sub-basins can be generated automatically on the basis of a digital terrain model (DTM), as shown in Figure 12.4. Such a topology forms the spatial structure for the hydrological modelling in MHMS. On the basis of available GIS functions, hydrological tools have been developed for the spatial interpolation of point data (for example, Thiessen polygons, inverse distance weighting, trend surface interpolation, kriging). Areal means may be estimated by similar geostatistical procedures. Figure 12.5 shows estimated areal precipitation depths for sub-basins which have been automatically delimited on a DTM basis.

To sum up, T_4HIM 1.0 offers a comprehensive hydrological GIS toolbox ready for use that should be completed gradually by additional tools relevant for hydrological modelling and spatial hydrological analyses.

12.5 Preview of forthcoming developments

In the next four years the hydrological toolbox T_4HIM is to be extended by additional tools. Conceptual studies for an interactive definition of homogeneous hydrological response units have already begun. The main focus is to be on the development of more sophisticated visualisation tools to support the analysis of hydrological data. Cartographic tools will be improved to support the presentation and interpretation of time-variant

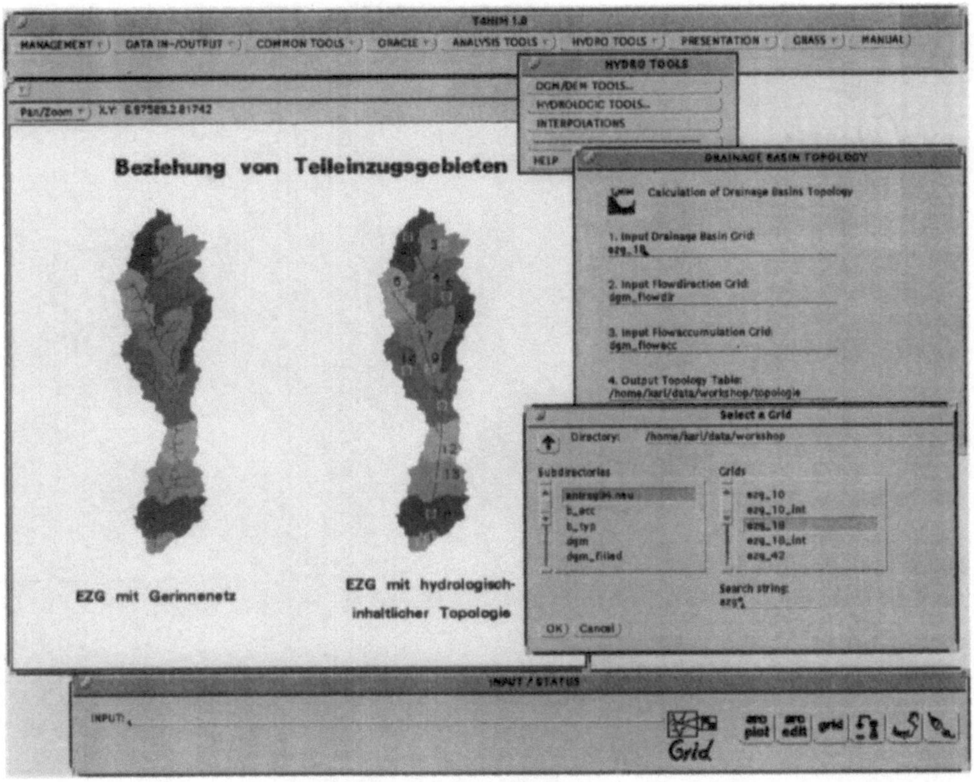

Figure 12.4 GIS tool of T_4HIM for automatic generation of topological relationships of river sub-basins used for hydrological modelling with MHMS.

hydrological processes. The analysis of the spatial distributions of measured data and hydrological parameters requires improved GIS functions as, for example, co-kriging and spatial filter operations. For the very important problem of changing the spatial scale of hydrological models, the available methods of spatial aggregation (for example, distribution functions instead of spatial means) and disaggregation (Streit and Paus, 1988) have to be improved and embedded into the GIS toolbox. All-in-all, the main focus of GIS application in hydrology should be shifted from a pure pre- and post-processing of the data to spatial analysis functions with the assistance of T_4HIM. The long-term aim is the full integration of hydrological models into the GIS platform. Additionally, two knowledge-based components are planned for T_4HIM to make the accumulated experience with hydrological GIS applications more easily available to users outside the special research initiative. For inexperienced GIS users in hydrology, an advisory system is to be developed which provides specific information about the technical use of GIS and T_4HIM. For this purpose, the PRO_PLANT expert system shell, developed at the Institute of Agricultural Informatics in Münster for the realisation and maintenance of the plant protection advisory system (Streit and Frahm, 1992), can be used. Alternatively, a graphically supported hypertext system may be considered. As a second step, it is intended to develop a decision support system for the proper choice of hydrological data, methods, tools and models.

If spatial analysis models and GIS tools are designed to meet the requirements of the planning practice, and if pragmatic solutions are found for actual problems, then reservations still existing about new methods and models can be quickly reduced. By the

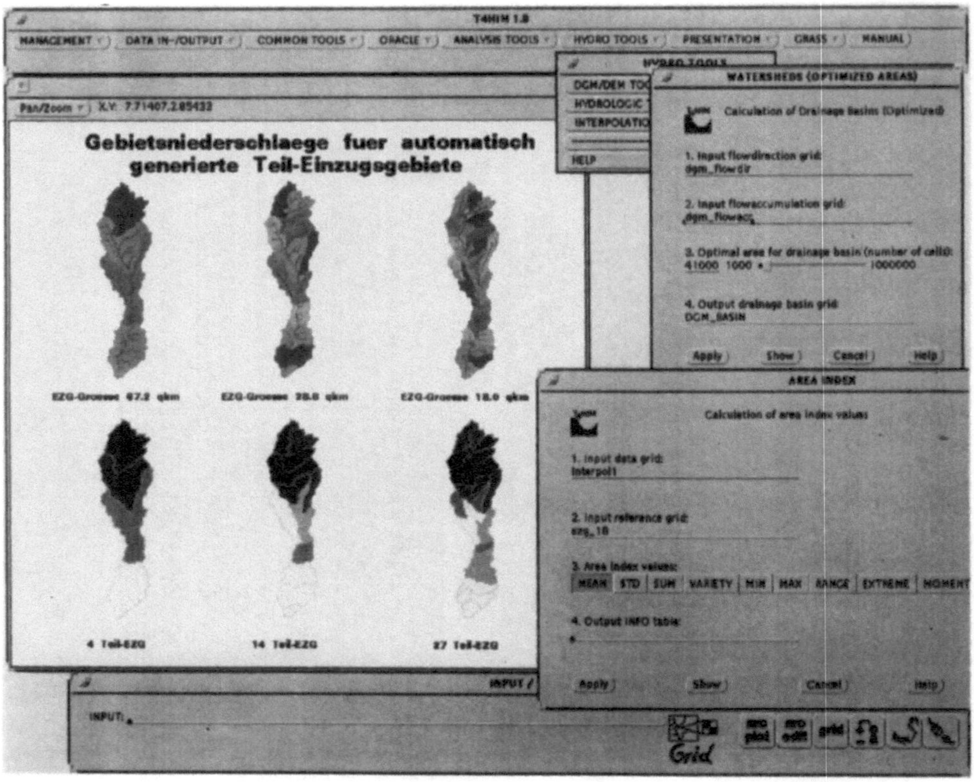

Figure 12.5 GIS tool of T_4HIM for automatic generation of river sub-basins and calculation of areal rainfall means.

example of a GIS-based system for the inventory and ecological evaluation of river systems, this fact has been demonstrated. For pen-top computers running under PEN-Windows, several GIS functions have been integrated with knowledge-based techniques (for example, rule-based tests of semantic plausibility or neural networks for classifying aquatic biocenoses) and digital image processing methods (for example, visualisation of multispectral satellite data and digital photography). An on-line operation of the differential global positioning system (DGPS) is in preparation. Meanwhile this portable 'GISpad' system (Remke *et al.*, 1993) with handwriting recognition is used for river-ecological investigations in the field by several environmental institutions and planning authorities in Germany. Based on these experiences, T_4HIM is to be developed to a complete GIS-supported 'workbench' for practical applications in hydrology and water resources management.

REFERENCES

BEVEN, K.J. and MOORE, I.D. (Eds), (1992). *Terrain Analysis and Distributed Modelling in Hydrology*, Chichester: John Wiley.

BRILLY, M., SMITH, M. and VIDMAR, A. (1993). Spatially oriented surface water hydrological modelling and GIS, in Kovar, K. and Nachtnebel, H.P. (Eds), *Application of Geographic Information Systems in Hydrology and Water Resources Management*, Proc. Hydro GIS '93, Vienna, pp. 547–75, IAHS Publ. no. 211, Wallingford: IAHS Press.

Büscher, K., Kirchhoff, Ch., Streit, U. and Wiesmann, K. (1992). Vergleich der Nutzbarkeit und Auswahl von GIS für die Regionalisierung in der Hydrologie, in Streit, U. (Ed) *Werkstattberichte Umweltinformatik – Agrarinformatik – Geoinformatik*, **1**, 37, Münster: Universität Münster.

Chou, H. and Ding, Y. (1992). Methodology of integrating spatial analysis/modeling and GIS, *Proceedings of the 5th International Symposium on Spatial Data Handling*, Charleston, South Carolina, **2**, pp. 514–23.

Fedra, K. (1993). Models, GIS, and expert systems: integrated water resources models, in Kovar, K. and Nachtnebel, H. P. (Eds), *Application of Geographic Information Systems in Hydrology and Water Resources Management*, Proc. HydroGIS '93, Vienna pp. 297–308, IAHS Publ. no. 211, Wallingford: IAHS Press.

Goodchild, M., Haining, R. and Wise, S. (1992). Integrating GIS and spatial data analysis: problems and possibilities, *International Journal of Geographical Information Systems*, **6**(5), 407–23, London: Taylor & Francis.

Grayson, R.B., Blöschl, G., Barling, R.D. and Moore, I.D. (1993). Process, scale and constraints to hydrological modelling in GIS, in Kovar, K. and Nachtnebel, H. P. (Eds), *Application of Geographic Information Systems in Hydrology and Water Resources Management*, Proc. HydroGIS '93, Vienna pp. 83–92, IAHS Publ. no. 211, Wallingford: IAHS Press.

Kemp, K.K. (1993). Environmental modelling and GIS: dealing with spatial continuity. in Kovar, K. and Nachtnebel, H. P. (Eds), *Application of Geographic Information Systems in Hydrology and Water Resources Management*, Proc. HydroGIS '93, Vienna pp. 107–15, IAHS Publ. no. 211, Wallingford: IAHS Press.

Leavesly, G.H., Restrepo, P., Stannard, L. and Dixon, M. (1992). A modular hydrologic modeling system – MHMS, *American Water Resources Association, 28th Annual Conference ans Syposium*, Reno, Nevada, 1.-5.11.1992.

Nyerges, T.L. (1992). Coupling GIS and spatial analytic models: Spatial Data Handling, *Proceedings of the 5th International Symposium on Spatial Data Handling*, Charleston, South Carolina, USA, 3.-7.8.1992, 2, pp. 534–43.

Remke, A., Pundt, H., Bluhm, M. and Streit, U. (1993). Wissensbasierte GIS-Werkzeuge zur Unterstützung fließgewässerökologischer Planungen, in Jaeschke, A. *et al.* (Eds), *Informatik für den Umweltschutz*, pp. 244–55, Heidelberg: Springer.

Streit, U. and Frahm, J. (1992). Entwicklung eines wissensbasierten Systems zur verbesserten Pflanzenschutzberatung, *Informatik-Fachberichte (Springer)*, **301**, 410–27.

Streit, U. and Paus, L. (1988). Construction of run-off maps derived from basin characteristics, *Geologisches Jahrbuch*, **A104**, 373–82.

Tuttle, M. (1991). Multiparticipant co-operation sparks success in river mapping project, *GIS World*, **5/6**, 43–5.

CHAPTER THIRTEEN

Spatial pattern of ecological processes: the role of similarity in GIS applications for landscape analysis

ENRICO FEOLI and VINCENZO ZUCCARELLO

13.1 Introduction

Space is where events may take place. Events are everything that could happen, they may be related to object movements or object functions or to both of them. Objects themselves are events: an object is a realised event, or a set of realised events. The space that we can observe and measure in some way is considered the 'real space' (geographical space or more generally phenetic space), while the space that we cannot see directly but only imagine and/or describe mathematically is abstract space. Both may be represented by maps. Abstract space is defined by the variables describing objects and/or phenomena and by objects and phenomena themselves. Variables and objects may be used as reference axes to represent and map the abstract space by ordination techniques (Orlóci, 1966, 1967, 1978; Gabriel, 1971; Legendre and Legendre, 1983; Jongman *et al.*, 1987; Whittaker, 1990; Feoli and Orlóci, 1991). Time may be also included as an axis of the abstract space.

Pattern is concerned with the location of events in both spaces. Orlóci (1988) considers pattern as the manner of arrangement in space/time of objects and relationships. This definition is very general; however, we would like to make it more general as follows: 'Pattern is the manner of arrangement of events in space/time'. In effect, any ecological study can result in a spatial pattern analysis of ecological events, and as such it uses abstract spaces with the aim of sampling the ecological space (Feoli and Orlóci, 1991). This sampling is done in order to describe and understand the relationships between and within living organisms (individuals, populations and communities) and between them and the chemical–physical environment. Abstract space is multidimensional, defined by all the factors (ecological factors), biotic and abiotic, controlling the biological processes that are the sources of ecological events. In this space each individual assumes during its life a precise set of co-ordinates that defines its life trajectory. The set of trajectories of individuals of the same species is included into a portion of the multidimensional space (hypervolume) that constitutes the niche of the species. The niche concept as formulated for species can be extended to any other kind of living category: taxonomic, ecomorphologic, ecophysiologic or functional and to vegetation types and ecological communities (Ganis, 1991a, b; Feoli *et al.*, 1991). Any event that takes place in the geographical space ('real space') is actually also taking place in the ecological space (abstract space). This chapter discusses what it means to describe the ecological processes by abstract spaces and what it means to transfer the results obtained by abstract spaces into maps representing the geographical space. This is done with the aim of illustrating the possibility of producing and representing hierarchical classifications of landscape based on ecological processes and measuring the predictivity of different descriptions by coupling the GIS technology with non-standard packages of multivariate data analysis based on probabilistic similarity functions (Goodall, 1964, 1966, 1968; Feoli and Lagonegro, 1983; Goodall *et al.*, 1991).

13.2 Ecological processes, spatial patterns and maps

Everywhere that there is a living organism, ecological processes take place since the organism is interacting with the environment. If there is a living organism in a place, it means that the environmental conditions let it survive there for at least some time. There are many ecological processes at all levels of biological organisation and scale from individuals, populations, communities, to sets of communities living in a given territory. They may be related to nutrition, growing, respiration, reproduction, death, decomposition, movement, and so on. The processes may be described by one variable at a time (for example, temperature, evapotranspiration, CO_2 absorption, biomass, energy, number of individuals born or dying or moving from site to site, single nutrients, pH and so on) or by several variables together. These are the chosen state variables of the system under study and they partially define the abstract space where the processes may be described by trajectories. The ecological processes can be divided into two rough categories: short term and long term. The first are those that must be recorded in a short time such as respiration and photosynthetic activity. The second may be recorded only during long-time observations, such as, for example, ecological succession, where the change of the community states is given by species substitution. This classification is evidently fuzzy because the speed of a process may be seen in relative terms and it may also be scale-dependent. Population dynamics is recordable in the short term for small organisms but it needs a long time-span for big organisms. However, the classification of processes into short and long time may be useful from an operational point of view.

The spatial pattern of ecological processes is related to the spatial pattern of living organisms and/or communities (Turner, 1989; Leps, 1990). These are the 'actors' of the processes. The spatial pattern of living organisms, however, is defined by the spatial pattern of combinations of environmental factors. These are controlling the physiological processes that make possible the life of the specific organisms in given sites. Ecological processes are strictly related to physiological processes of species. For this reason in ecology there is the need to group species into functional groups (carnivore, herbivore, euthrophic, oligotrophic, heliophilic and so on) irrespective the classic traditional taxonomy, as was firstly done by E. Warming (see *The History of Ecology* by Acot, 1989). In these terms, species that are taxonomically different may be ecologically similar or even so identical as to occupy the same ecological niche in the abstract space. A parallel taxonomy, however, is needed for ecological studies (Orlóci, 1991a) and the species concept remains the most operative one. Species are identifiable and may be described by ecophysiological characters. If a database of species description is available it is easy to 'translate' a taxonomic description of sites into other descriptions (Feoli, 1984; Feoli and Orlóci, 1991).

In ecology, spatial pattern analysis can be accomplished for the biological components (species, communities) and for the chemical–physical component (environmental factors). For the biological component the pattern analysis can be done by one species (monospecific pattern) and by many species (multispecific spatial pattern) at the same time (Bouxin, 1991). Podani (1991a) defines the first as autophenetic and the second as synphenetic. In autophenetic pattern analysis (APA) attention is mainly focused on departures from randomness. This is measured by different statistical models (Hill, 1973; Diggle, 1983; Greig-Smith, 1983; Bouxin, 1991; Podani, 1991a; Ver Hoef *et al.*, 1993). In synphenetic pattern analysis (SPA) attention is mainly given to similarity or resemblance between plots of different size and shape (quadrats, circles, rectangles, and so on) and to some concepts that are not familiar outside quantitative plant ecology, such as expected similarity (or resemblance), distinctiveness, associatum, and dissociatum (Juhasz-Nagy and Podani, 1983; Bartha, 1990; Podani, 1991a). We do not want to review these concepts in this chapter, but want to stress that SPA is the fundamental step not only to understanding the relationships between species and areas (or other spatial entities and area) but also to typifying ecological communities. This is an essential step for producing ecological maps of a given territory (for example, Feoli *et al.*, 1992) and maps of the community niches in the abstract spaces (for example, Banyikwa *et al.*, 1990; Feoli *et al.*, 1991). Spatial pattern analysis of chemical–physical variables can also be accomplished for single variables or for many variables together and in this case typification could be required (definition of environmental types: climatic types, pedological types, geological types, and so on). Sun and Feoli (1992) provide an example of typification based on climatic data and a climatic map of China based on such a typification.

Maps are necessary tools for spatial pattern analysis in both APA and SPA and in both real and abstract spaces. Maps can show the spatial distribution of individuals of the same species (autophenetic map) or different species (synphenetic map), and can show the distribution of one community (autophenetic map) or more communities (synphenetic map). Maps can show the distribution of one environmental type (autophenetic map) or more than one environmental type (synphenetic map). If the maps represent a partition of a given territory or a given abstract space by contiguous areas (or volumes) each one considered uniform (choropleth) and different from the adjacent ones they are called choropleth maps (Robinson *et al.*, 1984; Burrough, 1987). Choropleth maps may be obtained both in SPA and in APA, however, in the first instance choropleths represent in a discrete way the spatial variation of the 'assemblage' of variables between different types (ecological or environmental types); in the second instance, they represent, in a discrete way, the spatial variation of the intensity of a single variable (entity belonging to a taxonomic or ecomorphologic class or one environmental factor). In APA the choropleths are automatically given by the method used, for example, by trend surface analysis (Burrough, 1987; Mucina *et al.*, 1991; Hauser and Mucina, 1991). In SPA the uniformity (or homogeneity) has to be defined in some way, such as through the application of similarity functions and clustering algorithms.

The phenetic maps showing the distribution of biological and environmental variables in the 'real' space are indispensable tools for sampling ecological processes to obtain autofunctional maps and/or synfunctional maps. The processes may be long term or short term, continuous or discrete, cyclic or non-cyclic, trended or non-trended, but in all cases they have to be described as a sequence of maps at regular or irregular time intervals (Ord, 1979). Each map showing the pattern of a process in the 'real' space actually represents the effects of environmental factors on that process. In other words, it describes in a synthetic way the environmental landscape heterogeneity with respect to that process at time T_i irrespective the phenetic description.

13.3 OGUs, similarity functions, classification and choropleth maps

By studying the spatial pattern of the ecological components and processes it is necessary to define the operational geographic units (OGUs) that have to be described and compared. The concept of OGU has been introduced in biogeography by Crovello (1981), but it is useful also in landscape ecology. In APA, OGUs are points or small areas representing georeferenced individuals. In SPA, they are georeferenced areas. Computer programs are available for defining their optimal shape and size (Podani, 1984a, b; 1991a, b; Orlóci, 1991b; Orlóci and De Patta Pillar, 1991).

When the landscape is recorded by remote sensing techniques the OGUs are automatically given by pixels and described by the wavelength of the radiation used. Dobson (1993) discusses the possible integration between remote sensing techniques and GIS for building a 'better macroscope' for landscape analysis. Remotely sensed data may be used in both APA and SPA, provided a correct identification of species and community is available within OGUs given by pixels. When ecological processes are studied, new OGUs have to be sampled according to some sample design. In fact, it is impossible to record the functional variables everywhere within polygons or within all the pixels or individuals belonging to a choropleth. Sometimes it is assumed that within a phenetic choropleth the process is developing uniformly, following a random variation, but this has to be proved. Different organisms and/or different communities, or the same organism and/or community could respond differently or in the same way to different combinations of environmental variable states. This is well shown by the unimodal response curves of the coenocline model in direct or indirect gradient analysis (Whittaker, 1967, 1978; Orlóci, 1978).

The sampled OGUs can be used to obtain functional maps by many different methods. These may be grouped into two main approaches:

1. classificatory approach producing choropleth maps;
2. non-classificatory approach producing maps of isolines and/or splines.

A method not yet explicitly explored in landscape ecology is the comparison between the maps of 'real' and maps of abstract spaces. It is clear that proximity of OGUs in the abstract space means ecological similarity while distance in ecological space means ecological dissimilarity. If OGUs are distant in the abstract space and close in the 'real' space it means that there is a sharp environmental change in the landscape from one OGU to the other one.

There are very few examples of spatial pattern analysis of ecological processes. Kenkel (1991) presents an example of APA dealing with mortality of *Pinus banksiana* in Manitoba (Canada) by using a spatial tessellation model. Feoli *et al.* (1992) present an example of SPA by producing choropleth maps of the state of the ecological succession in the Karst region of NE Italy. This last example, based on the application of probabilistic similarity functions (Goodall, 1964, 1966, 1968; Goodall *et al.*, 1991), is one in which the similarity concept plays a fundamental role in producing maps of the 'real' space out of the abstract space. The abstract space has been used to measure the similarity between OGUs and it has been mapped to the 'real' space. In the maps, produced by a GIS, choropleths represent the probabilistic similarity that each OGU has with the vegetation type considered the climax of the area.

Although choropleths have several drawbacks (Burrough, 1987), they may be used in a process of successive approximation (Orlóci, 1991b) for their more precise definition, provided suitable operational geographic units (OGUs) are available. In this context techniques for edge detection should be used (Brunt and Conley, 1990; Orlóci and Orlóci, 1991).

The representation of spatial pattern by choropleth maps does not prevent the use of other techniques, for example on choropleth maps, isolines of variables (simple or composite) may be superimposed, and nor do they prevent the application of other spatial pattern analytical techniques, such as time series, spectral and autocorrelation analysis (Ord, 1979; Bennett, 1981; Cliff and Ord, 1981; Hepple 1981; Burrough 1987; Wildi, 1991a).

The classificatory approach is based on the concept of equivalence: objects considered similar enough are equivalent and grouped into the same class (class of equivalence). The similarity can be defined on the basis of one or many variables (monothetic and polythetic similarity respectively). There are several similarity functions. Dale (1991a) provides a review of them based on the 'Levenshtein distance' but, from a computational point of view, they may be roughly classified into four main categories:

1. based on the geometry of multidimensional spaces, such as Minkowski metrics and scalar products (for example, Euclidean distance and cosine of angles), or non-metric measures such as the Calhoun's distance which considers the number of points between two points (Orlóci and Kenkel, 1985);
2. based on set theory, such as the Jaccard's, Soerensen's, or Gower's similarity coefficients (Mueller-Dombois and Ellenberg, 1974; Dunn and Everitt, 1982; Legendre and Legendre, 1983) and Levenshtein's distance itself;
3. based on information theory, such as Rajski's metric, information divergence, mutual information (Feoli *et al.*, 1984);
4. based on probability theory, such as the probabilistic similarity index of Goodall (1964, 1966, 1968), Goodall and Feoli (1991) and Goodall *et al.* (1991).

In addition to the method of maximum likelihood, that may be included in the fourth category, the available GIS do not include similarity functions based on probability. The advantages of using such functions rely on the fact that similarity between OGUs may be tested in a probabilistic way within the data sets representing the system under study. With probabilistic similarity functions the similarity may be tested within and between time intervals. By analogy with phenetic comparisons, functional comparisons may also be made according to single-layer or multilayer processes. For example, an ecological succession may be described by using the vegetation layer, the pedological layer and also the animal community layer.

13.4 Probabilistic similarity functions

The probabilistic similarity functions we suggest for use in GIS applications have been proposed already by Goodall (1964, 1966, 1968) and they are included in a package described by Goodall *et al.* (1991). One is computed by program SIMIL, the other by program AFFIN. Both can deal with six categories of characters:

- binary
- qualitative with more than two alternative values
- ordinal
- small integer values
- quantitative grouped into classes
- quantitative not grouped into classes.

The output from SIMIL represents the probability that two particular OGUs would have a set of character values at least as similar as those actually observed if the two values for each character were a random selection from all the values of that character recorded in the set of OGUs (Goodall, 1964, 1966). The smaller the probability the more similar the OGUs. The one-complement of probability is a measure of similarity. The function of AFFIN gives the probability that a particular OGU in a set would have character values as close to those of a specific subset, if its character values were a random sample of those not in the subset (Goodall, 1968). Low probability values represent high affinity of a particular OGU of the set to the OGUs of the subset. Both SIMIL and AFFIN weight characters or character values by their rarity. High weight is given to rare events (joint occurrence of similar character values) as in information theory. The OGUs are compared two by two by one character at a time and the probability that two values of a character would be as alike as they are, if the two OGUs simply constituted a random sample of character values from the whole set, is combined with the probabilities of the other characters according to Fisher (1948):

$$\chi^2 = -2\sum_i \ln p_i \qquad i = 1, \ldots, s \qquad (s = \text{number of characters})$$

this is distributed as chi-square with $2s$ degrees of freedom. The probability of chi-square combines the s separate probabilities. Lancaster's (1949) correction is applied for discrete characters.

The function of SIMIL can be used for classification of OGUs and the function AFFIN for their identification when a classification is already available. The classification may be obtained by applying the clustering algorithms (Orlóci, 1978; Legendre and Legendre, 1983) to the similarity matrix obtained by SIMIL or by using the iterative process suggested by Goodall and Feoli (1991). If the iterative process is used, the probability criterion implicit in the method is used to define the clusters at higher hierarchical levels; if other methods are used this criterion has to be defined in some way (Dale, 1991b). In any case, the clusters or classes defined at different hierarchical levels must be defined by discriminant characters. Only after the finding the discriminant characters can we say that a classification has been established as a set of typologies of homogeneous OGUs that can be used to produce choropleth maps.

Choropleth maps may be seen not only as cartographic representations of hierarchical classification of the landscape at the end of an analytical process, but also as data sources for landscape spatial pattern analysis through the many different indices of the landscape structure such as shape, fragmentation, fractality, diversity and so on (Ebdon, 1977; Turner, 1989; Milne, 1991; Gardner and O'Neill, 1991; Fabbri, 1991; Baker and Yunming, 1992; Cullinam and Thomas, 1992; Gustafson and Parker, 1992; Olsen *et al.*, 1993).

The choropleths are themselves OGUs. On the basis of similarity between vegetation types and considering the areas of the corresponding choropleths, Feoli *et al.* (1992) have suggested the following measure of landscape biological diversity:

$$D = n\,H\,(1 - S)$$

in which H is the Shannon entropy calculated on the basis of the percentage of the area occupied by the community types of a territory, n is the average number of species per community type and S is the average probabilistic similarity between community types. Probabilistic similarity is also proposed by Feoli *et al.* (1992) to place a given state of an ecological system along a hypothetical dynamical axis ranging between 0 and 1. Given an ecological map of a territory it is possible to calculate its average similarity with a reference state as:

$$CS = \sum p_i S_i \qquad i = 1, \ldots, N$$

where p_i is the proportion of the area occupied by the state having S_i similarity with the reference state and N is the number of different similarity values.

13.5 Classification, ordination and prediction

GIS can manage different OGUs to obtain maps. In GIS, raster or vector, the OGUs are stored in databases generally structured according to the relational model (Wildi, 1991b). In the databases the OGUs may be described by different characters. The variables describing the ecological processes may be some of them. If a long- or a short-time process has to be recorded, the database must be dynamical. GIS may also have options for computing similarity between OGUs and for classifying and ordering them. The results of the proposed probabilistic similarity functions applied between OGUs within and between time intervals may be easily stored and displayed by new maps defined by the OGUs considered probabilistically similar. Such maps can show the similarity of OGUs at time T_0 with the corresponding OGUs at times T_i (i= 1, . . . , n). They should also be used to describe and quantify changes of ecological processes in the landscape.

Choropleth maps of processes may reproduce the pattern of the corresponding phenetic maps only if there is a strict correspondence between the phenetic choropleths and the functional choropleths. In this case the phenetic classification is highly predictive with respect to the functional classification. The functional map will be more fragmented then the original phenetic map only if more than one functional class is recognised within the same species, community, or environmental type. It will be less fragmented if different phenetic choropleths belong to the same functional class. The predictivity of phenetic maps with respect to functional maps, and the predictivity of functional maps with respect to other functional maps given by different characters or corresponding to different times, may be measured using a classification or an ordination approach. The first measures the predictivity between classifications by the chi-square, mutual information, or other related tests (Gokhale and Kullback, 1978; Feoli *et al.*, 1984, 1991) of the contingency tables given by the intersections between the classes of different classifications. The second measures the predictivity by the correlation between similarity matrices or between the ordination axes obtained from them. Some multivariate methods, such as *canonical correlation analysis* (Orlóci, 1978; Legendre and Legendre, 1983) and *Procrustes analysis* (Gower, 1971; Podani, 1991c), are suitable for the measurement of such a correlation. Feoli *et al.* (1993) provide an example of prediction between climatic and geographic data in China using the *Procrustes* method. The description of OGUs based on different sets of characters may be incorporated into the GIS database. Once the typologies are obtained, the GIS allows us to map them as choropleths by the *renumbering* function (Johnston, 1992). These maps represent the distribution of typologies (functional or phenetic) in the geographic space. The distribution of functional typologies in abstract spaces is defined by the interactions between environmental factors and the genetic pools of living organisms, so that knowledge of the distribution of environmental factors in geographic space is fundamental to making predictions about landscape changes at different territorial scales. In this respect, the example of Sun and Feoli (1992) shows how it is possible to produce maps of vegetation types based on changes of climatic patterns. A functional and/or phenetic choropleth corresponding to a given environmental type will shift to another one if the environment changes from that type to another. The probability of natural shifting or transition from one system state to another one can be estimated by the probability of similarity between the two states. This implies that it is easier to change to the one which is probabilistically closest in the abstract space rather than to one which is more distant. The consequence of this assumption is that the trajectories of an ecological system in the abstract space should not have big jumps, provided catastrophic events do not occur.

13.6 Conclusion

Similarity functions are useful tools for representing abstract spaces and measuring the proximity of choropleths (phenetic or functional) in such spaces. This is necessary to analyse and interpret the spatial pattern of ecological processes in given territories. The probabilistic similarity functions we suggest would improve this interpretation by introducing a probabilistic aspect to the prediction. In this way they add something new to GIS applications in landscape and territorial analysis.

REFERENCES

ACOT, P. (1989). Storia dell'ecologia, in Roma, L., *Historie de l'ecologie 1988*, Paris: Presse Universitaire de France.

BAKER, W.L. and YUNMING, C. (1992). The r.le programs for multiscale analysis of landscape structure using the GRASS geographical information system, *Landscape Ecology*, **7**(4), 291–302.

BANYIKWA, F.F., FEOLI, E. and ZUCCARELLO, V. (1990). Fuzzy set ordination and classification of Serengeti short grasslands, Tanzania, *Journal of Vegetation Science*, **1**, 97–104.

BARTHA, S. (1990). Spatial processes in developing plant communities: pattern formation detected using information theory, in Krahulec, F., Agnew, A.D.Q. and Willems H.J. (Eds). *Spatial Processes in Plant Communities*, pp. 31–47, Prague: Academia Prague.

BENNETT, R.J. (1981). Spatial and temporal analysis: spatial time series, in Wrigley, N. and Bennet, R.J. (Eds). *Quantitative Geography: a British View*, pp. 97–103, London: Routledge and Kegan Paul.

BOUXIN, G. (1991). The measurement of horizontal patterns in vegetation: a review and proposal for models, in Feoli, E. and Orlóci, L. (Eds). *Computer-Assisted Vegetation Analysis*, pp. 337–53, Dordrecht: Kluwer.

BRUNT, J.W. and CONLEY, W. (1990). Behaviour of a multivariate algorithm for ecological edge detection, *Ecological Modelling*, **49**, 179–203.

BURROUGH, P.A. (1987). Spatial aspects of ecological data, in Jongman, R.H.G., Ter Braak, C.J.F. and van Tongeren O.F.R. (Eds). *Data Analysis in Community and Landscape Ecology*, pp. 213–51, Wageningen: Pudoc.

CLIFF, A.D. and ORD, J.K. (1981). Spatial and temporal analysis: autocorrelation in space and time, in Wrigley, N. and Bennet, R.J. (Eds). *Quantitative Geography: a British View*, pp. 104–10, London: Routledge and Kegan Paul.

CROVELLO, T.J. (1981). Quantitative biogeography: an overview, *Taxonomy*, **30**, 563–75.

CULLINAM, V.I. and THOMAS J.M. (1992). A comparison of quantitative methods for examining landscape pattern and scale, *Landscape Ecology*, **7**(3), 211–27.

DALE M.B. (1991a). Mutational and nonmutational similarity measures: a preliminary examination, in Feoli, E. and Orlóci, L. (Eds). *Computer-Assisted Vegetation Analysis*, pp. 123–35, Dordrecht: Kluwer.

DALE M.B. (1991b). Knowing when to stop: cluster concept–concept cluster, in Feoli, E. and Orlóci, L. (Eds). *Computer-Assisted Vegetation Analysis*, pp. 149–71, Dordrecht: Kluwer.

DIGGLE, P.J. (1983). *Statistical Analysis of Spatial Point Patterns*, London: Academic Press.

DOBSON, J.E. (1993). Commentary: a conceptual framework for integrating remote sensing, GIS and geography. *PE&RS*, **59**(10), 1491–6.

DUNN, G. and EVERITT, B.S. (1982). *An Introduction to Mathematical Taxonomy*, Cambrdige: Cambridge University Press.

EBDON, D. (1977). *Statistics in Geography. A Practical Approach*, Oxford: Basil Blackwell.

FABBRI, A.G. (1991). Spatial data analysis in raster-based GIS: an introduction to geometric characterization, in Belward, A.S. and Valenzuela, C.R. (Eds). *Remote Sensing and Geographical Information Systems for Resource Management in Developing Countries*, pp. 357–88, Brussels and Luxembourg: ECSC, EEC, EAEC.

FEOLI, E. (1984). Some aspects of classification and ordination of vegetation data in perspective, *Studia Geobotanica*, **4**, 7–21.

FEOLI, E., GANIS, P., ORIOLO, G. and PATRONO, A. (1992). Modelli per il calcolo della diversità e loro applicabilità nella valutazione di impatto ambientale, *S.IT.E. Atti*, **14**, 29–34.

FEOLI, E., GANIS, P. and ZERIHUN WOLDU (1991). Community niche an effective concept to measure diversity of gradients and hyperspaces, in Feoli, E. and Orlóci, L. (Eds). *Computer-Assisted Vegetation Analysis*, pp. 273–77, Dordrecht: Kluwer.

FEOLI, E. and LAGONEGRO, M. (1983). A resemblance function based on probability: applications to field and simulated data, *Vegetatio*, **53**, 3–9.

FEOLI, E., LAGONEGRO, M. and ORLÓCI, L. (1984). *Information Analysis of Vegetation Data*, The Netherlands: Dr. W. Junk Publishers, The Hague.

FEOLI, E. and ORLÓCI, L. (1991). The properties and interpretation of observations in vegetation study, in Feoli, E. and Orlóci, L. (Eds). *Computer-Assisted Vegetation Analysis*, pp. 3–13, Dordrecht: Kluwer.

FEOLI, E., ORIOLO, G., PATRONO, A. and ZUCCARELLO, V. (1992). Phytosociology and G.I.S.: conceptual and technical tools to map landscape dynamics, *Documents Phytosociologiques*, **14**, 65–81.

FEOLI, E., PODANI, J. and SUN, C.Y. (1993). Correspondence between climatic and geographical spaces by Procrustes analysis, *Abstracta Botanica*, **17**, 141–6.

FISHER, R.H. (1948). *Statistical Methods for Research Workers* (10th Ed.), Edinburgh: Oliver and Boyd.

GABRIEL, K.R. (1971). The biplot graphic display of matrices with application to principal component analysis, *Biometrika*, **58**, 453–67.

GANIS, P. (1991a). PATT-Spatial autocorrelation analysis: Computer program and example of application with data sets of grassland vegetation under a natural reforestation process in the Karst near Trieste, in Feoli, E. and Orlóci, L. (Eds). *Computer-Assisted Vegetation Analysis*, pp. 489–96, Dordrecht: Kluwer.

GANIS, P. (1991). NICHE – Programs for niche breadth, overlap and hypervolumes, in Feoli, E. and Orlóci, L. (Eds). *Computer-Assisted Vegetation Analysis*, pp. 469–87, Dordrecht: Kluwer.

GARDNER, R.H. and O'NEILL, R.V. (1991). Pattern, Process, and Predictability: the use of neutral models for landscape analysis, in Turner, M.G. and Gardner, R.H. (Eds). *Quantitative Methods in Landscape Ecology*, pp. 289–307, New York: Springer-Verlag.

GOKHALE, D.V. and KULLBACK, S. (1978). *The Information in Contingency Tables*, New York: Marcel Dekker, Inc.

GOODALL, D.W. (1964). A probabilistic similarity index, *Nature*, **203**, 1098.

GOODALL, D.W. (1966). A new similarity index based on probability, *Biometrics*, **22**, 883–907.

GOODALL, D.W. (1968). Affinity between an individual and a cluster in numerical taxonomy. *Biometric Proximetrique*, **2**, 52–5.

GOODALL, D.W. and FEOLI, E. (1991). Application of probabilistic methods in the analysis of phytosociological data, in Feoli, E. and Orlóci, L. (Eds). *Computer-Assisted Vegetation Analysis*, pp. 137–46, Dordrecht: Kluwer.

GOODALL, D.W., GANIS, P. and FEOLI, E. (1991). Probabilistic methods in classification: a manual for seven computer programs, in Feoli, E. and Orlóci, L. (Eds). *Computer-Assisted Vegetation Analysis*, pp. 453–67, Dordrecht: Kluwer.

GOWER, J.C. (1971). Statistical methods of comparing different multivariate analyses of the same data, in Hodson, F.R., Kendall, D. G. and Tautu, P. (Eds). *Mathematics in the Archeological and Historical Sciences*, pp. 138–49, Edinburgh: Edinburgh University Press.

GREIG-SMITH, P. (1983). *Quantitative Plant Ecology* (3rd Ed.), Oxford: Blackwell.

GUSTAFSON, E.J. and PARKER, G.R. (1992). Relationships between landcover proportion and indices of landscape spatial pattern, *Landscape Ecology*, **7**(2), 101–10.

HAUSER, M. and MUCINA, L. (1991). Spatial interpolaton methods for interpretation of ordination diagrams, in Feoli, E. and Orlóci, L. (Eds). *Computer-Assisted Vegetation Analysis*, pp. 299–316, Dordrecht: Kluwer.

HEPPLE, L.W. (1981). Spatial and temporal analysis: time series analysis, in Wrigley, N. and Bennet, R.J. (Eds). *Quantitative Geography: a British View*, pp. 92–6, London: Routledge and Kegan Paul.

HILL, M.O. (1973). The intensity of spatial pattern in plant communities, *Journal of Ecology*, **61**, 225–35.

JOHNSTON, C.A. (1992). GIS technology in ecological research, *Encyclopedia of Earth System Science*, **1**, 329–46.

JONGMAN, R.H.G., TER BRAAK, C.J.F. and VAN TONGEREN O.F.R. (Eds). (1987). *Data Analysis in Community and Landscape Ecology*, Wageningen: Pudoc.

JUHASZ-NAGY, P. and PODANI, J. (1983). Information theory methods for the study of spatial processes and succession, *Vegetatio*, **51**, 129–40.

KENKEL, N.C. (1991). Spatial competition models for plant populations, in Feoli, E. and Orlóci, L. (Eds). *Computer-Assisted Vegetation Analysis*, pp. 387–97, Dordrecht: Kluwer.

LEGENDRE, L. and LEGENDRE, P. (1983). *Numerical Ecology*, New York: Elsevier.

LEPS, J. (1990). Can underlying mechanism be deduced from observed patterns?, in Krahulec, F., Agnew, A.D.Q. and Willems H.J. (Eds). *Spatial Processes in Plant Communities*, pp. 1–11, Prague: Academia Prague.

MILNE, B.T. (1991). Lessons from applying fractal models to landscape pattern, in Turner M.G. and Gardner R.H. (Eds). *Quantitative Methods in Landscape Ecology*, pp.199–235, New York: Springer-Verlag.

MUCINA, L., CIK, V. and SLAVKOVKY, P. (1991). Trend surface analysis and splines for pattern determination in plant communities, in Feoli, E. and Orlóci, L. (Eds). *Computer-Assisted Vegetation Analysis*, pp. 355–71, Dordrecht: Kluwer.

MUELLER-DOMBOIS, D. and ELLENBERG, H. (1974). *Aims and Methods of Vegetation Ecology*, New York: John Wiley.

OLSEN, E.R., RAMSEY, R.D. and WINN, D.S. (1993). A modified fractal dimension as a measure of landscape diversity, *PE&RS*, **59**(10), 1517–20.

ORD, J.K. (1979). Time-series and spatial patterns in ecology, in Cormack, R.M. and Ord, J.K. (Eds). *Spatial and Temporal Analysis in Ecology*, pp. 1–94, Fairland, Maryland: International Cooperative Publishing House.

ORLÓCI, L. (1966). Geometric models in ecology. I. The theory and application of some ordination methods, *Journal of Ecology*, **54**, 193–215.

ORLÓCI, L. (1967). Data centering: a review and evaluation with reference to component analysis, *Systematic Zoology*, **16**, 208–12.

ORLÓCI, L. (1978). *Multivariate Analysis in Vegetation Research*, The Hague: Junk.

ORLÓCI, L. (1988). Detecting vegetation patterns, *ISI Atlas of Sciences, Plants and Animals*, **1**, 173–7.

ORLÓCI, L. (1991a). On character-based plant community analysis: choice, arrangement, comparison, in Feoli, E. and Orlóci, L. (Eds). *Computer-Assisted Vegetation Analysis*, pp. 81–6, Dordrecht: Kluwer.

ORLÓCI, L. (1991b). Statistics in ecosystem survey: computer support for process-based sample stability tests and entropy/information inferece, in Feoli, E. and Orlóci, L. (Eds). *Computer-Assisted Vegetation Analysis*, pp. 47–57, Dordrecht: Kluwer.

ORLÓCI, L. (1991c). Poorean approximation and Fisherian inference in bioenvironmental analysis, *Advances in Ecology*, **1**, 65–71.

ORLÓCI, L. and DE PATTA PILLAR, V. (1991). On sample size optimality in ecosystem survey, in Feoli, E. and Orlóci, L. (Eds). *Computer-Assisted Vegetation Analysis*, pp. 41–6, Dordrecht: Kluwer.

ORLÓCI, L. and KENKEL, N.C. (1985). *Introduction to Data Analysis with Examples from Population and Community Ecology*, Fairland, Maryland: Int. Coop. Publishing House.

ORLÓCI, L. and ORLÓCI, M. (1991). Edge detection in vegetation: Jornada revisited, in Feoli, E. and Orlóci, L. (Eds). *Computer-Assisted Vegetation Analysis*, pp. 373–85, Dordrecht: Kluwer.

PODANI, J. (1984a). Spatial processes in the analysis of vegetation: theory and review, *Acta Bot. Hung.*, **30**, 75–118.

PODANI, J. (1984b). Analysis of mapped and simulated vegetation pattern by means of computerized sampling techniques, *Acta Bot. Hung.*, **30**, 403–25.

PODANI, J. (1991a). Computerized sampling in vegetation studies, in Feoli, E. and Orlóci, L. (Eds). *Computer-Assisted Vegetation Analysis*, pp. 17–28, Dordrecht: Kluwer.

PODANI, J. (1991b). SYN-TAX IV. Computer programs for data analysis in ecology and systematics, in Feoli, E. and Orlóci, L. (Eds). *Computer-Assisted Vegetation Analysis*, pp. 437–52, Dordrecht: Kluwer.

PODANI, J. (1991c). On the standardization of Procrustes statistics for the comparison of ordinations, *Abstracta Botanica*, **15**, 43–6.

ROBINSON, A.H., SALE, R.D., MORRISON, J.L. and MUEHRCKE, P.C. (1984). *Elements of Cartography*, pp. 337–66, New York: John Wiley.

SUN, C.Y. and FEOLI, E. (1992). Trajectory analysis of Chinese vegetation types in a multidimensional climatic space, *Journal of Vegetation Science*, **3**, 587–94.

TURNER, M.G. (1989). Landscape ecology: the effect of pattern on process, *Annual Review Ecological Systematics*, **20**, 171–97.

VER HOEF, J.M., CRESSIE, N.A.C. and GLENN-LEWIN, D.C. (1993). Spatial models for spatial statistics: some unifications, *Journal of Vegetation Science*, **4**, 441–52.

WHITTAKER, J. (1990). *Graphical Models in Applied Multivariate Statistics*, Chichester: John Wiley and Sons.

WHITTAKER, R.H. (1967). Gradient analysis of vegetation, *Biology Review*, **42**, 207–64.

WHITTAKER, R.H. (1978). Direct gradient analysis, in Whittaker, R.H. (Ed.), *Ordination of Plant Communities*, pp. 7–50, The Hague, The Netherlands: W. Junk Publishers.

WILDI, O. (1991a). Sampling with multiple objectives and the role of spatial autocorrelation, in Feoli, E. and Orlóci, L. (Eds). *Computer-Assisted Vegetation Analysis*, pp. 29–39, Dordrecht: Kluwer.

WILDI, O. (1991b). The relational model for data bases in community studies, in Feoli, E. and Orlóci, L. (Eds). *Computer-Assisted Vegetation Analysis*, pp. 71–7, Dordrecht: Kluwer.

CHAPTER FOURTEEN

Integrating GIS into the planning process

T. A. ARENTZE, A. W. J. BORGERS and H. J. P. TIMMERMAN

14.1 Introduction

Commercially available geographical information systems essentially remain systems for storing, retrieving and mapping spatial information. Geographical information systems contain a spatial database, a series of functions to manipulate the database and a series of modules to display the information in this database. Although the observation unit differs between geographical information systems, typically co-ordinates or grids are used as the carrier of the information. Accordingly, where geographical information systems are used in planning, their use is not strongly related to particular stages of the planning process. Instead, they are considered as useful media for storing and displaying spatial data (Harris and Batty, 1992). To the extent that analytical or modelling capabilities are added to geographical information systems, they are often limited to the most simple models that were often developed in the 1960s. Some geographical information systems constitute an exception in this respect, at least to some extent. For example, the network allocation algorithms included in TRANSCAD belong to the most advanced algorithms available. In other areas, however, this system has much less to offer.

Hence, it is no surprise that many scholars have recently advocated to give the improvement of the spatial analyses and spatial modelling capabilities of geographical information systems a high priority on the GIS research agenda (see, for example, Clarke, 1990; Douven *et al.*, 1993; Fischer and Nijkamp, 1992; Goodchild, 1991a; Openshaw, 1990). Probably most researchers working on geographical information systems can sympathise with this viewpoint. However, the area of spatial analysis and spatial modelling itself has rapidly grown in the past. Much progress has been made in areas such as exploratory data analysis, spatial econometrics, non-linear simulation models, activity scheduling modelling, strategic choice analysis, and data envelope analysis. One should realise that most techniques and models serve different purposes. Incorporating all possible techniques and models, or even only those receiving most attention, would probably have adverse effects on the system's speed, and would at least have dramatic effects on memory requirements, which raises the issue of how integrated the envisioned systems should be. Perhaps more important is the observation that different models require different data structures which are not necessarily consistent with the structures of current geographical information systems. Hence, the problem of improving the functionality of geographical information systems for planning and design requires answers to questions such as which methods, techniques and models to include? Which data structures would be most efficient for the various models? Which strategy to use to link the models to geographical information systems.

The aim of this chapter is to provide some thoughts on these questions. We will give an overview of some important developments in spatial analysis and spatial modelling and their applications to socio-economic problems. We wish to emphasise though that this review is necessarily biased as the number and types of models in various areas has really exploded in the last decades. We will occasionally use classifications of techniques and models as a means of providing some structure to our arguments. We will also argue that the integration of some more advanced modelling tools may require substantial changes in typical database structures and illustrate this using examples from our own ongoing or completed work.

From the outset, it should be emphasised that in the present discussion we adopt a broad view of the terms 'spatial analysis' and 'spatial modelling'. Some authors seem to restrict the discussion to geostatistical analyses. We adopt the view that all models and techniques relevant to spatial problems should potentially be considered. Hence, in our view the discussion on the integration of techniques and models into geographical information systems should also include for example conventional non-spatial statistical analyses, tools for policy development and evaluation, models of individual choice behaviour, activity patterns analyses, and project management algorithms.

This chapter is organised as follows. First, we will discuss some of the recent developments in spatial data analysis and spatial modelling and their application to socio-economic planning. In Section 14.3, we will place these analyses and models in the context of geographical information systems. This is followed, in Section 14.4, by a discussion of some of the implications for developing geographical information systems. Finally, we summarise our viewpoints and discuss some avenues for future research endeavours.

14.2 Spatial analysis and modelling

14.2.1 Statistical analysis

If one considers all available techniques and models that have been used in the spatial sciences in the past or that could potentially be used, one realises it is virtually impossible to find a classification of low dimension that would encompass all of them. Keeping this in mind, we have chosen to present a simple classification that suits our purpose. Table 14.1 presents a classification that is based on two dimensions. First, it is relevant to distinguish between the type of data available. Most data used in the spatial sciences involve a spatial component, which can either represent a point, a line or an area; a temporal component which may refer to distinct time periods, or be coded in terms of events, and the other attributes of the observations, which may either be of nominal, ordinal, interval or ratio nature. It should be noted that this table is neither mutually exclusive nor exhaustive. For example, while LISREL is given as an example of a statistical technique for interval data, it can be used for lower order data as well. The table is not exhaustive simply because many more techniques exist. Secondly, we distinguish between descriptive analysis, associative analysis and data reduction. The

Table 14.1 Examples of techniques for spatial analysis

	Data	Type of Analysis		
		Description	Association	Data Reduction
X-Y	Point	Geostatistics, point pattern analysis	Pattern analysis	Spatial filtering
	Line	Graph theory	Pattern matching	
	Area		Spatial correlation	Contingency analysis, regionalisation
T	Time	Statistics	(Spatial) time series analysis	(Spatial) smoothing analysis
	Events	Event history analysis	Event history analysis	
Z	Nominal	Statistics, exploratory data analysis	Loglinear analysis	Correspondence and latent class analysis
	Ordinal	Statistics, exploratory data analysis	Ordered logit analysis	Preference mapping
	Interval	Statistics, exploratory data analysis	Regression analysis, LISREL	Cluster analysis, factor analysis
	Ratio	Statistics, exploratory data analysis	Regression analysis, LISREL	Cluster analysis, factor analysis

purpose of descriptive analysis is to characterise the data in terms of their distribution. Hence, we will typically have some measure of central tendency and some measure of dispersion. The specific measure will depend on the type of data and their properties. Associative analyses aim at finding and expressing the functional relationships between two or more variables. Again, techniques differ in terms of their underlying assumptions which are often related to the properties of the data. Most of these techniques are usually used for explanation, but they can be used for prediction as well if one is willing to assume that the estimated functional forms are invariant across time. Assumptions regarding the behaviour of the system under investigation or policy scenarios can then be translated into values of the explanatory variables and these variables, together with the estimated parameters, then used to predict the dependent variable of interest. Finally, the aim of data reduction is either to reduce the number of variables into a smaller number of underlying (latent) dimensions, or to group the observational units into a smaller number of clusters.

Evidently, the techniques discussed here are the conventional statistical analysis techniques, which in many cases do have their spatial equivalents. It may be a concern whether non-spatial statistical analysis should be included into geographical information systems. We feel this is important simply because the non-spatial techniques are more frequently used than their spatial analogues. Advances in statistical analyses during the last decade concentrated on techniques for data of lower measurement levels (for example, Tukey's exploratory data analyses) and improving methods for identifying latent dimensions. To our knowledge, spatial equivalents of such techniques are still largely absent in the literature.

14.2.2 Design and planning tools

The techniques discussed in the previous section typically allow one to process the data in the database of the geographical information system and derive summary measures, associations and/or classifications. In most cases, the purpose underlying such usage is data analysis or information processing. In the context of socio-economic planning the findings of such analyses may be useful in that the policy-maker can develop an understanding of the spatial structure, functional organisation and dynamics of the area of interest. While this is important information, it is also highly limited information that is often used only in the very beginning of any planning process.

Table 14.2 gives an overview of a very simple planning process. The planner may use different sets of techniques to support his activities. In the work we have been conducting in our group, we differentiate between those techniques and models that are external to the planning activities and those that support these activities. External techniques and models typically provide information that planners can use to base their decisions upon. Examples are population forecasting models, input/output analysis, and spatial choice models. Other techniques and methods support these activities directly. Examples are design methods, multicriteria evaluation, scheduling and time management algorithms. If we examine currently available geographical information systems, they may include some modules for models of the first type; options for methods of the second kind are rare. Let us now elaborate a little on the kinds of methods and models that would be useful to support socio-economic and spatial planning processes.

The first step in any planning process is that of problem identification. Currently available geographical information systems are well suited for this task. Search, retrieval and overlay options allow the planner to test if certain areas meet the conditions he or she thinks are indicative of problems. The analytical capabilities of geographical information systems will certainly improve significantly if more statistical modules are added to them. Geographical information systems are very powerful for such diagnostics and, together with their graphical capabilities, this functionality alone should probably already suffice to switch from traditional information systems to geographical information systems.

The second phase concerns goal analysis and specification, and this constitutes a different matter altogether. Geographical information systems do not allow one to perform such analysis. Still, developing goals for certain areas, given analytical results, constitutes a central element of a planning process and it makes sense to use the same environment for this as one can immediately define goals seeing the results of the analysis. Useful techniques in this area are tree structures that allow one to specify the

Table 14.2 Stages of a planning process

Problem identification
Goal analysis and specification
Generating alternatives
Information collection
Evaluation
Decision
Plan implementation
Monitoring and early warning

vertical and horizontal relationships between goals, graphical tools that support these activities, goal compatibility analysis and so on.

The next step involves generating alternatives. Again, to date this is still largely a manual exercise. Potentially, geographical information systems provide a suitable environment for performing this task if the problem is to find suitable locations for developing or reorganising activities. Map-overlay techniques are generally useful for identifying possible options or narrowing down the space to be searched. To identify interesting or promising options, the technique of potential measures is helpful (Breheney, 1988). Using this technique, potential maps displaying zones of equal potential can be produced to assist spatial search. Alternatively, one could develop intelligent algorithms based on heuristic knowledge that would generate all feasible or promising options, giving the goals specified in the previous phase of the planning process. One can also think of more interactive routines. Both numerical and graphical approaches are relevant as some planners are inclined to favour quantitative approaches, whereas others are more visually oriented. General combinatorial algorithms are useful for generating alternative spatial configurations.

Once the alternatives are defined, one collects information on how these alternatives perform on the articulated goals or objectives. To support this activity, forecasting models and methods for measuring system performance are available. Forecasting might involve many different types of model, such as population forecasting models, economic forecasting models, demand forecasting systems, some kind of spatial choice/spatial interaction forecasting system, traffic assignment and routing models, to name a few. In principle, most of these models are not difficult to incorporate into any geographical information system, as they rely on the same observational units as most geographical information. However, it is interesting to note that recent advances in some of these types of model dramatically shifted in conceptualisation so that they have significant implications for the inclusion of spatial models in geographical information systems. We wish to illustrate this point using developments in spatial choice/spatial interaction theory as an example. As is well known, planners' interest in spatial interaction modelling was well developed in the 1960s. Spatial interaction models or their entropy-maximising version were frequently used for demand forecasting, trip generation forecasting, and predicting interaction flows for all kinds of purposes such as recreation, shopping, trip to work, and so on. The traditional models were based on data on observed trips between zones and attraction and distance functions were fitted to such data. The specification of the attractiveness and distance decay function may differ, but all these models were based on data on zonal flows. Most geographical information systems, if they do include any of such models, offer versions of the original spatial interaction models developed in the 1960s.

Conventional spatial interaction models have been seriously criticised though, and this has led to the development of three new types of models/approaches: discrete choice models, stated preference models, and trip-chaining/activity pattern models. Discrete choice models are derived from assumptions about individual choice behaviour, and although some authors would still estimate such models using zonal flow data, they strictly speaking require data on individual choices. While it is still possible to include individuals in geographical information systems, it will of course affect the required resolution of the system. Stated preference and choice models carry this argument one step further. These models have in common with discrete choice models the assumption that individuals combine their evaluations of the attributes of choice alternatives into some overall utility measure and use utility-maximising rules to arrive at some choice. However, these models are not derived from data on observed choice, but rather on experimental design data. Hence, the estimation of such models requires data structures that are typically not used in geographical information systems. These are not very difficult and often do not require much memory, but if one wishes to include stated preference and choice models, additional data structures and links with the other databases have to be created. Substantial progress has also been made in relaxing some of the highly restricted assumptions underlying conventional discrete choice models. Conventional models such as the multinomial logit model were based on the assumption of independently and identically distributed double exponentially distributed error terms. This assumption was required to derive a closed form model, at the cost of the independence from irrelevant alternatives property which states that the introduction of any new choice alternative will draw market shares from the existing alternatives in direct proportion to their original market share, an assumption these models share with traditional spatial interaction models. Ideally, one would wish to have a model that would allow one to incorporate every imaginable variance–covariance matrix of error terms, heterogeneity, taste variation, state dependence and so on. Recently, the method of simulated moments allows one to estimate such less rigorous models. However, the estimation may still take a day for reasonably-sized problems using high power number crunching machines. In fact this statement applies to most simulation models, and this raises the question whether one should use the most advanced models outside a GIS environment or be satisfied with less sophisticated models that are part of a geographical information system? Trip-chaining and activity-scheduling models are not that demanding, but they do cause other interesting problems. Scholars have realised that interaction flows are often part of a more general trip structure or even an activity pattern. Our predictive ability may therefore improve if we would not model, for example, shopping as a single-purpose, single-stop trip, but rather as a multipurpose, multi-stop trip. Hence, in order to estimate such a model, one requires data on where, and for what purpose, individuals stopped at the various stages that make up a journey. Geographical information systems are not well suited for such data structures, and hence again the incorporation of such models requires additional steps and programming in database management.

The next step of the planning process involves evaluating the alternatives on the basis of their evaluation scores. Techniques such as decision analysis, multicriteria evaluation and data envelope analysis are helpful to support this stage of the planning process. Note that we have adopted a viewpoint that the planning style is subjective in that the planner generates the alternatives and sets the weights in the evaluation process. Alternatively, one can choose a more rational style. In that case, the task of the planner would more typically be to formulate objectives and conditions, and choose the alternative that maximises the formulated objectives subject to a set of conditions. Techniques such as multiple objective programming have been developed for such purposes. In some sense, location-allocation models may serve a similar purpose for spatial problems in that they allow the planner to find the best location or location pattern in a plane or along a network that optimises some objective function such as minimum travel times, maximum coverage, and so on.

Once a decision has been made, a plan is implemented. Perhaps the most powerful techniques in this phase are techniques such as PERT, and methods for efficient time management, project management and costing. These are typically methods developed for business applications, but they are generally useful for socio-economic planning processes as well.

Finally, we have identified the aspect of monitoring the process. This allows the planners to assess whether the actual evolution of the system is consistent with the predictions underlying the plan. Closely related to monitoring is early warning. In this case, the system warns in advance that the evolution of the system would not be according to plan if the present trend continues. Models for early warning are often the same as the ones used for predictive and evaluation purposes. Hence, the same modules can be used for this purpose. However, they should be accompanied by rules and/or performance indicators that define when an early warning is in order. Too often this is simply seen as some direct measure of the difference between predictions and observations. More sophisticated systems would try to identify the causes underlying such differences. This poses a complicated modelling problem in itself.

We have now discussed several tools that may be helpful in a particular stage of the planning process. However, alternatively, algorithms could be developed that would involve several of these stages. For example, the authors have recently developed a generic expert system for spatial search that can be used in a GIS environment (Arentze *et al.*, 1996). The system is based on an algorithm that searches an area for sites that best match a specified ideal profile. Compared with the dominant procedure in current methods, this algorithm improves the efficiency of search and probably corresponds more closely to the typical human way of solving these problems. Typically, current methods involve testing all possible sites against a set of requirements to select the set of feasible options. Then, in the second phase, each of the selected options is evaluated in-depth to identify the best alternatives or to rank alternatives, possibly, by using MCE methods. In contrast, the algorithm we propose adopts an iterative procedure of selection and evaluation. The best options are found through cycles of selecting the most promising option based on currently available evaluation results and evaluating additional attributes of that option. The process stops when all attribute scores of the option selected are known. A hierarchical top-down procedure is used to select optional sites. In this procedure, the area is searched by searching successively smaller areas. Because of the iterative character of the procedure, the search does not proceed irreversibly from top to bottom. When the evaluation of sublocations reveals shortcomings of an area, the process may return to a higher level to select the currently best alternative area for evaluation and so on. So, the principles of iterative and hierarchical search results in a tentative process of zooming in on promising groups (areas) of options. The algorithm reduces the expected evaluation costs needed to identify best sites to a minimum, essentially by evaluating in each stage of the process the most informative (high level) attribute of the most promising options. Thanks to the efficiency of the procedure, a large number of options can be taken into consideration. In this way the algorithm may improve the quality of the outcome. The user of the system specifies a problem-specific knowledge base including knowledge on the area to be searched, the ideal profile and, optionally, methods for evaluating attributes. The system uses this input knowledge to identify the best locations in the area following the search procedure outlined above. Options are evaluated in various degrees of user interaction dependent on the specification of the input knowledge base.

14.3 GIS functionality

GIS provides a suitable environment for performing the kind of analysis and modelling described in the previous section. First, GIS data management and display capabilities are useful for storing, retrieving, manipulating and visualising input and output data.

Secondly, GIS technology has the potential of enabling or facilitating spatial analysis and modelling (Ding and Fotheringham, 1992). Analyses such as, for example, calibrating spatial models, may be easily conducted in a GIS environment. Goodchild (1991b) argues that the power of GIS lies in its ability to link objects of different classes (point, line and area) based on their relative geographic location. He derives from this ability several classes of unique functions. We mention the following functions that are often involved in generating, selecting and organising data for spatial analysis and modelling:

1. Define the objects of analysis such as buffer zones around line objects, proximal areas of point objects, and overlaps of overlaid polygon objects,
2. Generate object attributes such as the distance to nearest service centres, the location of representative centroids of zones, and the number of points contained in areas,
3. Select relevant objects such as zones within a distance band, points within an area, and objects that meet certain attribute and location characteristics,
4. Aggregate data to the appropriate level of scale such as from points to census tracts or from zones to superzones.

Furthermore, generating distance data is often needed to analyse or model interactions between objects. By storing data and providing tools for these functions, GIS makes a powerful environment for such analyses. This is particularly important in cases where modelling is an iterative or interactive process. For example, optimising model fit may be based on successive cycles of refining data and model (for example, Batty and Xie, 1993). Such a process requires efficient tools for handling spatial data which may be provided by GIS.

14.4 Implications for developing GIS and DSS systems

The question of the integration of techniques and models into geographical information systems involves various aspects. First, it relates to how much time and effort would be required for such a task. Secondly, decisions have to be made regarding the nature of this integration. Should one develop fully integrated systems, or should one focus on general user-interfaces allowing one to easily export and import relevant files (Batty and Xie, 1993).

The amount of time and effort required should not be an issue. Most of the senior people currently working on geographical information systems were also among those experimenting with and developing spatial analysis techniques and spatial models. Consequently, it is fair to assume that a wide variety of software should be available for those techniques developed in the early years. For example, in Eindhoven we developed a package for geostatistical analysis which included techniques for descriptive analyses, different variants of nearest-neighbour analyses, quadrat sampling methods, and so on, simply because at the time no commercial package for geostatistical analysis was available. Similarly, models such as Monte Carlo simulations, spatial interaction models, population forecasting techniques and input–output analyses have received considerable attention; hence access to software should not be the issue. It is much more difficult to decide which techniques and models to include, as, obviously, demands for such software is likely to differ considerably among users. For example, examining the recent literature on geographical information systems, many scholars seem to advocate incorporating geostatistics, ranging from descriptive geostatistics to point pattern analysis and autocorrelation. This seems to make sense in that such techniques indeed represent basic tools for spatial analysis. Moreover, such software may appeal to scholars working in this area. However, we doubt whether there is a huge demand for such software among practitioners. Several years ago we saw applications of geostatistical analyses in applied research, but maybe our experiences and observations are biased.

Another tendency that raises doubts is that, in order to demonstrate the potential applications of geographical information systems, scholars with only a limited background and expertise in spatial modelling start using and incorporating the simplest version of some modelling approach, often without knowing such models have been heavily criticised and not taken very seriously by specialist modellers. Illustrations of this tendency are the use of the original Huff model to predict spatial shopping behaviour and the use of simple graph-theoretic indices for measuring accessibility. One could argue that such options should be included into a GIS for the sake of completeness, but we feel one should allow this only if more advanced models are included as well. Unfortunately, with a few exceptions, most of the techniques and especially models included in geographical information systems reflect the state-of-the art in modelling in the mid-1960s and early 1970s.

Another aspect of integration concerns the problem of how to integrate. Two years ago, the Dutch Working Group on Mathematical Geography and Planning organised a conference on the topic of 'Modelling and Geographical Information Systems'. It became clear at that conference that the vast majority of scholars and research institutes used a GIS to generate the data required for subsequent analysis. Selection, retrieval, overlay and shortest route routines were typically used to prepare data matrices which were exported as stand-alone files. Dedicated software for spatial analysis and spatial modelling was then used to analyse these data. The results of the analyses were saved as separate files and imported into their geographical information system for display. To be fair, this was hardly ever a conscious decision; it simply reflected the fact that most commercially available geographical information systems have a closed architecture and did not allow them to use an integrated system. One could ask if this strategy is really problematic. Of course, the stand-alone files require more memory, but then again it is very likely that one will save such files anyway for subsequent analyses, also within a GIS environment.

As the other extreme, one could develop fully integrated systems that would encompass a variety of techniques and models that could be activated within the GIS environment. Such an approach would be similar to recent advances in software development in general where increasingly more options become available within a single package. Similar to concerns with respect to such software, one could argue that geographical information systems incorporating many different options for analyses and modelling are not very efficient in that most users will only use a small subset of all possible options. This holds true for stand-alone software as well; however, this is not really an issue, especially if one were to allow users to buy or install only those modules they wish to use.

However, even in this case, the overall system may not be very memory-efficient in that complete modules for highly similar techniques and modules are incorporated in the system (Densham and Goodchild, 1989). Efficiency can be obtained if one realises that most spatial analysis techniques and models can be broken down into a small set of constituent components. For example, models of spatial interaction and spatial choice basically involve an attractiveness function and a distance decay function. The attractiveness function of a wide variety of spatial interaction and choice models can be classified into additive, multiplicative and exponential types. Similarly, all distance decay functions ever used can be classified into a small set of basic types. Similar functions are used also in all kinds of system performance measures, goodness-of-fit statistics, multicriteria evaluation methods, and so on. Hence, substantial savings in memory requirements can be achieved by programming modules for such basic operations, rather than separate fully-fledged modules for individual techniques or modules. Users could then interactively program their own models and save them as macros. Alternatively, one could pre-program models using the same structure. No doubt the more advanced models currently available for spatial analysis are probably too difficult for the average user, but this problem can be circumvented by including small expert systems that prevent users from making mistakes or translate verbal specifications into mathematical structures that are required to apply a particular model (Armstrong *et al.*, 1986; Elam and Konsynski, 1987).

The authors have recently developed a prototype of such an approach for the measurement of various aspects of demand–supply relationships, such as for example accessibility (see Arentze *et al.*, 1993). Measurement methods of this kind involve a series of successive elementary steps. In the first step the subsets of demand (residential zones) and supply points (facility centres) that are relevant for the given evaluation problem are selected. The relevance of points may be defined in terms of their location, attributes, group membership or combinations of these factors. The next step is to measure distances between selected points by determining in some way the costs of travelling across a network, for example, the shortest-path travel time. Based on the distance relationships, each selected demand point is then allocated to one or more supply points. Allocation may be based on a rule, such as centres within a distance band, or a behavioural model, such as a discrete choice model. Rules typically result in all-or-nothing allocations, whereas models typically return probabilities of interaction across destination alternatives. Then, an aspect of the allocation relationships is measured using an evaluation function. The evaluation function defines the aspect that is measured and the measurement method used. For example, the availability of supply from each demand point may be measured by summing the amount of supply in destinations, while taking the allocation intensities (for example, interaction probabilities) into account. Optionally, the obtained scores are then aggregated across the points of predefined groups, to evaluate system performance at group level, for example by calculating a weighted mean.

In each step a choice is made among a number of alternative methods or function specifications. Therefore, a specific indicator is defined as a specific combination of methods chosen. The prototype system we have developed incorporates for each step a set of optional functions. The system is based on the GIS software TRANSCAD, which provides optional functions for the selection of points and the measurement of distances. Functions for performing the remaining steps are added to the GIS. The system includes various kinds of deterministic rules and behavioural models for allocation. Furthermore, a range of evaluation functions are built in to calculate accessibility, availability, welfare, efficiency and effectiveness measures in various ways. Finally, several central tendency and dispersion measures are available to obtain scores at higher levels of aggregation. The system enables the user to define a specific measure by selecting for each step an optional function. Definitions can be saved in the form of macros or formulas, which can be used to analyse other cases. Furthermore, the system provides a base of standard definitions from which the user can choose. Besides savings in memory requirements, as mentioned above, this modular approach has the advantage of improving the flexibility of the system. The user is enabled to customise performance measures to specific information needs on the one hand and available data resources on the other hand. The task of specifying adequate performance measures could be automated or supported by incorporating analytic knowledge in the system. Then, the lay user would be able to maximally benefit from the capabilities of the system.

Integrating constituent elements of models rather than the models themselves results in a multilayered system. At the level of the model elements (for example, tools for network analysis, allocation, evaluation and so on) the geographical information system is general for different application areas (for example, transportation analysis, location analysis and so on.). Model base management facilities should be available to build models out of the constituent elements. Thus, the system can be customised to specific application areas, resulting in a dedicated geographical information system. Database management tools are available to build a database for a certain application. Therefore, at the database level the system is application specific.

14.5 Conclusion and discussion

The aim of this chapter has been to discuss some issues related to improving the spatial analysis and spatial modelling capabilities of geographical information systems. We have argued that considerable progress still has to be made in this area, and we have suggested some techniques and models we feel would be important in this respect. Some colleagues who may question our suggestions may argue that the systems we envision would be better called spatial decision support and expert systems rather than GIS. Of course, this is primarily an issue of terminology. In fact, we ourselves have advocated in the past a very strict use of these terms, 'geographical information system' being reserved for systems that excel in their database management and graphical capability and offer less or nothing at all in terms of modelling, direct design, or decision support functions. However, at the same time, we believe that such GIS have less to offer to the planner and have advocated and developed 'decision support systems' for the last decade.

In analysing and discussing some recent developments in spatial analysis and spatial modelling, we have pinpointed some elements that, in our opinion, require serious consideration and discussion when developing GIS that incorporate modelling capabilities. First, we have indicated that most models included in current GIS stem from the 1960s. We feel this is bad practice as people will use these simple models often not knowing they have been heavily criticised and that much better alternatives are available. Hence, every attempt should be made to include state-of-the-art models, rather than the most simple version that only suffices to demonstrate that particular types of analysis can be conducted. Secondly, we have argued that in trying to incorporate more advanced models, the database structures currently used are not necessarily appropriate any more. Hence, an attempt to improve the modelling capabilities of geographical information systems may in some cases imply that one has to reconsider the database architecture of the system as well. Thirdly, we have indicated that some recently developed models are too demanding in terms of computer time to be included in any GIS. Hence, we feel there are also limits to attempts to improve the modelling capabilities of GIS. Fourthly, even this brief summary of techniques and models for spatial analysis already suggests there are far more techniques than one can reasonably include. Moreover, in some research areas progress is so fast that one can be pretty sure the models included in a GIS are likely to be outdated once the package is shipped. Therefore, we have advocated the development of modules for the basic common elements of most techniques and models that allow users to program their own techniques and models within a GIS environment, if necessary helped by simple expert systems.

Finally, we disagree with those arguing that GIS will benefit most by extending their geostatistical functionality. We feel this set of techniques may have some appeal to some academics working in this area, but other, often non-spatial methods, techniques and models are potentially more important or useful to address socio-economic and spatial planning problems.

REFERENCES

ARENTZE, T.A., BORGERS, A.W.J. and TIMMERMAN, H.J.P. (1993). Geographical information systems and the evaluation of the location of facilities, in Harts, J.J., Ottens, H.F.L. and Scholten, H.J. (Eds). *Proceedings of the Fourth European Conference on Geographical Information Systems Vol. 2*, pp. 962–72, Utrecht: EGIS Foundation.

ARENTZE, T.A., BORGERS, A.W.J. and TIMMERMAN, H.J.P. (1996). An efficient search strategy for site-selection decisions in an expert system, *Geographical Analysis*, **28**, 126–47.

ARMSTRONG, M.P., DENSHAM, P.J. and RUSHTON, G.F. (1986). Architecture for microcomputer-based decision support systems, *Proceedings of the Second International Symposium on Spatial Data Handling*, pp. 120–31, Williamsville, New York: International Geographical Union.

BATTY, M. and XIE, Y. (1993). 'Modelling inside GIS'. Paper presented at the Third International Conference on Computers in Urban Planning and Urban Management, Atlanta, Georgia.

BREHENEY, M.J. (1988). Practical methods of retail location analysis: a review, in Wrigley, N. (Ed.), *Store Choice, Store Location and Market Analysis*, pp. 39–105, London: Routledge.

CLARKE, M. (1990). Geographic information systems and model based analysis: towards effective decision support systems, in Scholten, H. and Stillwell, J.C.M. (Eds). *Geographical Information Systems for Urban and Regional Planning*, pp. 165–75, Dordrecht: Kluwer.

DENSHAM, P.J. and GOODCHILD, M.F. (1989). Spatial decision support systems: a research agenda. *Proceedings GIS/LIS '89*, pp. 707–16, Bethesda, MD: ASPRS/ACSM.

DING, Y. and FOTHERINGHAM, A.S. (1992). The integration of spatial analysis and GIS, *Computers, Environment and Urban Systems*, **16**, 3–19.

DOUVEN, W., FISCHER, M.M. and SCHOLTEN, H.J. (1993). 'Environmental and socio-economic problems, GIS and spatial analysis'. Paper prepared for the Expert Meeting on Exploratory and Explanatory Spatial Analysis, European Science Foundation, Amsterdam.

ELAM, J.J. and KONSYNSKI, B. (1987). Using artificial intelligence techniques to enhance the capabilities of model management systems, *Decision Sciences*, **18**, 487–501.

FISCHER, M.M. and NIJKAMP, P. (1992). Geographic information systems and spatial analysis, *Annals of Regional Science*, **26**, 3–17.

GOODCHILD, M. (1991a). Progress on the GIS research agenda, in Harts, J., Ottens, H.F.L. and Scholten, H. (Eds). *Proceedings of the Second European Conference on Geographical Information Systems Vol.1*, pp. 342–50, Utrecht: EGIS Foundation.

GOODCHILD, M. (1991b). Geographic information systems, *Journal of Retailing*, **67**, 3–15.

OPENSHAW, S. (1990). Spatial analysis and geographical information systems: a review of progress and possibilities, in Scholten, H. and Stillwell, J.C.M. (Eds). *Geographical Information Systems for Urban and Regional Planning*, pp. 153–63, Dordrecht: Kluwer.

PART THREE

Spatial Dynamic Modelling

CHAPTER FIFTEEN

Spatial simulation modelling

ANTÓNIO S. CÂMARA, FRANCISCO FERREIRA and PAULO CASTRO

15.1 Introduction

Geographic information systems (GIS) provide rich spatial databases but have been traditionally static. The coupling of dynamic models to GIS provide an insight to the evolution of spatial phenomena as discussed by Grossman and Eberhardt (1992).

These authors identified three separate categories of combinations of dynamic models with GIS. In the first category, dynamic models use spatially aggregated variables but their results may be modified, for each site, by locally varying parameters. The second category includes the 'active area dynamic' models. In these models, the development of each area is modelled separately taking into account, however, the changes caused by adjacent areas. The third category includes the classical transport models, most of them with governing equations taken from physics.

Traditional dynamic simulation models represented in these three categories are based on a process view of the world. However, many natural systems may be perceived as a set of interacting entities and not as a set of processes representing empirical or theoretical principles. Thus, a fourth category of dynamic simulation models based on object-oriented representations have emerged to accommodate this entity view proposed by among others Larkin *et al.* (1988).

Differential equations and partial differential equations have been the mathematical tools of choice for most of the models that have been developed under these four categories. Toffoli (1984) and Toffoli and Margolus (1987) proposed cellular automata models to replace differential equation based models. Cellular automata models can be used both for process view models belonging to the first three categories and for entity view models.

However, as Phipps (1992) mentioned, there is a need to test cellular automata models concurrently with well-known models. This is the main purpose of the research work reported here. Cellular automata were used to model process-based surface water quality and fire propagation problems and predator–prey relationships viewed from an entity standpoint. The results obtained with these cellular automata models were then compared with the results derived from traditional models. Issues such as computer implementation with raster GIS and model evaluation were also discussed.

15.2 Cellular automata modelling

15.2.1 Cellular automata and differential equations

Computational resources have suggested new approaches to the modelling of systems. Digital computing devices are finite and discrete in nature, and their potential can be best realised when applied to discrete dynamic systems (Fogelman *et al.*, 1987). Cellular automaton is a modelling approach with an increasingly important role for conceptual and practical modelling of discrete dynamic systems (Toffoli and Margolus, 1987).

The theory of cellular automata was first introduced by John von Neumann (1966) in the late forties. This concept gained popularity three decades later through John Conway's work in the *Game of Life* (Fogelman *et al.*, 1987; Toffoli and Margolus, 1987).

Most spatial models are initially based on mass balances. These mass balances are written as differential or partial differential equations. Initially, these were manipulated and solved analytically. The emergence of digital processing made numerical techniques based on finite difference and finite element formulations highly suitable to integrate them. However, these algorithms truncate the power series and are solved with round-off using limited precision machines.

Cellular automata models of dynamic systems consider a lattice of cells on a line (one-dimensional cellular automata) or a uniform grid (two- or three-dimensional cellular automata), with each cell containing a few bits of data. Time advances in discrete steps Δ, and the laws are represented as transition rules, through which each cell determines its state at the next time $(t+t)$ from that of its neighbours and itself at the previous time (t). Since they are based on microscopic behaviour, the transition rules are generally quite simple. However, the resulting overall behaviour of the system can appear to be quite complex.

One- and two-dimensional cellular automata examples are shown in Figure 15.1. Each cell has two possible states (black and white), and the local neighbourhood of a cell is defined by two adjacent neighbour cells. The transition rules simply specify that the state of a cell at time $t+\Delta t$ is equal to the state of its two neighbours at time t if these have the same state; otherwise the state of a cell will remain unchanged.

In terms of structure, this computational scheme is similar to the ones employed in the numerical manipulation of partial differential equations. The difference is that the state variable at each cell of the lattice is only allowed to assume a small set of values, typically two states per cell, and that the transition functions do not assume an algebraic form (Hogeweg, 1988) but may be deterministic or stochastic.

This simplicity coupled to their inherently locality or parallelism (one can associate every N cells with a processor, without a significant overhead from the splitting of the problem among several processors), makes it possible to realise cellular automata models exactly on digital hardware. With appropriate architectures, performance gains are at least 10 000 in the simulation of cellular automata (Toffoli and Margolus, 1987). These gains offset the costs resulting from the increased number of variables (one per cell) in cellular automata models compared with the relatively few variables considered in typical differential equations. This improvement also opens the possibility of a more detailed and efficient dynamic modelling of spatial systems.

In summary, the cellular automata approach provides a model representation closer to the real physical system by bridging the differences between macroscopic and microscopic representations; it does not involve any power series truncation; it is not subject to round-off; it can be easily extended from a one-dimensional to a two- and

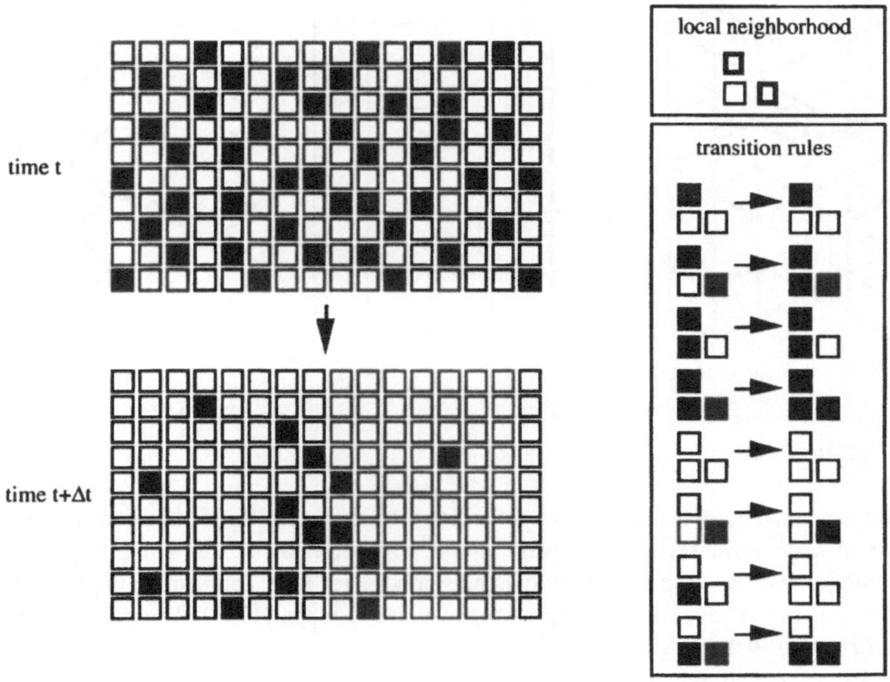

Figure 15.1 Cellular automata examples.

three-dimensional representation; and, by focusing on simple, microscopic behaviour, the emphasis of spatial modelling is placed on spatial mechanisms and not on the numerical solution techniques.

15.2.2 Cellular representations and boundary conditions

Cellular automata models assume a discretisation of space in mosaics of cells. Traditionally, these mosaics or tessellations in cellular automata models are mostly based on cells with identical size and shape to simplify computations (regular tessellations). The three simplest regular tessellations rely upon triangle, square or hexagonal cells as in Figure 15.2. However, it is possible to consider tessellations with cells with varying size and shape.

The geometry of regular tessellations refers to the shape, adjacency, connectivity, orientation, self-similarity, decomposability and packing properties of the tessellation. Square tessellations present significant advantages due to the equality of sides, decomposability, and stability of orientation and aggregation and they are dominant as they appear in the form of arrays of picture elements, grid cell data recording and electronic data capture and display (Laurini and Thompson, 1992).

In cellular automata models, one may look at transition rules as a method of propagation or elimination of a finite number of tokens (the possible state values).The behaviour of the transition rules at the boundaries of the cell mosaics depends on boundary conditions that may be of three dominant types: cyclic – the tokens crossing a boundary reappear in a cell at the opposite boundary; finite – the tokens are not able to transpose the boundary so they tend to accumulate near it, and infinite – the entities that cross the boundary disappear from the simulation.

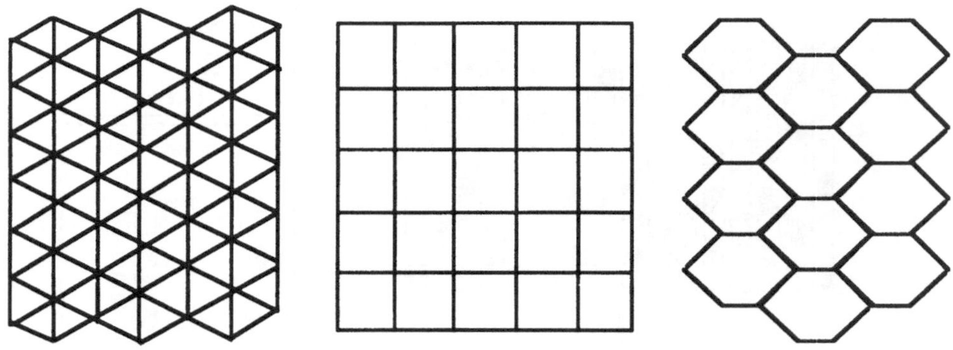

Figure 15.2 Cellular representations.

15.3 Applications

Cellular automata models are ideally applied for diffusion modelling, considering a small number of state variables and relatively simple transition rules. This is the case of the water quality and fire propagation models presented here. They can also be implemented to individual-based spatial modelling as suggested by Phipps (1992). The predator–prey model described below belongs to this category.

The fire propagation model is the only one that was implemented coupling a dynamic model to a raster GIS. However, both the implementations of the water quality and the predator–prey model are more interesting from a research standpoint. The water quality model was implemented on a parallel-processing machine and performance results were assessed taking into account a varying number of processors. The predator–prey model was implemented with an application that may include a map as an interacting background for the predators and preys.

The water quality and predator–prey models may be calibrated and verified using traditional statistical tests. Similar procedures for a fire propagation model require options based on image analysis.

15.3.1 Water quality modelling

Castro *et al.* (1993) have developed a river Biological Oxygen Demand (BOD)/Dissolved Oxygen (DO) model based on cellular automata. The basic partial differential equation for such a model is (Thomann and Mueller, 1987):

$$\partial C/\partial t + u\, \partial C/\partial x = E\, \partial^2 C/\partial x^2 - k_d C + k_a C \qquad (15.1)$$

where C is the pollutant concentration, t is the time, x is the distance in the longitudinal direction, u is the advective velocity, E is the longitudinal dispersion coefficient, k_d is the first-order decay rate constant and k_a the first-order reaeration constant.

This equation considers four essential processes affecting the pollutant distribution: advection (represented by the velocity term), longitudinal dispersion, decay and aeration. It is usually solved using finite difference or finite element schemes.

The cellular automaton model is based on a line of cells (a one-dimensional cellular automaton). The amount of pollutant is represented by the number of particles, with an arbitrarily defined mass, present in each cell. Pollutant concentration at a given cell is derived by dividing the product of the respective number of particles and the particle mass by the water volume of that cell.

Each cell has no defined limit on the number of particles it can contain. This does not mean however that an unlimited number of states per cell are considered. The transition rules are iteratively applied to each particle. This process corresponds to a two-state cellular automaton (particle or 1, and no particle or 0).

The cellular automaton cells are grouped in subsets, each subset representing a reach of uniform characteristics in the river. In the upstream cell, fixed concentrations of BOD and DO are defined. Infinite boundary conditions are assumed in the downstream end. BOD and DO particles are simply permanently removed from the system as they cross the last downstream cell. All cells of a certain reach follow the same transition rules. The rules account for advection, dispersion, decay and aeration.

The advection process is considered through a probability of particles to move to the next cell in the downstream direction during a simulation time step. The advection probability, P_{adv}, is defined as:

$$P_{adv} = (u\Delta t)\Delta x \qquad 0 \leq P_{adv} \leq 1 \tag{15.2}$$

where Δt is the simulation time step and Δx is the cell length. A careful choice of Δt and Δx values assures that P_{adv} is never greater than one.

The dispersion process is modelled following a random walk approach (Bear and Verruijt, 1987). These authors, using a particle-tracking formulation, considered that a particle moves by a deterministic amount due to advection, and by a random amount of a maximum magnitude as a result of dispersion. Assuming a uniform distribution of the random component of the movement, the probability distribution of the particle movement for a great number of independent steps should follow a Gaussian distribution.

Castro *et al.*, (1993) showed that the probability distribution representing a random walk can be shown to be identical to the analytical solution of the one-dimensional advection–dispersion differential equation for an instantaneous spill. Thus, the dispersion probability P_{dis} was defined as:

$$P_{dis} = (2s - 1) \quad (6\,E\,\Delta t)^{0.5}/\Delta x \qquad -1 \leq P_{dis} \leq 1 \tag{15.3}$$

where s is a uniformly distributed random number between 0 and 1. Again, a careful selection of Δt and Δx guarantees that the absolute value of P_{dis} is never greater than 1. When P_{dis} is negative the particle moves to the adjacent upstream cell; when P_{dis} is positive to the adjacent downstream cell; otherwise the particle stays in the original cell.

A relationship between decay probability, P_{dec}, and the decay rate constant k_d was derived by Toffoli and Margolus (1987):

$$P_{dec} = k_d.\Delta t \qquad 0 \leq P_{dec} \leq 1 \tag{15.4}$$

The formulation for the reaeration constant is similar:

$$P_{aer} = k_a.\Delta t \qquad 0 \leq P_{aer} \leq 1 \tag{15.5}$$

Through a careful choice of the appropriate value for Δt, P_{dec} and P_{aer} are typically much less than one.

To define the appropriate simulation time step, a time step is independently calculated based on each process. These values are then compared to determine the minimum value which will be the time step to be used in the simulation. The following expressions are used to calculate the time step for each process:

$$t_{\mathrm{adv}} = \Delta x/u \tag{15.6}$$

$$t_{\mathrm{dis}} = (\Delta x)^2/6E \tag{15.7}$$

$$t_{\mathrm{dec}} = 1/k_{\mathrm{dec}} \tag{15.8}$$

$$t_{\mathrm{aer}} = 1/k_{\mathrm{a}} \tag{15.9}$$

The actual time step used in the simulation is given by:

$$t = \min\left(t_{\mathrm{adv}}, t_{\mathrm{dis}}, t_{\mathrm{dec}}, t_{\mathrm{aer}}\right) \tag{15.10}$$

This cellular automaton model was implemented on a Intel personal supercomputer iPSC/1 with a MIMD (multiple instruction, multiple data) architecture, distributed memory and a hypercube topology (Fox, 1991).

Figure 15.3 shows the results obtained with the cellular automata models and the traditional formulations for a set of synthetic data. These results show that the approach based on cellular automata is capable of describing BOD and DO behaviour as well as the established differential equation models.

To illustrate the impact on the total simulation time of a cellular automata model implemented in parallel processors, several simulations were performed in the iPSC/1 using a different number of processors. The result is shown in Figure 15.4. Although it is possible to significantly decrease the simulation time by increasing the number of processors, such improvement tends to level off due to the communication overhead.

15.3.2 Predator–prey modelling

Câmara *et al.* (1993) have modelled a predator–prey relationship using a cellular automata formulation and compared the results with a traditional differential equation-based model. Regular time steps were considered for both models.

Predators and preys were assigned locations in cells in a mosaic representing a territory (Figure 15.5). Cyclic boundary conditions were considered. To simulate species growth, random reproduction and death rules were assumed. Probabilities for these rules

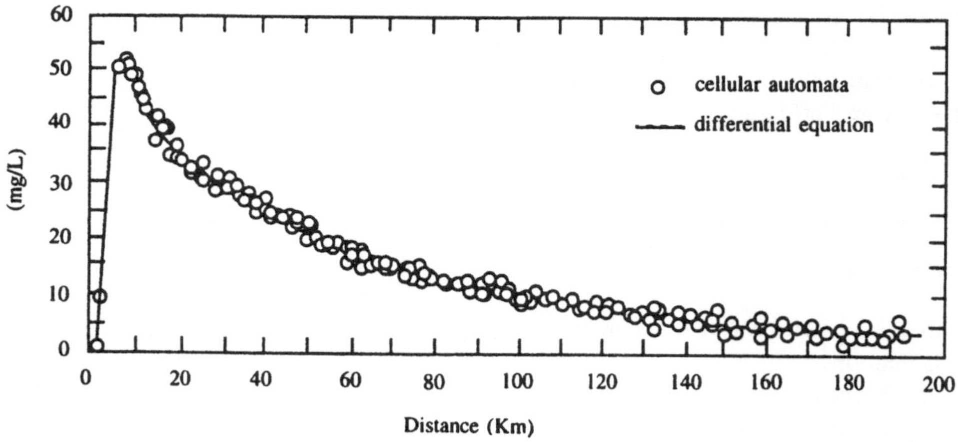

Figure 15.3 BOD/DO cellular automata model and Streeter and Phelps model results for a steady-state scenario involving a single discharge in a river with uniform characteristics.

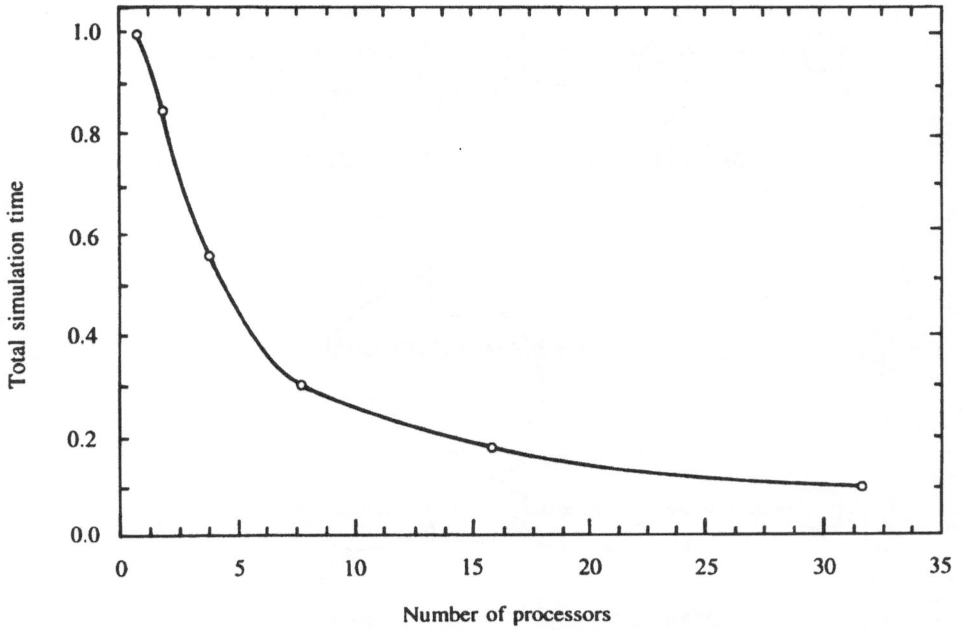

Figure 15.4 Total simulation time as a function of the number of processors used in water quality cellular automata simulation.

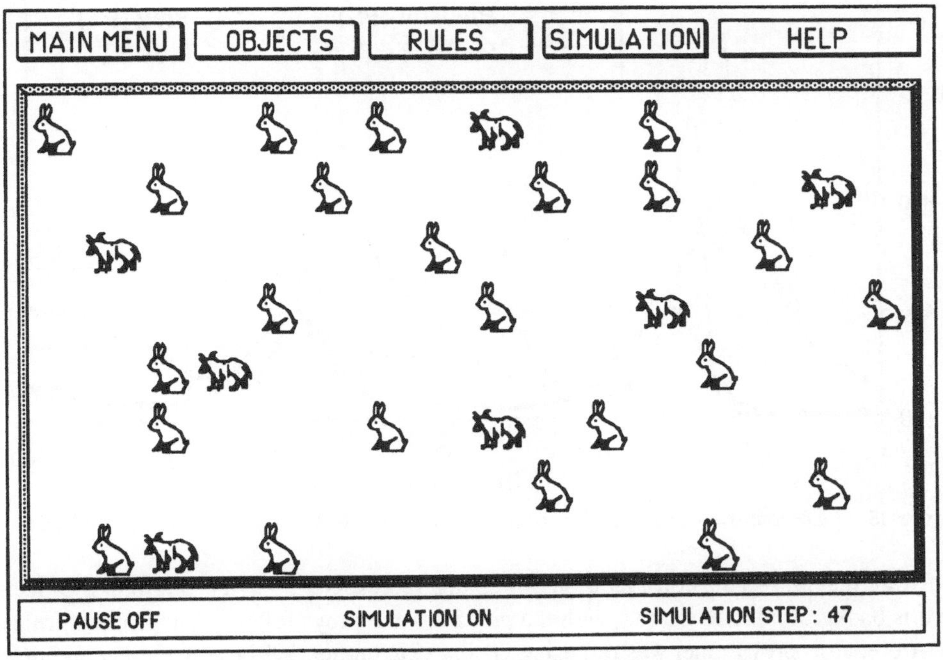

Figure 15.5 Object representation of a predator–prey cellular automata model.

for both the predators and the preys were determined based on birth and death rate constants following derivations similar to the ones used in Eqs (15.4) and (15.5). Predators and preys could meet in the same cell. Then, for a certain probability, the prey could die. This cellular automata model was implemented on a personal computer using

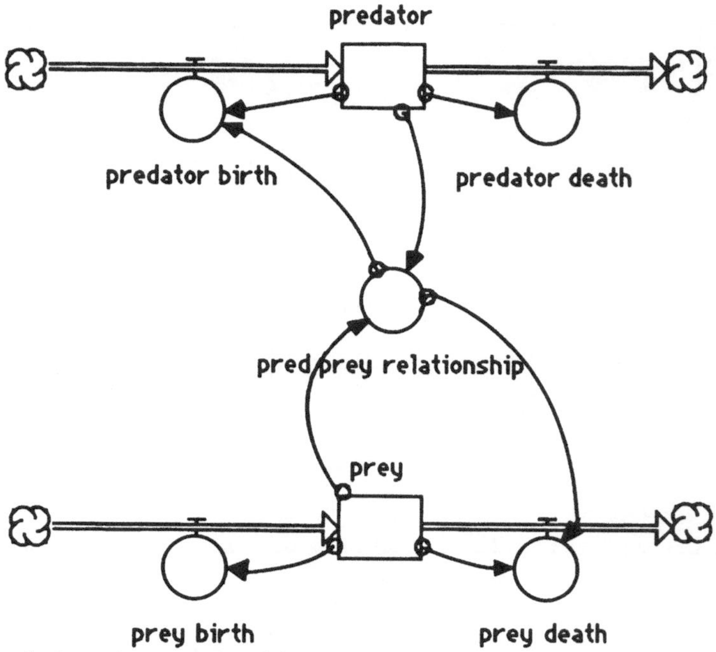

Figure 15.6 iThink predator–prey model.

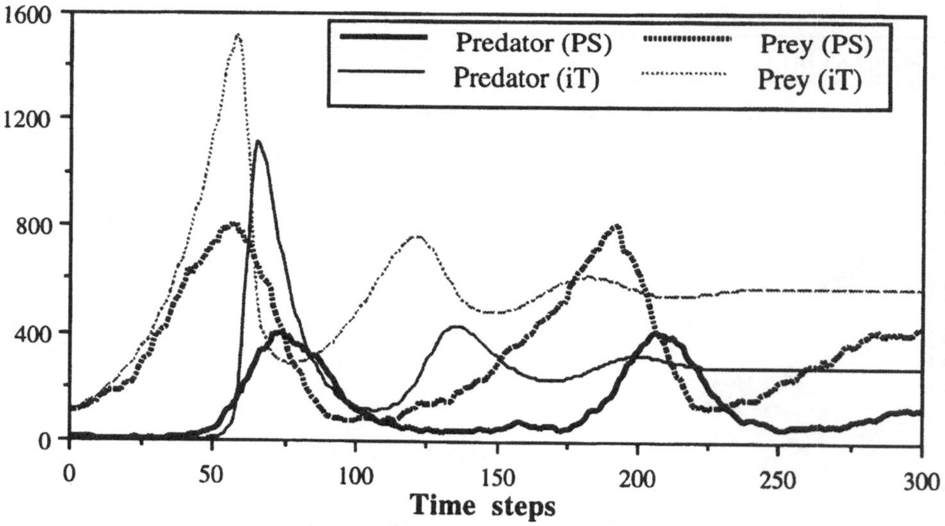

Figure 15.7 Comparison of the cellular automata model run (CA) and the iThink (iT) model run.

software that allows the consideration of a spatial representation as a background. Objects in this background may interact with the predators and preys following appropriate rules.

A predator–prey model was run using iThink (Richmond, 1991), considering an initial number of individuals, birth and death rates for both species identical to the ones used in the cellular automata model (Figure 15.6). Figure 15.7 shows the difference between both approaches: one that takes into account the spatial characteristics of the individuals (the cellular automata approach); and another approach that only considers their number. Note that while the cellular automata model continues to perform oscillations, the iThink model stabilises.

15.3.3 Fire propagation modelling

Landscapes can be represented as cellular automata (Green *et al.*, 1989). Each cell represents a fixed surface area and has attributes that correspond to environmental features such as vegetation cover and topography. Thus, it is possible to apply cellular automata formulations to a number of a landscape diffusion process such as forest fires.

Conventional forest-fire modelling relies upon empirical models. The most successful of those is the one proposed by Rothermel (1972). Rothermel's model enables the forecasting of the fire's speed of propagation and intensity, based on data on the following parameters: fuel characteristics such as moisture, proportion of live and dead material, density, surface volume coefficients and mineral content; wind speed and direction; and topographic factors such as altitude and slope.

Diogo *et al.* (1993) developed a cellular automata model for forest fire propagation that combines Rothermel's model with a cellular automata view of the landscape. The proposed model is based on the following assumptions:

- Rothermel's model is representative of the fire's speed of propagation.
- Vegetation is homogeneous within each cell.
- There are eight independent wind directions.
- Slope is null in each cell.
- Only cells with the fire already extinct or without vegetation are not subject to fire propagation.
- In each time step, each cell can propagate the fire to only one of the eight adjacent cells.

Fire ignition may start in one or more cells. Each cell may then have one of four possible states: (1) unburned; (2) fire with no capacity to propagate (first burn); (3) fire with capacity to propagate (burning); and (4) burned.

One of the most difficult issues in cellular automata modelling is the definition of a relationship between real time and the simulation model time step. This definition is essential to compute the real time associated with fire propagation and to take into account the temporal differences in the fire evolution on different vegetation covers.

There are a number of time steps associated with each of the cell states. This means that each cell assumes a given state for a number of time steps of the model. Each step corresponds in this model to one minute in real time.

Note that a untouched cell (state 1) will move to state 2 when burned; and an extinct cell (state 4) will always remain in that state. Thus, the definition of the number of time steps is only relevant for states 2 and 3. For state 2, this number is defined as

$$\text{number of steps} = \text{rate of spread (slope} = 0\text{, average wind intensity)} \times \text{cell length} \qquad (15.11)$$

For state 3, the number of steps is determined by

$$\text{number of steps} = 10 \times \text{time of permanence in state 2} \qquad (15.12)$$

Factor 10 was obtained by calibration with the Rothermel model.

For each model interaction, there is a propagating cell and eight adjacent cells as in Figure 15.8. To determine the rate of spread (RS) associated to each cell one has to apply the Rothermel based expression. In this expression, wind is considered to have eight wind directions with intensities varying for each direction. To consider the slope variable, only

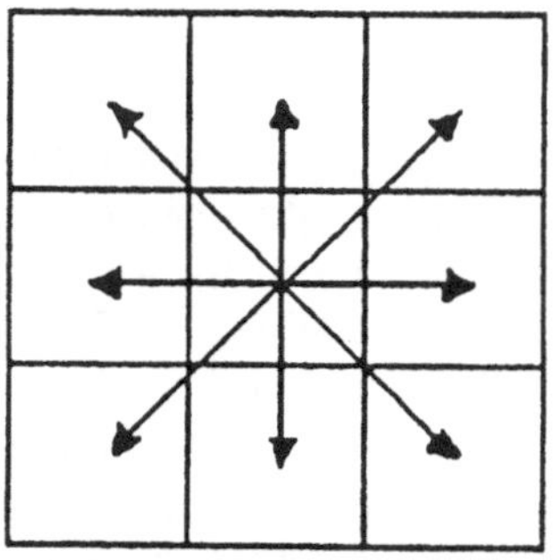

Figure 15.8 Possible directions for fire propogation.

slopes between cells are considered, as slope in each cell is assumed to be null. This is achieved using the empirical expression (Ritsperis, 1992):

$$\text{RS}_i \text{ final} = \text{RS} \times \text{sen}(\arctan(\text{slope between cells}))\ i = 1, \ldots, 8 \tag{15.13}$$

The rule of fire propagation from one cell to another may be now defined according to the following steps.

- Square the rate of spread RS_i of each of the adjacent cell.
- Add all the RS_i.
- Assign probabilities of propagation, defining intervals for the probability of fire ignition P_i by dividing the square of RS_i with the square of the summation of the RS_i. The squaring amplifies the differences among the RS_i of each cell.
- Generate a random number between 0 and 1.
- The contaminated cell will be the one where P_i corresponds to the generated random number.

Expanding this methodology to the whole mosaic, one obtains the global evolution of the fire spread (Figure 15.9).

To evaluate the cellular automata model performance, the traditional Rothermel model was used as images of burned landscapes are difficult to obtain. For both models, shape and areas indices were computed for a number of runs (Turner *et al.*, 1989). For this purpose, fractal dimensions (Burrough, 1986; Mandelbrot, 1983) and the burned areas interface index (Turner *et al*, 1989) were calculated. The results obtained with both models were similar. However, the jagged lines obtained with the cellular automata model seem to be more realistic (Figure 15.9).

The cellular model and the evaluation procedures were implemented on a personal computer as an application under the name FireGIS (Diogo *et al.*, 1993). This application enables the visualisation of the model's results in raster format and in a digital terrain model. FireGIS accepts data in several formats, namely IMG from IDRISI.

15.4 Conclusions and future developments

Cellular automata appears as a promising approach to simulate spatial phenomena. It was shown that for water quality and forest fire propagation modelling the results obtained with cellular automata were similar to the ones obtained with the traditional models.

However, these efforts showed that for process-based models, the definition of

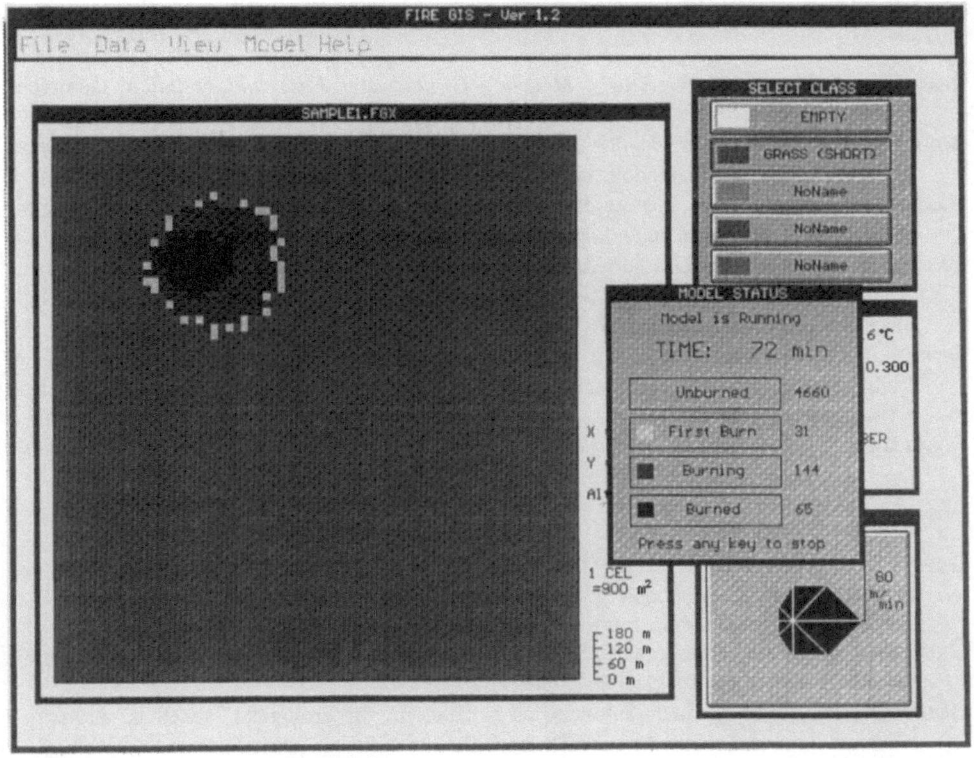

Figure 15.9 FireGIS application results

cellular automata transition rules may be cumbersome. Future practical application of cellular automata in these cases will certainly require built-in transition rules.

For a predator–prey problem, the cellular automata entity-based model used simple transition rules. For this problem, transition rules such as attraction and repulsion will have to be developed. Naturally, the spatial nature of the cellular automaton model made it produce different results from the conventional differential equation lumped model.

Future developments should not be centred only on transition rules. The modelling of more complex systems will require the handling of a larger number of interacting bit planes. Extensions to three dimensions may be also relevant for a number of environmental applications.

Finally, cellular automata concepts may be extended to simulate real or symbolic multidimensional objects. Câmara *et al.*'s (1993) work points towards that direction by simulating pictorial objects defined by position, size and colour. Future work should also take into account the shape of objects (basically a set of connected cells).

Acknowledgements

The authors acknowledge the contributions to this work of Daniel Gallagher at Virginia Tech and Pedro Gonçalves, Paulo Diogo, João Pedro Rodrigues, Edmundo Nobre and José Eduardo Fialho at the New University of Lisbon. This research has been partially supported by Direcção Geral da Qualidade do Ambiente under contract no. 86/91/J, a fellowship from JNICT awarded to the third author and a grant from the US National Science Foundation.

REFERENCES

Bear, J. and Verruijt, A. (1987). *Modeling Groundwater Flow and Pollution.* Dordrecht, Holland: D. Reidel.

Burrough, P.A. (1986). *Principles of Geographic Information Systems for Land Resources Assessment.* Oxford: Clarendon Press.

Câmara, A., Ferreira, F., Nobre, E. and Fialho, J.E. (1994). Pictorial modeling of dynamic systems, in *Systems Dynamics Review*, **10** (4), 361–73.

Castro, P., Gallagher, D. and Câmara, A. (1993). 'Dynamic water quality modeling using cellular automata'. Department of Civil and Environmental Engineering, Virginia Tech, Blacksburg, Va. (not-published).

Diogo, P., Gonçalves, P. and Rodrigues, J.P. (1993). 'FireGIS'. Department of Environmental Sciences and Engineering, New University of Lisbon, Monte de Caparica, Portugal (unpublished).

Fogelman, F., Robert, Y. and Tchuente, M. (1987). *Automata Networks in Computer Science: Theory and Applications.* Princeton, NJ: Princeton University Press.

Fox, G.C. (1991). Achievements and prospects for parallel computing, *Concurrency: Practice and Experience*, **3**(6), 725–39.

Green, D.G., Reichelt, R.E. van der Laan, J. and Macdonald, B.W. (1989). A generic approach to landscape modeling, in *Proceedings of Eight Biennial Conference*, pp. 451–6, Canberra, Australia: Simulation Society of Australia.

Grossman, W.D. and Eberhardt, S. (1992). Geographical information systems and dynamic modeling, *The Annals of Regional Science*, **26**, 53–66.

Hogeweg, P. (1988). Cellular automata as a paradigm for ecological modeling, in *Applied Mathematics and Computation*, **27**(1), 81–100.

Larkin, T., Carruthers, R. and Roper, R.S. (1988). Simulation and object-oriented programming: the development of SERB, *Simulation*, **52**(3), 93–9.

Laurini, R. and Thompson, D. (1992). *Fundamentals of Spatial Information Systems.* London: Academic Press.

Mandelbrot, B. (1983). *The Fractal Geometry of Nature.* San Francisco, CA: Freeman.

Phipps, M.J. (1992). From local to global: the lesson of cellular automata, in DeAngelis, D.L. and Gross, L.J. (Eds). *Individual-Based Models and Approaches in Ecology*, pp. 165–87, London: Chapman and Hall.

Richmond, B. (1991). *iThink: User's Manual.* Lyme, NH: High Performance Systems.

Ritsperis, A. (1992). *Evaluating a New Methodology for Prediction of Fire Behaviour, with Particular Reference to the Eucalyptus Plantation of Portugal*, MSc thesis, London: Imperial College.

Rothermel, R. (1972). 'A mathematical model for predicting fire spread in wildland fuels'. Res. Paper INT-115, US Department of Agriculture, Forest Service, Intermountain Forest and Range Experiment Station, Odgen, UT.

Rothermel, R. (1983). 'How to predict the spread and intensity of forest fire and range fires'. Rep. INT-143, US Department of Agriculture, Forest Service, Intermountain Forest and Range Experiment Station, Odgen, UT.

Thomann, R.V. and Mueller, J.A. (1987). *Principles of Surface Water Quality Modeling and Control.* New York, NY: Harper and Row.

Toffoli, T. (1984). Cellular automata as an alternative to (rather than an approximation of) differential equations in modeling physics, *Physica*, **10D**, 117–27.

Toffoli, T. and Margolus, N. (1987). *Cellular Automata: A New Environment for Modeling.* Cambridge, MA: MIT Press.

Turner, M.G., Constanza, R. and Skalar, F. (1989). Methods to evaluate spatial simulation models, *Ecological Modeling*, **48**, 1–18.

Von Neumann, J. (1966). *Theory of Self-Reproducing Automata*, Burks, A.W. (Ed.), Urbana, Illinois: University of Illinois Press.

CHAPTER SIXTEEN

CityLife: a study of cellular automata in urban dynamics

GEOFFREY G. ROY and FOLKE SNICKARS

16.1 Introduction

The study of interacting dynamic systems has traditionally been approached from a detailed analysis and modelling of the interconnections and interactions amongst the system's component parts. There are many complex systems, however, where this approach has not been particularly productive, especially in those which comprise natural or ecological phenomena such as the growth of natural organisms, the spread of fire and the emergence of weather patterns.

These systems can be said to be complex in the sense that they function through a multitude of interacting mechanisms which define them and control their change and evolution. On the other hand, recent theories propose that the underlying forces which shape these systems arise from much more primitive interactions among individual components which are now often described as representing their natural chaotic behaviour.

One of the best known and pioneering studies in this area was done by John Conway *et al.* (1982), Gardener (1974), which emerged in Conway's *Game of Life*. While this work was essentially abstract, it demonstrated that the repeated application of very simple rules to some random initial state could generate interesting, and recurring patterns as the state of the system evolved. The *Game of Life* is played on a grid of cells, some of which are said to be alive and some dead. Given a random initial state, at each generation some new cells are born, and some existing living cells die. A live cell dies if it has two or three neighbouring live cells, while a dead cell with three live neighbours will be reborn.

Earlier work by computing pioneer John Von Neumann (1966), introduced the concept of cellular automata in an attempt to represent the idea that a computing machine could reproduce itself, or even more interestingly produce a more complex machine. More recently, similar ideas have been applied to a range of natural systems from biology to economics (McGlade, 1993; Arthur, 1993).

In most of these approaches the essential continuity of nature and natural systems is approximated by a discrete set of cells which interact with each other using rather simple rules. This type of modelling is thus conceived as a cellular automaton. Our interest in this chapter is to apply a similar approach to the study of the dynamics of an urban system. Many contributions to the study of urban systems have been made, generally by attempting to apply various modelling strategies to capture the underlying relationships among the components which constitute an urban system. Our approach is initially much more primitive in some classical regards while placing the emphasis on the dynamics of change.

The literature on urban and regional planning abounds with examples of simulation models in which the evolution of land use is traced out by, more or less, sophisticated mechanisms of change. Some of these models have a thorough theoretical underpinning, showing the consequences of guiding economic principles as competition for land among a set of activities. Several models in this tradition have been developed by Anas (1982) and Fujita (1989). In these models, space is either represented in the form of a grid or as a set of geographically delimited zones. The ambition is to show the equilibrium distribution of land use when the demand for space has been matched with supply. These models are normally static with the dynamic model of Fujita and Kashiwadani (1989) as a prime exception. In some models, markets are cleared via prices throughout, whereas different rationing schemes are adopted for some submarkets, as in the model developed by Anas *et al.* (1987). Most of the models in this category are theoretical constructs with limited connection to the practical world of urban and regional planning.

There are also applied urban simulation models where the mechanisms of change, and the market-clearing schemes, are not derived directly from economic theory but represent ways of mimicking the behaviour of agents in a historical perspective. Examples of such applied urban simulation models are Roy and Snickars (1992) and the micro-simulation model of Wegener (1980). In these models, which most often model dynamics in a recursive fashion, the emphasis is on consistency among submodels rather than on adherence to theoretical principles.

A third tradition of urban and regional modelling which has a bearing on the current problem is the multiobjective optimisation models for land use and transport interaction developed by, for instance, Lundqvist (1977). In these models, regions are subdivided into zones of unequal size with limited capacity, the problem being to allocate activities over these zones to maximise one or several criteria, basically expressed as the accessibility to other urban activities.

There are also a number of theoretical models where space is treated as a homogeneous entity, thus forecasting distribution functions as the ultimate outcome of spatial competition. These models most often deal with the use of land for residential purposes, aiming to show how land use and land value is affected by changes in household incomes and preferences, as well as transport technologies. An analysis of the variety of such urban economic models is given in Fujita (1989). This research has been extended to games between landowners and landlords by Asami *et al.* (1990). These latter game-theoretical analyses employ discrete space and attempt to forecast the distribution of the returns to land from the strategic interdependence between landowners and landlords. A classical example of modelling in this tradition is the Hotelling game of spatial competition (for example, Rasmussen, 1989).

Game-theoretical analyses are of particular interest as a background to the current CityLife model. The linking factor is the representation of space in discrete cells, the explicit treatment of actors who co-operate or compete with one another about locations in space, and the explicit treatment of the utility for each actor associated with a pattern of land use. The aim of the CityLife model is to show how complex patterns of land use will evolve from simple behavioural rules for the strategically interdependent actors in combination with simple representations of land-use restrictions imposed by public policy agents.

CityLife is a modelling framework based on a grid of cells, just like the Game of Life. Each cell is intended to represent a unit of space which can contain some particular urban activity, typically dwellings, workplaces, transport infrastructure, or green space. The spatial arrangement of cells is intended to reflect the spatial organisation of an urban system. From some initial state, and through the imposition of set of behavioural rules and selection criteria, the system will grow and evolve spatially. Our interest here is to study these changes in land use patterns and to contemplate whether such a simple approach to modelling can make a contribution to the understanding of the emergence of complexity in urban systems.

16.2 Modelling the urban system in CityLife

Urban systems exhibit a substantial degree of variability over time. Our approach here makes many simplifications as regards the properties of activities, the temporal mechanisms of change, the transport network structures and the behavioural rules. First, we will assume that the spatial structure can be approximated by a two-dimensional regular grid of cells. Distances among cells are measured either in terms of straight lines or in terms of horizontal and vertical inter-cell distances. Each cell contains a single activity type. While, in general, there are no special restrictions on the variety of activity types, we limit our interest to a small number which might characterise the dominant urban activities. In CityLife we characterise the activity classes into the following types:

empty cells	where no activity exists, thus available for urban use;
green cells	which have been set aside for green urban purposes;
dwelling cells	which contain essentially residential activities;
workplace cells	which contain essentially employment activities;
reserved cells	which are reserved for transport or other urban or rural use.

Other cell types can also be included, for example cells which exhibit water-front features or terrain unsuitable for building. Such cells are fixed in space but may influence the relative attractiveness of nearby cells.

Each cell in the model has a computable 'attractivity' for each type of urban activity based on some prescribed function. In our case we have adopted a simple accessibility function of the form:

$$a_i(k) = \sum_j \exp(-\mu(k) * d_{ij}(k)) * x_j(k)/N(k) \tag{16.1}$$

$x_j(k)$ = 1 if cell j is used for an activity of type k, = 0 otherwise;
$d_{ij}(k)$ = distance from cell i to cell j for an activity of type k;
$\mu(k)$ = accessibility coefficient for an activity of type k;
$N(k)$ = number of cells containing an activity of type k, where $\Sigma_j\, x_j(k) = N(k)$;
$a_i(k)$ = accessibility from cell i to cells containing an activity of type k, a measure of the attractivity of cell i for activity k.

The attractiveness of any cell for the location of an activity thus depends on the accessibility to a range of other activities. To model this we have adopted a simple weighted function to estimate the average attractiveness:

$$A_i(k) = \sum_l b(k,l) * a_i(l) \tag{16.2}$$

In Eq. (16.2), the factors $b(k,l)$ are weights which describe the relative importance of the accessibility profile of activity of type l for activity of type k. $A_i(k)$ is the composite attractiveness of cell i for activity of type k.

Given an initial state, and the desire to allocate additional activity of type k, then the cell that would be considered most attractive would be the one having the largest value of $A_i(k)$. We thus have a simple mechanism to study the dynamic behaviour of an urban system. Although it is simple it exhibits properties which can be interpreted in a wide variety of ways, from willingness to pay to planning norms.

To study some simple systems we initially assume that the only imposed constraints on the dynamic process are that once a cell is occupied, it remains occupied by the same activity type. Starting from an arbitrary state, and following some growth strategy, for instance, allowing a specified number of new cells of each type to be allocated at each generation of CityLife, we can follow the spatial growth of the system. The growth pattern will be evolutionary in the sense that the future development of the spatial structure will depend on the trajectory followed up to the present.

In the simple model, different land uses will compete with one another for attractive locations in view of the mutual benefits from interaction, at the same time affecting each other as a result of the assumption that a chosen land-use is never altered. The rule of dynamics of the coupled land-use processes is that a certain number of new cells are to be allocated at each time point. When a full allocation has been made at a point in time, attractivities are recomputed to form the basis for the allocation process at the next time point. The process described for the simple illustrative model differs from the one in the *Game of Life* in the sense that in that model only growth is treated, not substitution.

To demonstrate the model in operation we take the initial state as shown in part (a) of Figure 16.1. Here we have two cells used as green-space, four cells as housing activities and two cells containing workplace activities. At each generation, two green-space activity units, four dwelling activity units and two workplace activity units are added. The system is simulated for a number of planning generations. Part (b) of Figure 16.1 shows the result after 10 generations. Part (c) of Figure 16.1 shows the spatial pattern after 20 generations. It may be noted even after 10 generations the patterns become tightly mixed with one another, thus causing continued spatial trajectories which are restrained by one another.

For the simple dynamic model presented above the weighting coefficients $b(k,l)$ are given in Table 16.1.

The figures in Table 16.1 refer to the case when the attractiveness of a cell for a particular activity is only determined by the accessibility to cells housing activities of its own kind. The interpretation is that agglomeration is favoured in the allocation of new urban green space and when additional dwelling and workplace activities are choosing to locate among cells. Thus, the benefits for the locating activities are supposed to be enhanced when green spaces, dwelling areas and workplace districts in the city each form contiguous patterns.

Table 16.1 Relative weighting coefficients for the interaction among activities in model version 1

From	Green space	Dwelling	Workplace
Green space	1.0	0.0	0.0
Dwelling	0.0	1.0	0.0
Workplace	0.0	0.0	1.0

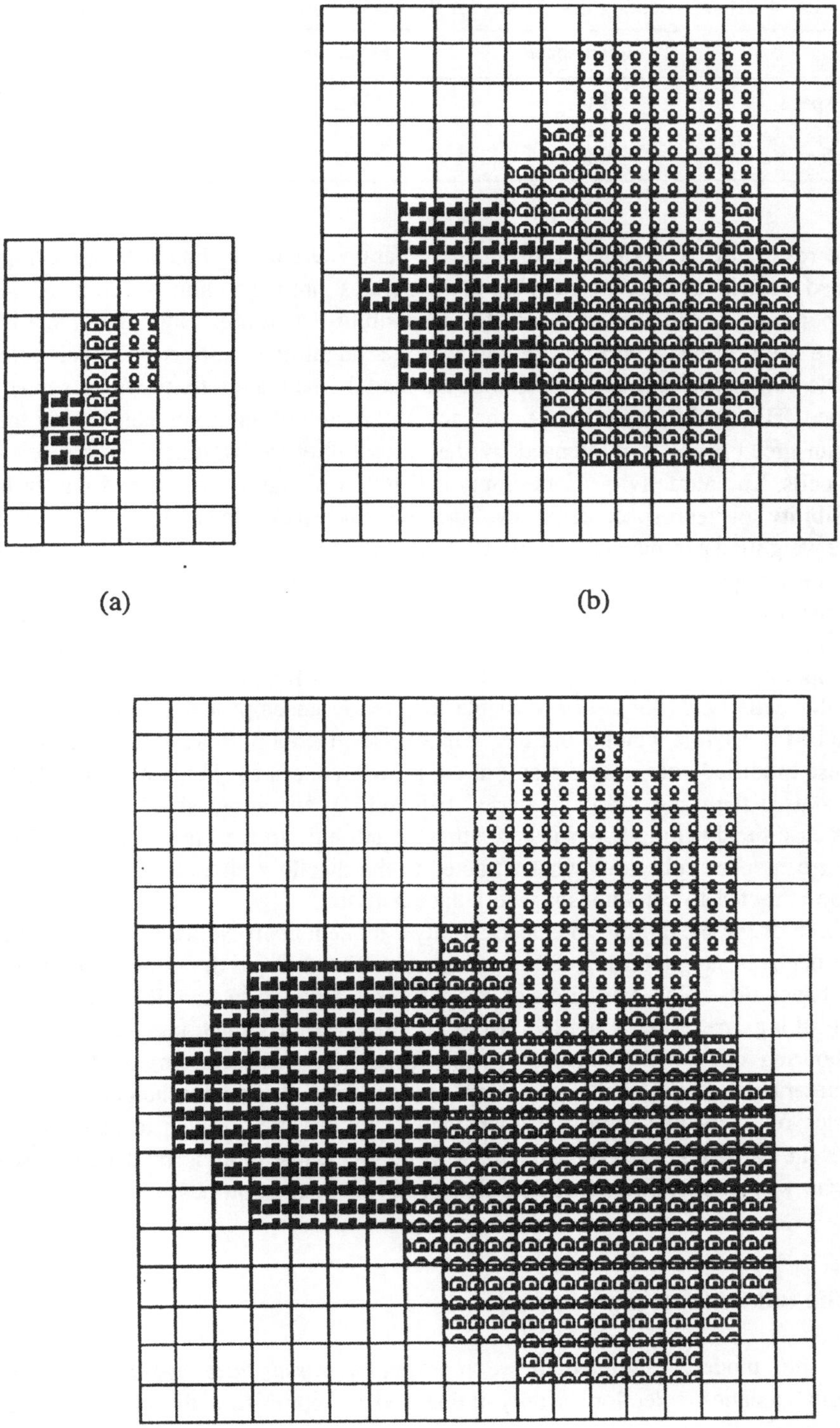

Figure 16.1 Simulation with simple model showing (a) initial state, (b) resulting spatial pattern after 10 generations, and (c) after 20 generations.

Table 16.2 Relative weighting coefficients for the interaction among activities in model version 2

From	Green space	Dwelling	Workplace
Green space	1.0	0.5	0.0
Dwelling	1.0	1.0	0.5
Workplace	0.0	0.0	1.0

The results of the simulation using these parameters are in line with what might be expected. Each activity gradually agglomerates into an almost circular pattern corresponding to the influence of the accessibility measure. Note that the distance measure used in this version of the model is a straight line, cell-centre to cell-centre.

Table 16.2 shows a second version of the model where attractivity weights have been modified. The modification implies that the attractivity of an empty cell having the role of urban green space is influenced by the accessibility to dwelling cells besides green space cells. The attractivity of an empty cell for dwellings is determined jointly by the accessibility to green-space cells, dwelling cells and workplace cells.

The weighting parameters in Table 16.2 sum to different values for each urban land-use. Since the composite attractivity measure is computed for green spaces, dwellings and workplaces separately, the fact that the weighting factors have different sum totals does not affect the allocation mechanism. The weights should be interpreted as coarse indicators of how important the accessibility to a certain activity is for the attractivity of a cell for a particular activity. It should be possible to provide an empirical underpinning for the choice of the relative weights but they can also be given a normative interpretation.

These modified coefficients result in the growth pattern as shown in Figure 16.2. The initial state is the same, but the patterns at 10 and 20 generations show some differences. In this case the dwelling activities are attracted strongly to the green space cells and the urban green space cells somewhat attracted to the dwelling districts. Dwelling activities are also attracted to cells housing workplace activities.

In both of these examples the accessibility exponent coefficients, which describe how the influence of neighbouring cells declines with distance, are the same for each activity type. One could explore variations of this assumption and show how the effects influence the resulting patterns. Rapidly declining interaction functions mean that only neighbouring cells will contribute to the attractivity measures. In the *Game of Life* these interactions are even more simple since only the closest neighbours have an influence on each other. On the other hand, activities are allowed to outcompete one another in particular cells in the *Game of Life*. Competition is modelled in a more elaborate way in CityLife as will be clear from the discussion below.

16.3 The influence of planning rules

The simple models examined above represent a *laissez-faire* approach where each individual location selection is only influenced by optimising the choice of the best available cell. Planning constraints, or planning norms and rules, can be introduced to reflect not only formal urban development restrictions as zoning, but also other planning goals and objectives. These restrictions can be looked upon as legal or pragmatic rules which have to be fulfilled by any spatial pattern which can occur during the development of the city.

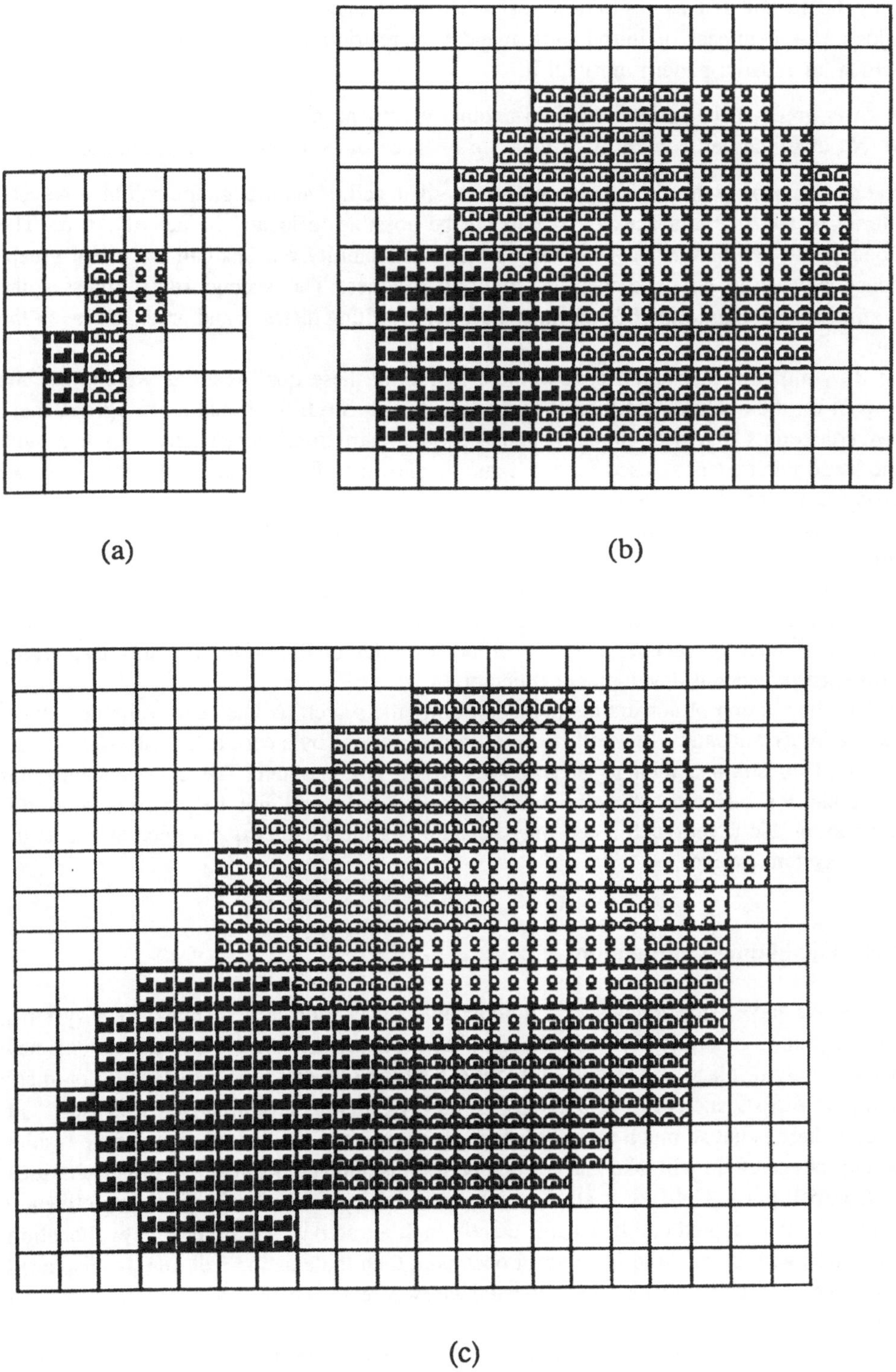

Figure 16.2 Simulation with simple model with modified weighting factors showing (a) initial state (b) results after 10 generations, and (c) after 20 generations.

There is a wide variety of options for the formulation of such planning rules. They may be formulated as restrictions or in terms of uncertainty and risk. Here we will illustrate the impact of introducing mandatory restrictions via two examples of rules which any spatial pattern must fulfil.

- Any green-space cell must adjoin another green-space cell.
- No dwelling cell can be more that two distance units away from a green-space cell.

Such rules qualify the choice prospects for a given cell. Even though the cell may be very attractive for a certain activity, it will not be possible to locate the activity there. The conceptual idea behind the first rule is that sustainability reasons call for urban green-space always to be developed in contiguous zones. The second rule expresses the common planning notion of proximity between dwelling districts and green spaces in the city.

By simulating the urban development including these qualifications we can see the impact on the evolution of the urban system. In the CityLife simulation program, these two constraints have been implemented as formal restrictions. When activating them with the same parameters as used for the example shown in Figure 16.1 and Table 16.1, we arrive at the results in Figure 16.3.

In this case we see that the basic circular agglomeration trends indicated above are now heavily modified. The continuity of the green-space cell area is ensured under the assumption that adjacency is interpreted as being any one of the eight adjoining cells to a particular cell. The dwelling type cells are forced to follow the corridor of green-space cells. It is therefore apparent that by imposing rules of this kind we can considerably influence the spatial development trajectories.

The imposition of constraints is made during the search for the best available cell for each activity allocation. If a cell being examined for use by a certain activity satisfies the imposed constraints, then its accessibility value is computed. The cell then becomes available for selection. A cell which violates a rule will not be considered in this generation. However, such a cell may become eligible in subsequent generations as the urban system evolves.

16.4 Facilitating competition

So far we have considered a cell untouchable once allocated. In real urban systems, competition is generally allowed. That is, if a location is sufficiently attractive then a new activity may displace an old one. In reality this type of behaviour is dependent on many complex factors, including land prices, zoning regulations and urban growth itself. Our interest here is not so much to model these processes, but to illustrate in a simple fashion what types of effect will take place if competition is allowed between the urban activities.

Competition in CityLife is allowed for each activity on the basis of the attractiveness values. If an occupied cell has a sufficiently high attractivity value for a new allocation, compared with its value to the current occupant, then the existing cell may be displaced. The displaced cell will be then allocated elsewhere to the best available cell for that activity type.

This description must be qualified by a practical problem. During the process described, a deadlock may be created where two cells simply keep replacing each other. It will then not be possible to proceed along the current spatial trajectory without changing the parameter structure of the model simulated.

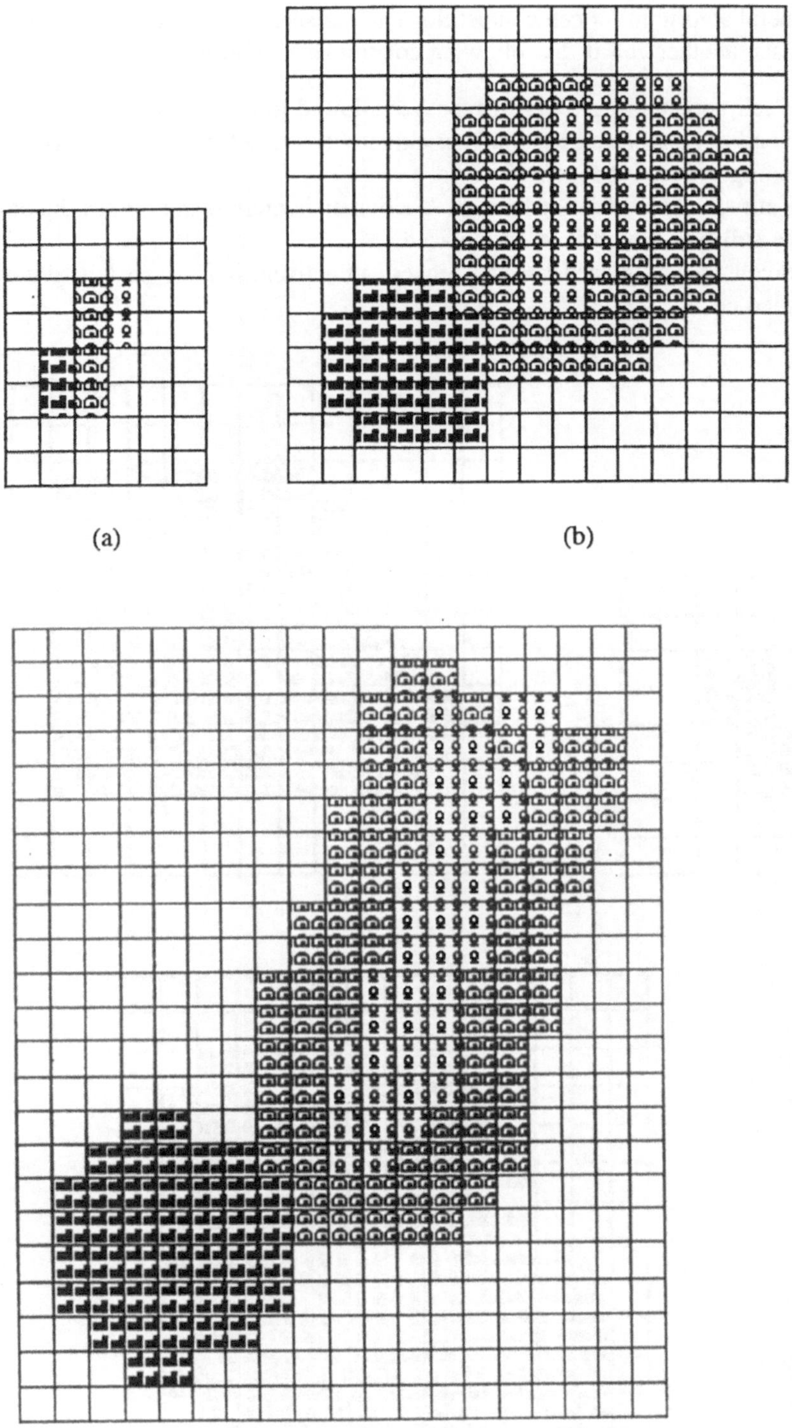

(a) (b)

(c)

Figure 16.3 Simulation with simple model using parameters according to version 1 with rules on green space and dwelling type cells, showing (a) initial state, (b) results after 10 generations, and (c) after 20 generations.

A general algorithm for cell competition and replacement might be that a cell of type *k* will replace another cell if the following conditions are fulfilled.

- If the cell containing activity of type *k* is allowed to be exposed to competition,
- if the cell to be replaced has a higher attractivity value than any empty cell for a cell containing activity of type *k*,
- if the attractivity of the cell for type *k* activities is more than Ω times the attractivity for the cell type which is to be replaced,
- if the cell to be replaced has the highest attractiveness for type *k* activities for all occupied cells.

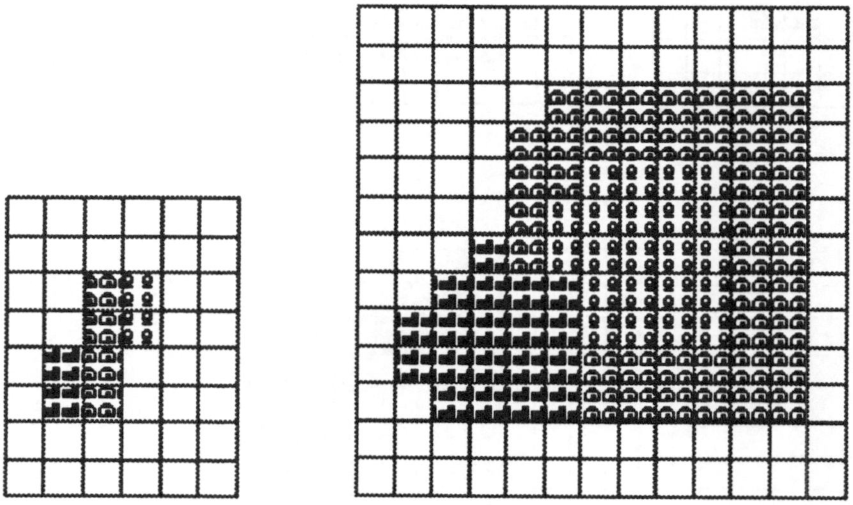

(a) (b)

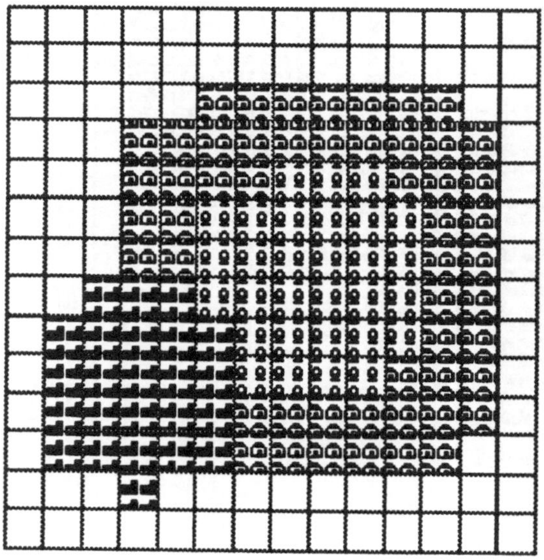

(c)

Figure 16.4 The simple model allowing competition after 10 and (almost) 14 generations.

Figure 16.4 shows the results where all of the activities can compete with each other. In this case the weighting factors are those in Table 16.1, the constraints from Figure 16.3 are also imposed. Each of the three activities can compete for an already allocated cell if its attractivity value for them is higher than the attractivity of the cell for the existing activity. In CityLife the level of attractiveness at which an activity can displace another one in a cell can be set for each combination of cells as a percentage of the value for the existing occupant.

The value of Ω is the value set for each combination of activity types. A very high value, say, Ω=1000 per cent would effectively prevent displacement, while a value of Ω=0 per cent would ensure displacement. A value of Ω=100 per cent means that a new activity will compete with an existing one for a particular cell if its attractiveness value is the same.

The results in Figure 16.4 show that, since the green-space activity can compete, land use tends to agglomerate in a circular region. The constraints on the dwelling type cells force them to locate around the green-space cells. As a result, after 14 generations we reach an impasse where no further dwelling cells can be allocated.

In CityLife it is perfectly permissible to allow green-space activity to compete with dwelling activity in a particular cell. It simply means that dwelling districts in the long run can be theoretically transformed into a green-space urban state. Running this same model, but preventing the green-space activity from competing, produces the result given in Figure 16.5 after 10 generations.

The results show that dwellings successively take over the centrally located green space. The reason for this is that the attractivity measure encourages the dwelling districts to be spatially contiguous. This tends to push the urban green space outwards while keeping the location pattern of that activity as concentrated as possible.

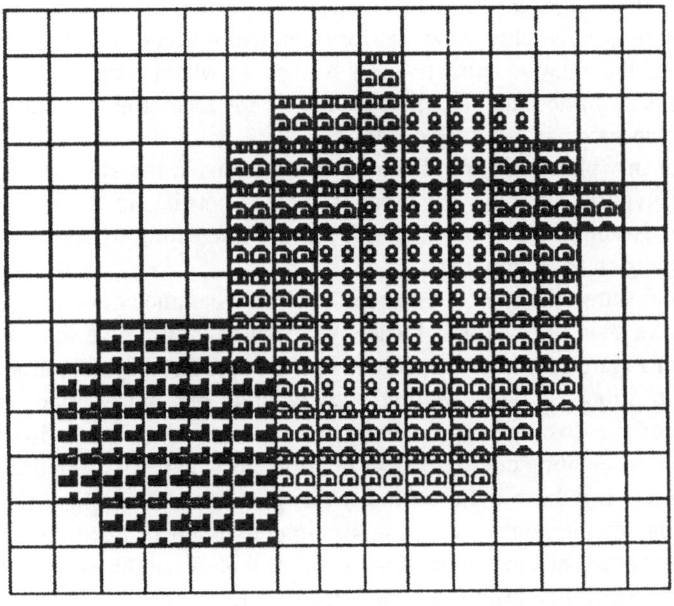

Figure 16.5 Model simulation as in Figure 16.4 but with no competition from green space, showing the resulting spatial pattern after 10 generations.

16.5 Implementation

CityLife has been implemented for Apple Macintosh® as a test-bed tool for the experimental development and evaluation of various low level modelling strategies of the cellular automata kind. The basic cell grid is limited to a 100×100 size. For each new allocation various options are available to guide the principles by which the transformation of patterns takes place over time:

- to allocate a cell of a selected type;
- to allocate a single cell of a selected type;
- to allocate a specified number of cells of a selected type;
- to allocate a specified group of cells (set numbers of one or more types).

The initial state is typically constructed by a manual allocation of appropriate activity types to the cell grid. The automatic allocation process is preceded by a computation of the attractivity values for all available cells. For practical reasons of computation time only that part of the complete grid which is in distance terms close to occupied cells is included in this computation. The computation of the attractivity values is quite time consuming and the time needed increases rapidly with the size of the part of the grid which is being analysed.

The set number of activity units is then allocated to those cells which are allowed and which have the highest attractivity values. Normally, this would be a small number since the actual attractivity values will change with each cell allocated. We have implemented the allocation process in this way to avoid having to recompute the complete set of attractivity values after each allocation. For the group options above, the user can specify a number of generations, after each of which the attractivities are completely recomputed.

Various parameters can be specified and adjusted by the user to characterise a model, for example, the weight given to each accessibility component in the attractivity value, the value of the accessibility exponent for each activity type, whether each activity type can compete, the relative attractiveness values at which displacement may occur if competition is permitted, and the distance metric for accessibility computations in terms of direct distances or as city block measures.

The order in which cells are allocated is important in the simulation process. In the current prototype of CityLife the order is fixed according to the designated activities green space, dwellings and workplaces. It may be interesting to explore alternative, user selected, strategies for comparison.

Figure 16.6 shows the effect of changing selected parameters in another simple model where we have a set of dwellings located near a highly attractive seaside location. Parts (a) and (b) of Figure 16.6 show the initial state and the development after 50 generations of block allocations of two dwelling activity units per time period. In this case the attractivity for the dwelling type cells is only influenced by other dwelling type cells, using the same weighting coefficients as in Table 16.1. Part (c) of Figure 16.6 shows the results from altering the weights so that water proximity is weighted five times that for dwelling cells among themselves. In this example we have forced the distance effect of neighbouring water cells to diminish much faster that of neighbouring dwelling cells, by making the accessibility exponent five times larger.

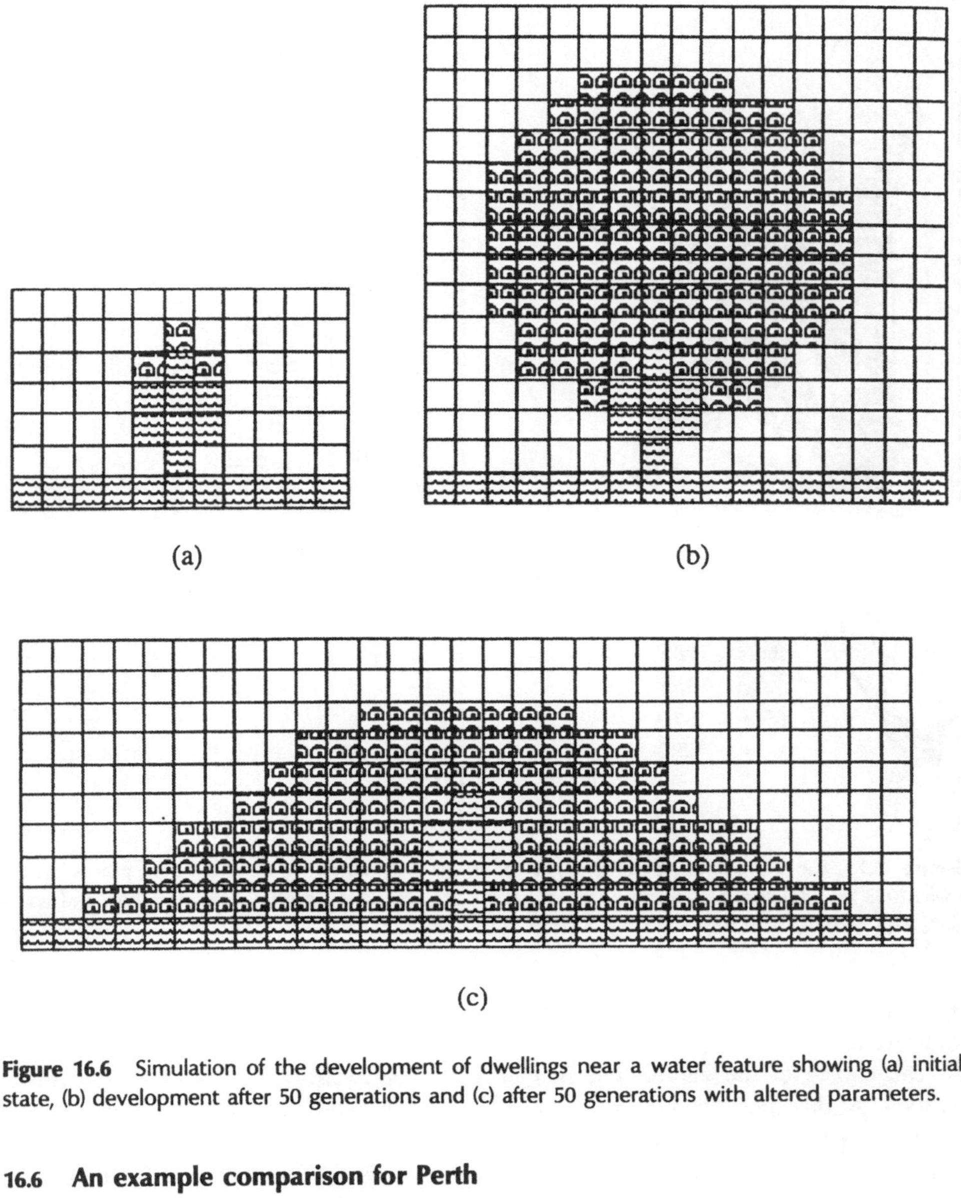

(a) (b)

(c)

Figure 16.6 Simulation of the development of dwellings near a water feature showing (a) initial state, (b) development after 50 generations and (c) after 50 generations with altered parameters.

16.6 An example comparison for Perth

Ultimately, modelling processes of this kind depend on their relevance to reality. While it is too early to claim this approach produces realistic forecasts, we can produce some preliminary results for the urban development of the city of Perth in Western Australia. Clearly the development of a large urban system as Perth, which has a population of about 1 million, is a much more complex urban system that what we will be able model here. It might still be interesting to see whether we can capture some of this complexity, even with the current simple modelling approach.

Perth is a city with a southern Mediterranean climate and so the ocean beaches and a significant estuarine river system dominate preferences in the choice of residential location. Our simple model has only two activity types to allocate over cells, urban development and water. The model thus develops through the allocation of urban

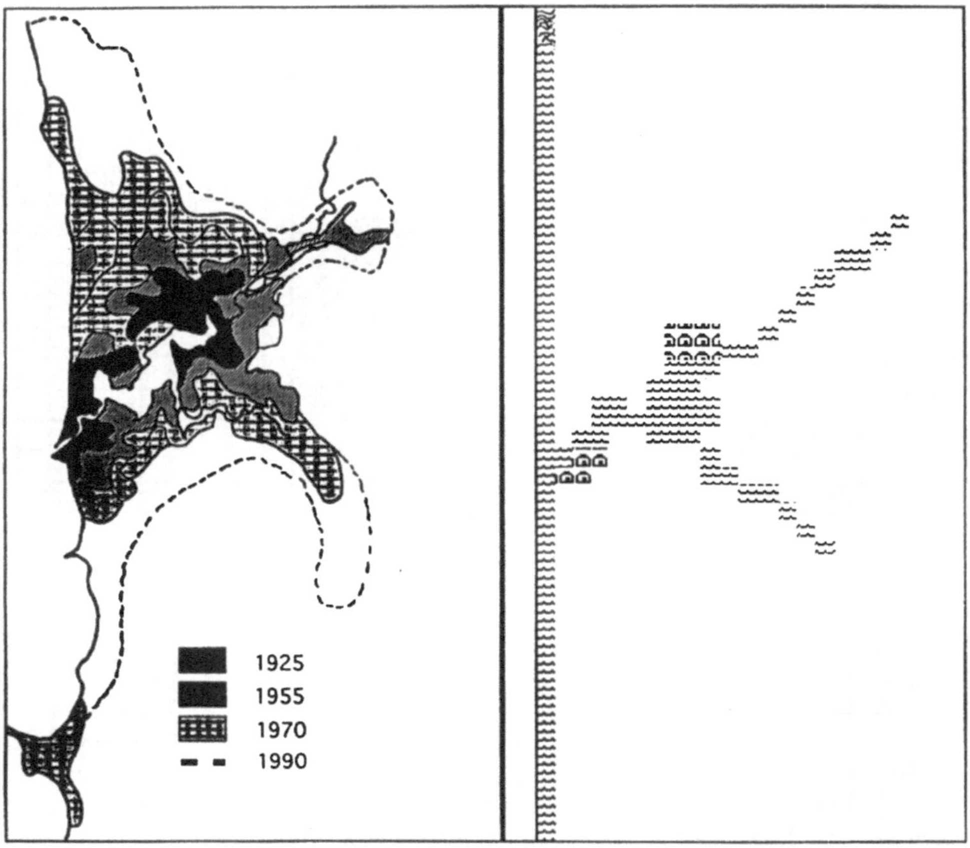

Figure 16.7 The approximate extents of the urban development in Perth at different points in time from 1925 to 1990 (left) and starting state for the Perth simulation model showing urban and water cells only (right).

activity to cells exhibiting different distances between one another and to the waterfront.

The choice of the various coefficients is of course critical to the way the model evolves. At this stage we have no definitive way of deriving the parameter values except through experimentation. For this model we have explored a range of values. The relative attractiveness weights for adjacent urban and water cells have been assumed to be 1.0 and 2.0 respectively. This means that for an urban cell, an adjacent water type cell has twice the value of an adjacent urban type cell. These values were chosen after some experimentation. In addition, the exponent coefficients for the accessibility functions were taken to be 1.0 for both urban and water cells. This means that the attractiveness of cells falls off rather quickly as the distance between cells increases.

The actual spatial development of Perth is represented in Figure 16.7. This diagram shows the approximate extents of the urban development from 1925 to the early 1990s in the left part of the figure. The starting state for the model is shown in the right part of the figure which represents the extent of the urban development in the 1880s and 1890s. Figure 16.8 shows the state of the system after allocating urban activity to 100 cells on the left side, and the results after allocating urban activities to 200 cells on the right side.

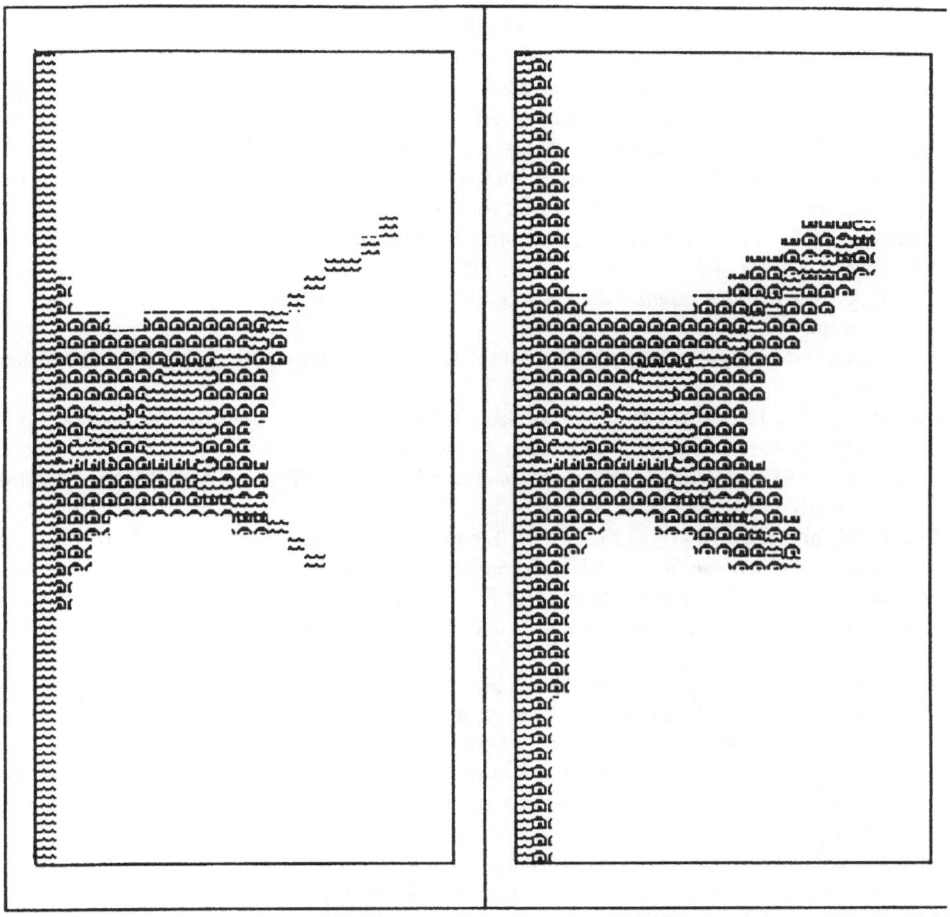

Figure 16.8 The Perth model state after allocation of urban activities to 100 cells (left) and 200 cells (right).

While we cannot claim an exact simulation the results appear to indicate that, given the assumed coefficients, the resulting states bear some resemblance to the actual states. This result is interesting given the simplicity of the spatial allocation model.

16.7 Conclusion

CityLife is a modelling environment in which various allocation and logical rule models can be developed and studied. The intention is to create a tool in which we can study various types of low-level relationship between the land-use activities and then follow the dynamics of the urban spatial system. We need to gain some confidence that the behaviour of the resulting systems is consistent with what we might expect from known behaviours and more comprehensive modelling methods before we can proceed to more elaborate variants of the modelling approach in analogy with cellular automata.

The next task in the exploration of the use of cellular automata is to further examine the spatial development within some urban systems, and to attempt to create planning and economic models which can follow their evolution.

REFERENCES

ANAS, A. (1982). *Residential Location Markets and Urban Transportation. Economic Theory, Econometrics, and Policy Analysis with Discrete Choice Models*. New York: Academic Press.

ANAS, A., CARS, G., HÅRSMAN, B., JIRLOW, U., RHAE CHO, J. and SNICKARS, F. (1987). *The Economics of a Regulated Housing Market–Modelling Principles and Policy Perspectives*. Document D1987:37, Swedish Council for Building Research.

ARTHUR, W.B. (1993). Complexity: Pandora's Marketplace, *New Scientist*, February 6.

ASAMI, Y., FUJITA, M. and THISSE, J-F. (1990). *On the Design of Non-cooperative Games Supporting Optimal Spatial Allocations*. CORE Discussion Paper 9019, Catholic University, Louvain.

BERLEKAMP, E., CONWAY, J. and GUY, R. (1982). *Winning Ways, Vol. 2*, New York: Academic Press.

CÂMARA, A.S., FERREIRA, F. and CASTRO, P. (1996). Spatial simulation modelling, *Spatial Analytical Perspective on GIS*, London: Taylor & Francis.

FUJITA M. (1989). *Urban Economic Theory, Land Use and City Size*, Cambridge: Cambridge University Press.

FUJITA M. and KASHIWADANI M. (1989). Testing the efficiency of urban spatial growth–a case study of Tokyo. *Journal of Urban Economics*, **25** (2), 156–92.

GARDENER, M. (1974). The Game of Life. *Time*, January 21.

HOLLAND, J. (1992). *Adaption in Natural and Artificial Systems*. Cambridge, MA and London: MIT Press.

KAUFFMAN, S. (1984). Emergent properties in random complex automata, *Physica D*, **10**, 145–56.

KAUFFMAN, S. (1990). Requirements for evolvability in complex systems: Orderly dynamics and frozen components, *Physica D*, **42**, 135–52.

LUNDQVIST L. (1977). Planning for freedom of action, in Karlqvist, A., Lundqvist, L., Weibull, J.W. and Snickars, F. (Eds), *Spatial Interaction Theory and Planning Models*, Amsterdam: North-Holland.

MACLELLAN, B.J. (1990). *Continuous spatial automata*. Working Paper CS-90-201, Department of Computer Science, University of Tennessee, Knoxville.

MCGLADE, J. (1993). Complexity: alternate ecologies, *New Scientist*, February 13.

RASMUSSEN, E. (1989). *Games and Information. An Introduction to Game Theory*. Oxford: Basil Blackwell.

ROY, G.G. and SNICKARS, F. (1992). Computer-aided regional planning. Applications for the Perth and Helsinki regions, *The Annals of Regional Science*, **26**, 79–95.

SANDERS, L. (1996). Dynamic modelling of urban systems, *Spatial Analytical Perspectives on GIS*, London: Taylor & Francis.

VON NEUMANN, J. (1966). *Theory of Self-Producing Automata*. Urbana and Chicago: University of Illinois Press.

WEGENER, M. (1980). *A multilevel economic-demographic model for the Dortmund region*. Institute of Spatial Planning, University of Dortmund.

WHITE, R. and ENGELEN, G. (1993). Cellular automata and fractal urban form: a cellular modelling approach to the evolution of urban land-use patterns, *Environment and Planning A*, **25**(8), 1175–99.

WUENSCHE, A. (1993). *The ghost in the machine basins of attraction of random Boolean networks*. Paper 281, May, School of Cognitive and Computing Sciences, University of Brighton.

CHAPTER SEVENTEEN

Dynamic modelling of urban systems

LENA SANDERS

17.1 Introduction

Dynamic models are of great utility in understanding the evolution of urban systems. Theoretical and technical improvements in the field have given rise to the development of more and more complete tools. This chapter aims to compare different applications of dynamic models of urban systems, to show their respective advantages and disadvantages, and to analyse how they can be coupled with GIS. Although there has been a lot of research in dynamic systems, most of it focuses on theoretical developments and most models have been transferred into geography from other disciplines, such as mathematics, physics, artificial intelligence and biology. Improvements in the treatment of complex systems in these fields have led to useful applications in urban geography where sets of spatial units such as cities, at an inter-urban scale, or districts, at an intra-urban level, often have a systemic organisation. Geographers have successively incorporated mathematical tools such as non-linear differential equations, cellular automata and distributed artificial intelligence into their models. Parallel improvements in GIS, particularly the growing possibility of integrating spatial analysis within them, can give a new impetus to the development and application of dynamic models and increase their usefulness as learning and prediction tools.

First, the problems associated with applications of dynamic systems models will be investigated and a number of approaches compared. This step is essential to validate the use of this conceptual framework in geography. A constant coming and going at several different levels between the real, observed behaviour of a system and its modelling is necessary. There are two levels in the transfer, that of the theoretical framework and that of the methodological tools, with different kinds of contributions and validations. The type of empirical work that has led to the use of concepts and methods developed in other fields will be examined. Next, the different ways of introducing geographical theories in the models which have been transferred from other fields will be discussed. Validation methods will be compared and the kinds of understanding that can be obtained by applying such methods to observed geographical data will be outlined. Finally, methods of coupling dynamic models and GIS will be discussed.

17.2 Modelling transfer

Empirical research has shown that the functioning of a city system cannot easily be described with linear tools and that feedback effects often structure the evolution of such systems. In French cities, for example, it has been shown that evolution is the result of two complementary forces. One, the more deterministic, expresses the regularities of the evolution and relates to a slow dynamic. The other, more irregular force, reflects fluctuation effects and refers to a faster dynamic. Depending on the general situation of the system, these fluctuations can sometimes lead to structural changes. For example, although in total the French urban system remains stable, the constant decline of northern industrial cities since 1962 has deformed the structure of the urban mesh. This kind of trend has been observed in the evolution of city size (Robson, 1973; Pumain, 1982; Guerin-Pace, 1993) and change in the economic and social profiles of cities (Pumain and Saint-Julien, 1989).

As these observations were made, new theories and tools for analysing the dynamics of complex systems were introduced into geography. Associating determinism and random fluctuations, continuity and bifurcations, stability and dramatic structural change, they offered an ideal conceptual framework for analysing the evolution of urban systems. In particular, Allen has adapted concepts and methods based on the auto-organisation theories developed at the 'Brussels school' by Prigogine (Prigogine and Stengers, 1979) and others for modelling regional and urban dynamics (Allen, 1978; Allen and Sanglier, 1979; Allen *et al.*, 1982). The field of synergetics developed by Haken in Stuttgart relates to the same conceptual framework. Using a master equation approach, Weidlich and Haag have built a family of models to analyse the dynamics of social systems (Haag, 1989; Haag *et al.*, 1992; Weidlich and Haag, 1983, 1988). Even if from a physical and mathematical point of view they refer to the same theoretical framework, when applied to geography these two approaches give rise to conceptual differences. They appear to be methodologically complementary and each of them has advantages corresponding to eventual constraints and simplifications of the other one. These reciprocal properties will be examined.

A quite different approach derives from the use of artificial intelligence where many improvements have recently been made, especially in the analysis of self-organising systems with properties of autonomy and the emergence of new structures. These models are more intrinsically spatial as the position of the spatial units and their contiguities are central to the description of their evolution. The formalisation of models of this kind involves a set of rules which are globally defined but executed at the level of each spatial unit. Interest in using cellular automata modelling in geography was expressed several years ago (Tobler, 1979; Couclelis, 1985) and a few applications have been made at the urban scale (White and Engelen, 1994, 1995; Roy and Snickars, 1996). Multiagent systems, part of distributed artificial intelligence, have many applications in biology (Ferber, 1994, 1995) and have recently been used in geography (Bura *et al.*, 1996). From the point of view of the geographer or the planner needing an operational tool to understand and predict the evolution of a complex urban system, these newer methods are complementary to more traditional, differential equations based approaches but do not replace them.

17.3 The integration of theory

When a model which has been developed in another discipline is used in geography, attention must be paid to two problems associated with the transfer.

- What are the implications for geography of the theoretical background of the original discipline?
- What are the possibilities to incorporate in that model theories or laws which refer directly to the modelled spatial system?

A transfer is only complete when at the same time the concepts, methods and techniques of the original discipline have a meaning in geography. For example, when auto-organisation theories are used to analyse the spatial structure of an agglomeration, this involves assumptions about the auto-organising capacity of urban units. Qualitative observations are generally used to build up the analogy and to justify the use of the associated methodologies, for example the use of non-linear systems of differential equations to describe the dynamics of a city or a neighbourhood. There is then a contrast between the qualitative justification of the conceptual transfer and the quantitative use of the associated methodologies. In fact, these two levels are strongly linked. The properties of systems of complex differential equations reflect what can be observed empirically and qualitatively in spatial systems. In most situations this is continuous change with the possibility of bifurcations in specific configurations due to particular values of some parameters or to a change in the initial situation which may have, further on, important effects. This very general level of the transfer, that of the concepts, does not depend on the geographical scale of the application or the nature of the problem to be treated. The conceptual framework is, in fact, the same for all applications in the field of complex dynamic modelling and many spatial systems can be described in these terms.

If this general level of transfer is accepted and justified, theoretical assumptions directly related to the spatial system which is analysed have to be taken into account in the model-building. Economies of scale, economic base theory, the friction of distance at an intra-urban level, central place theory and mechanisms of competition at the inter-urban level are some obvious examples of geographical regularity to be considered. How these well known ideas are introduced into a dynamic model depends to a large extent on the chosen methodological framework. This is examined in relation to four such frameworks associated with differential equations, synergistic models, cellular automata and multiagent systems.

17.3.1 Differential equations

First, such theories can be converted into mathematical expressions and combined in a general differential equation. Allen's intra-urban model is an example of such an approach. It is based on six interacting variables, with two kinds of population and four kinds of activity. The model is described by a system of $6 \times N$ differential equations, where N is the total number of spatial units in the agglomeration. An example of a single equation, referring to the dynamics of one activity, the non-basic tertiary sector, is given in Figure 17.1. This equation makes many assumptions about the evolution of an activity in a zone of an agglomeration.

- There is a logistic growth where the maximal capacity of a zone is measured by the induced demand.
- This demand for a given zone is a function of its relative attractivity and the distribution of the population in the agglomeration.
- The attractivity of a zone over another results from the combination of three components. These are agglomeration and saturation effects related to the importance

$$\frac{dS_j^b}{dt} = \epsilon^b S_j^b \left(1 - \frac{S_j^b}{\sum_{j'} (\beta^b \sum_k P_j^k \frac{A_{jj'}^{\alpha}}{\sum_{j''} A_{j''j}^{\alpha}})} \right) \quad (1)$$

with

$$A_{jj'} = \left(\frac{1 + \rho^b S_j^b (1 - \Psi^b S_j^b)}{1 + \Phi^b d_{jj'}} \right) \left(\frac{\tau^b}{\tau^b + \sum_k \gamma^k P_j^k + \sum_l \gamma^l S_j^l} \right) \quad (2)$$

where

S_j^b is the number of jobs of activity b (non basic tertiary sector) in zone j, the other activities S_j^l are industry, fundamental tertiary and regional tertiary activities)

P_j^k is the population of type k in zone j

$A_{jj'}^{\alpha}$ is the attractivity of zone j on zone j'

α is the sensitivity of the entrepreneurs for attractivity differences

ϵ^b is the speed of reaction of the system for activity b

ρ^b is the propensity of activity b to agglomerate

Ψ^b is the saturation effect for activity b

Φ^b is the sensitivity for distance to the consumer

τ^b is a measure for the maximum surface of activity b

γ^k, γ^l are the average occupation of respectively population k and activity l.

Figure 17.1 Example of equation used in Allen's model.

of activities which already exist in the zone, competition for space within the zone by all activities, and its location relative to other zones, with decreasing exchange as distance increases.

This attractivity, A_{jj}, of zone j on another zone j' includes the importance of the non-basic tertiary sector, the interrelations between this sector and the other variables (residential populations and other activities), competition, and the interactions between zone j and the other zones of the agglomeration. If each of the assumptions has a single and plausible transcription through a mathematical expression, the form of the combination in the global equation is more difficult to justify. This combination, which reflects a certain degree of complexity in reality, carries with it a high level of mathematical complexity. The difficult problem is to get a realistic link between these two levels of complexity.

All these mechanisms refer to a meso or macro scale and the micro level appears only through the parameters of the model which express, for example, the sensitivity of different actors to economies of scale or of the workers to distance to work.

17.3.2 Synergistic models

Models developed in synergetics are built up in a quite different way and most of those applied in social science are based on a master equation approach which focuses on the integration of the individual and the aggregate levels. These models refer in general to the choice process of the actors and to theories developed at the individual scale such as

The master equation

$$\frac{dP(n,t)}{dt} = \sum_i \sum_j w_{ij}(n'_{ij})P(n'_{ij},t) - \sum_i \sum_j w_{ij}(n)P(n,t) \qquad (3)$$

where:

$\mathbf{n} = (n_1, n_2, \ldots, n_L)$ is a vector describing the repartition of the total population of the system between L locations (regions, quarters, cities for example),
$P(\mathbf{n}, t)$ is the probability to observe a certain distribution $\mathbf{n}$ at time t,
$\mathbf{n}'_{ij}$ is a configuration different from $\mathbf{n}$ only by the position of one single individual, it means $\mathbf{n}'_{ij} = (n_1, n_2, \ldots, n_{i+1}, \ldots, n_{j-1}, \ldots, n_L)$,
$W_{ij}(n'_{ij})$ is the configurational transition rate from a configuration $\mathbf{n}$ to a neighbouring configuration $\mathbf{n}'_{ij}$.

The mean value equation

$$\frac{dn_k(t)}{dt} = \sum_{<n>} n_k \frac{dP(n,t)}{dt} = \sum_i n_i p_{ki} - \sum_j n_k p_{jk} \qquad \text{for } k = 1, \ldots, L$$

where:

p_{ki} is the individual transition rate for migrating from i to k. It has the following expression:

$$p_{ki}(t) = v_0(t) f_{ki} \exp(A_i(t) - A_k(t))$$

where:

$v_0(t)$ measures the global level of mobility of the individuals; it is constant through space but it varies through time.
f_{ki} is a parameter which takes into account all symmetric effects existing between two places and influencing the migration between them. Geographical distance of course but also measures of all kinds of cultural or social proximity
$A_i(t), i = 1, \ldots, L$ is a parameter which measures the attractivity of place j.

Figure 17.2 A master equation approach, the model of Weidlich and Haag.

utility maximisation. Dynamic models derived from synergetics have been applied in geography to describe the evolution of the population in inter-regional and inter-urban systems. The master equation approach is used with two steps, a general theoretical one describing the dynamics of the probability of observing a certain distribution, and a more applicable one based on the mean value equation. An example of the kinds of equation used is given in Figure 17.2.

This corresponds to an application of the model to the evolution between 1954 and 1982 of the French system of cities, which focuses on the redistribution of the urban population between the cities and reflects on the mechanisms of competition between them (Sanders, 1992). Change due to births, deaths, or migration to and from the rural hinterland could also be considered in a more global approach.

In building this model, the term p_{ki} is interpreted at a micro-level and in particular, $\exp(A_i - A_k)$ is formalised in the same way as a utility difference. Meanwhile, using the model at an aggregated level implies a change of perspective. The parameters f_{ki} and A_i then represent the average behaviour of the individuals and have to be interpreted with a

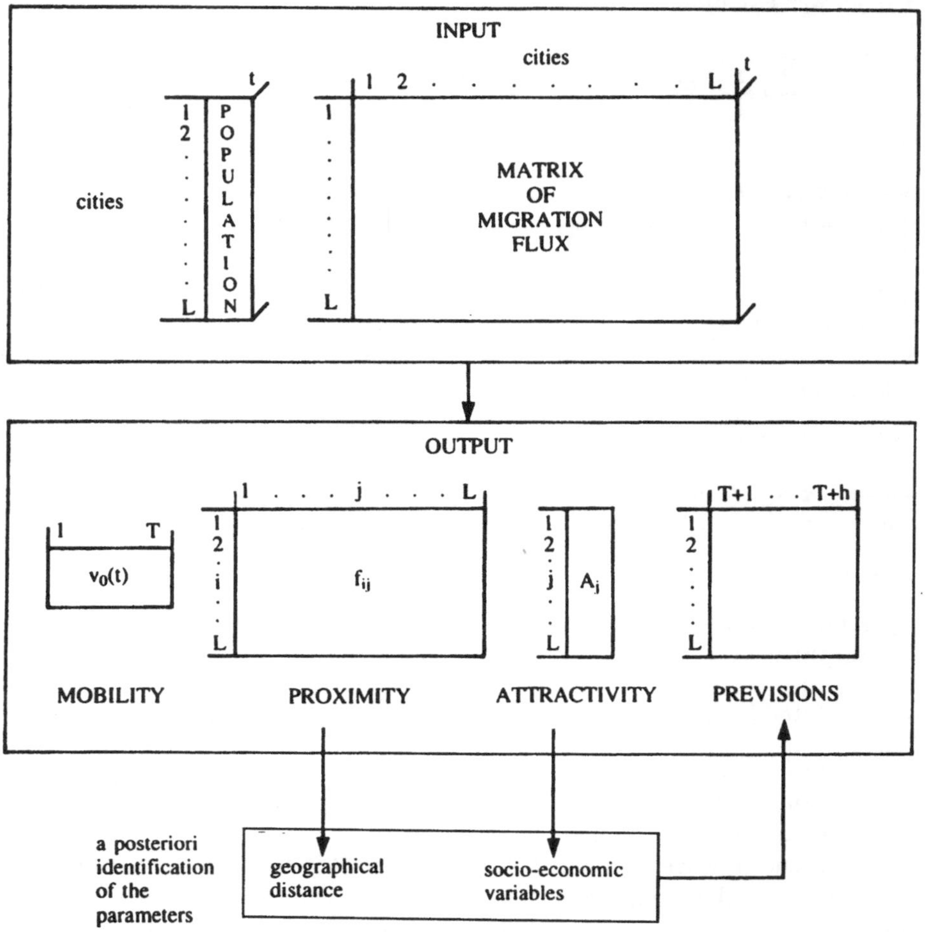

Figure 17.3 A schematic representation of the working of the model.

macro-scale logic. In fact, when these models are used at an aggregated level, they turn out to function in an inverse way to the preceding types. They assume as few as possible hypotheses *a priori* in order to find out the forces of change *a posteriori*. This is not inherent in the synergetic approach but it is in its application to spatial systems. Figure 17.3 gives a schematic representation of the working of the model for an application to the French system of cities. It can be seen that the original theoretical background loses its meaning at the scale of spatial aggregates, but analysis of the outputs of the model makes it possible to identify *a posteriori* the meso- and macro-level components which explain differential dynamics from one city to another. In that way the model is extremely useful as an exploratory tool.

17.3.3 Qualitative methods and cellular automata

In a third family of models, the theory behind the models can be expressed through a set of rules based on a combination of qualitative laws and quantitative criteria. This kind of approach has old roots and was first used by Hägerstrand (1953) in his well-known

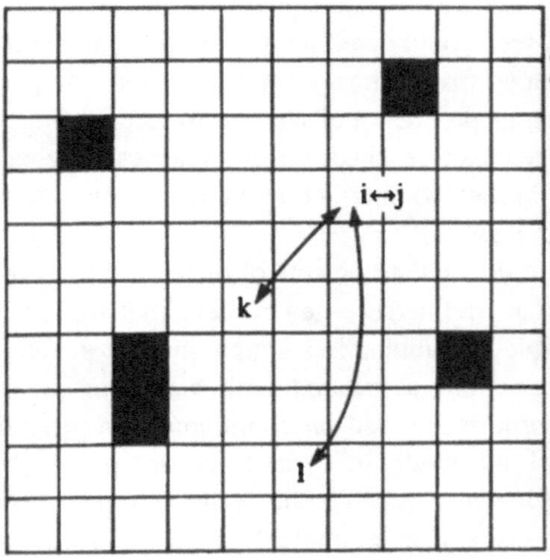

■ cell with predefined characteristics, unable to change. This kind of cell can be used to express constraints: for example, uninhabitable mountainous cells, or heavy transport infrastructures unchangeable except in the long term.

↔ expression of an interaction in form of information (for example about the state of cell j, k or l, of which knowledge is necessary to determine the change of i) or of goods or people. Only three examples corresponding to different ranges are represented here in order to keep the figure legible.

Figure 17.4 Cellular automata and multiagent systems: example of a grid.

diffusion models. More sophisticated tools such as cellular automata and multiagent systems are nowadays available to handle a much greater number of interacting spatial units.

In such modelling, geographical space is often represented by use of a grid, and Figure 17.4 presents some examples. Imagine, for example, that this grid represents land use in an agglomeration. Each zone is defined by a state (industry, services, residential) and any change of state depends on the states of contiguous cells. If a cell is occupied by industry, a very simple rule to describe change would be that it changes from industry to residential if three out of four contiguous cells are also residential. More sophisticated rules, based on larger neighbourhoods, could be used and are generally defined by a transition function. This kind of approach using cellular automata gives a central role to hypotheses about geographical pattern involving spatial interactions, spacing, and the relative position of different kinds of object. To simulate the occupational evolution of a given place, one refers on the one hand to the actual occupation and its degree of inertia, and on the other to the state of all surrounding places. In an application to the city of Cincinnati, White and Engelen (1994) used a neighbourhood of level 6, with a distance decreasing effect. Only a few hypotheses are introduced in this kind of modelling but they are all fundamentally geographical. Cellular automata can usefully be coupled with other tools such as classical dynamic (White and Engelen, 1995), or network models (Roy and Snickars, 1996).

17.3.4 Multiagent systems

Multiagent systems are a part of distributed artificial intelligence. This kind of modelling is based on mechanisms of communication and co-operation between agents and is especially useful for simulating the emergence of new structures and for analysing adaptation and auto-organisation. Each spatial entity is represented by an agent and Figure 17.4 can be used again to present a simple example, at the inter-urban level this time. Each spatial unit is characterised by its status, that is its qualitative state (uninhabited, agricultural village or a city with a certain functional level) and some variables (population, income, resources). Let a cell be occupied by a city. The evolution of this city will depend on a set of rules which could combine quantitative and qualitative criteria. For example, the city will change size with a growth rate depending on its economic situation, that is its ability to sell its products and services to surroundings cells. If some conditions of size and economic level are reached, the city reaches a higher level in the hierarchy and so acquires possibilities to interact at larger ranges. A whole hierarchy of cities can be built up in this way.

Multiagent systems are well adapted to manipulate objects of different scales or kind and to handle their overlapping. The rules are defined at a general level but are handled at the local level of each cell. So, it is possible to combine in a simple manner general laws of evolution, more specific types of behaviour associated with particular subsets of objects, and local exceptions. As the approach is based on communication possibilities between cells, it is possible to model all kinds of spatial interaction, exchange, competition and so on. Different kinds of hypotheses can be introduced, some concerning the evolution of the variables at each place (for example hypotheses about wealth

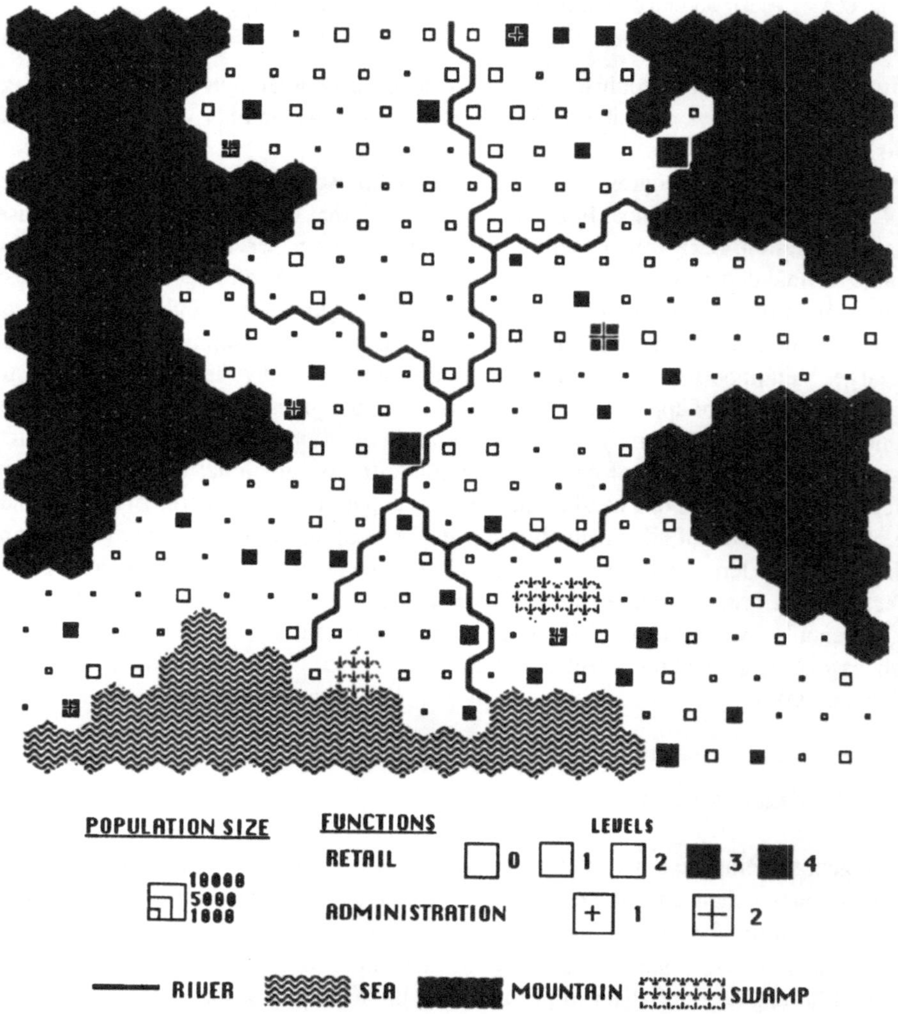

Figure 17.5 The settlement system after a 1000 years' simulation of SIMPOP (source: Bura *et al.*, 1996).

accumulation process, productivity evolution and population growth processes) and hypotheses about the relationships between different places (for example, exchanges through mechanisms of supply and demand, competition, and diffusion). These two levels of hypotheses are easy to link in this approach and in this way give better modelling possibilities than do cellular automata. In the SIMPOP model Bura *et al.* (1996) have used this environment to model the evolution of a settlement system over a period of 2000 years. Starting from a relatively regular pattern of small and dispersed agricultural settlements, SIMPOP models the aggregation process including the emergence of cities and their progressive evolution. Some stagnate, some decline and some grow and get more and more high-level functions and a larger spatial influence. Using this approach, it has been possible to simulate the progressive establishment of a hierarchical system of cities. Figure 17.5 shows an example of a settlement system obtained after a 1000 years' simulation.

Cities are distinguished by four levels of trade and two of administration. The process of forming a hierarchy is well illustrated in Figure 17.6 which shows the rank size distribution at different periods of the simulation.

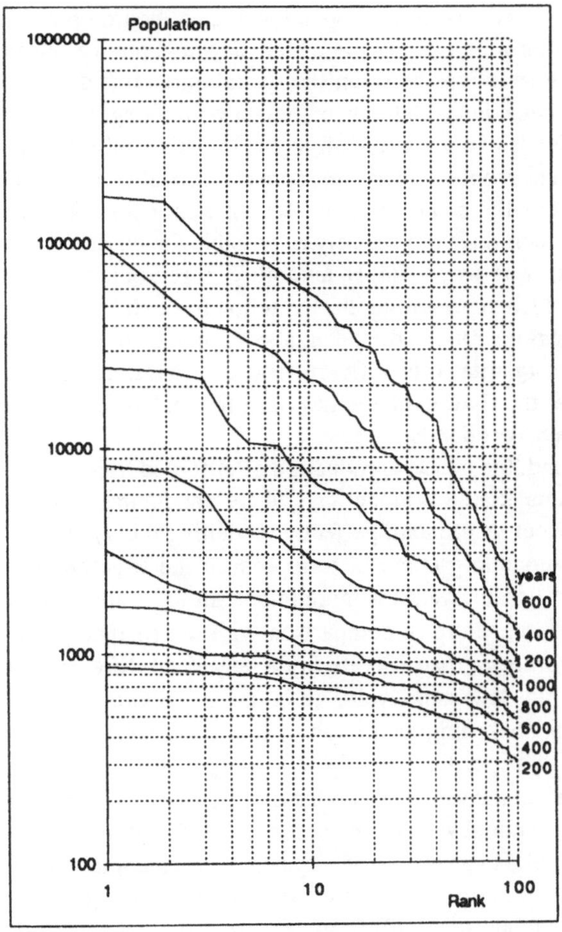

Figure 17.6 Rank size distribution of the settlement pattern at the different periods of 1 600 year simulation of SIMPOP (source: Bura *et al.*, 1996).

17.4 Calibration and validation

Calibration and validation are important aspects of model-building. If the model is able to correctly reproduce the observed evolution of a system, there is a good chance that one has managed to identify the mechanisms involved. A validated model can eventually be used as a prospective tool, for example to test the effects of different planning policies. One can also simulate what could have happened if initial conditions had been different or the possible futures of the system according to different hypotheses. If this step is needed, it raises many operational problems that arise in the main from the dynamic context. Two approaches can be distinguished, using trial and error or using analytical calibration.

17.4.1 Calibration by trial and error

Trial and error is often used for calibration and at first the contrast between the empiricism of this approach and the degree of elaboration of models using system dynamics or artificial intelligence seems surprising. Meanwhile, the more sophisticated the model, the less easy it is to develop automatic calibration procedures. For systems of differential equations which express in a realistic way both the interrelations between the variables and the interactions between the spatial units, it is difficult to get an analytical insight into their behaviour. Cellular automata or multiagent systems based mostly on rules are also difficult to associate with any automatic calibration procedure. As well as this kind of technical problem, there are more theoretical ones. What, for example, does a good calibration actually mean? In the context of classical static models, it usually means that one is able to reproduce as precisely as possible the observed zone by zone values. In a dynamic context, this would imply an ability to reproduce the observed evolution of each zone. Figure 17.7, representing the evolution through four dates of the 17 communes constituting the agglomeration of Rouen, allows one to see the difficulty of such a task. All trajectories are qualitatively different and a calibration simply using the final data does not mean that the right dynamic has been found. There are multiple possibilities of trajectories between two points.

On the other hand, the exact reproduction of observed trajectories does not correspond to the spirit of this kind of modelling where randomness is introduced and where bifurcations can occur. More than perfectly modelling the dynamics of each local spatial unit, this kind of modelling is meant to reproduce the structural evolution of the whole system. At an intra-urban level, it is more important to be able to reproduce, for example, a reinforcement of the centre in comparison with the periphery and to understand what has led to such a process, than it is to model any local differences between two similar peripheral zones. In the same way, it is more important to be able to predict a reinforcement or a slackening in the top of the urban hierarchy than it is to predict the change in rank of two specific cities. When emphasis is placed on global structure rather than local fit, classical measures of goodness of fit based on some function of the zone by zone differences, are not always the most useful. Different indicators can be used in order to characterise the global structure of a system. For example, the fractal dimension of a specific land-use distribution can be used to characterise the global spatial structure of an agglomeration (Frankhauser, 1990) or to validate a simulated spatial organisation (White and Engelen, 1993). At the scale of a system of cities, the parameters of the rank size distribution can be used to assess the truth of a simulated evolution.

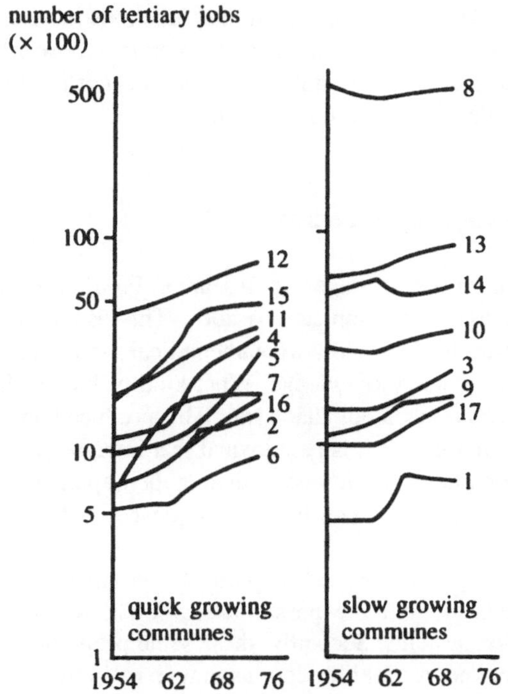

Figure 17.7 Diversity of communal evolutions (source: Pumain *et al.*, 1989)

In dynamic models, local fit can only be obtained by introducing some exogeneous attractivity or repulsion indicator. In Allen's model, such indicators, varying from zone to zone, are introduced in some equations and they are supposed to reflect the global effects of physical constraints such as slopes or riversides, on the location of basic activities. In the cellular automata approach of White and Engelen, constraints of the same kind have been introduced in a 'suitability' measure defined for each cell for each function. In multiagent modelling a similar procedure is used to introduce different kinds of specificities of one particular settlement unit. They then intervene as a constraint when applying the general rules associated with that kind of unit.

The existence of these two levels, local and global, and their interrelationships have to be considered in the calibration of complex systems. Trial and error methods are then the only way to proceed.

17.4.2 Analytic calibration

If adequate data are available, the models derived from synergetics can be calibrated analytically. The model described in Figure 17.2 is built up around a parameter decomposition with f_{ij} $(i, j=1, \ldots, n)$, which is symmetrical and which expresses the proximity between all pairs of places and A_j $(j=1, \ldots, n)$ which represents the attractivity of each place. These parameters are estimated through a maximum likelihood method, comparing the observed inter-urban migratory flows with the simulated ones. With these models, the quality of the fit is excellent, which, given the large number of parameters, is not surprising. However, even if the fit is good, the processes underlying the change are not identified. The analysis of the values of the parameters gives, *a posteriori*, an insight into the mechanisms of change. This approach thus functions more as an exploratory tool than as a theoretical model. In conclusion, it can be seen that these differences in building and validating dynamic models also implies that they do not produce the same kind of knowledge and it is to this that attention now turns.

17.5 Validation and knowledge creation

Modelling is useful at different stages in research. First, it is useful as an aid in conceptualisation and as a communications tool. This is common to all kinds of modelling. Secondly, and this is specific to mathematical modelling, it is useful as a tool for prediction and for producing knowledge. What kind of knowledge can emerge from the application of a dynamic model that cannot be produced in an other way? This question is not easy to answer and it is paradoxical that interest often seems to be greater in the mathematical properties of such models than in their application. The mathematical properties of dynamic models have two consequences which are in some ways contradictory.

First, they are of major conceptual importance as they enable us to reproduce multiequilibria, bifurcations, and the possibilities for chaos that are all fundamental properties of all complex systems. Secondly, these same properties lead to difficulties in real applications. If one knows that bifurcations may exist in the analysed system, and if the formalisation of the model makes it possible to produce bifurcations, then it is almost impossible to adjust for such events. It is rarely possible to interpret the threshold of a parameter which corresponds to a bifurcation in the model and it is just as difficult to reproduce an observed bifurcation. There is a paradox. A non-linear system of differential equations is often chosen because of its ability to produce discontinuous change, yet in practice such a model appears to be most useful for the stable periods of the system's evolution through configurations far away from a transition phase where change is almost continuous and the trajectory is towards a single attractor. So there is a clear contrast between the realism of the conceptual framework and the risk of artefact at the methodological level. Such contrast has to be taken into account in the interpretation of any applications.

Even if the models referred to here derive from a same theoretical framework they produce very different kinds of knowledge:

- The application of a model like Allen's can produce global measures of the characteristics of a city's dynamics and provide a basis for comparison between different cities. If one already knows that the evolution of the intra-urban spatial organisation refers both to general laws (residential functions being pushed towards the periphery, for example) and to local specificities (a firm settling down near the childhood home of its manager), only the application of a model makes it possible to measure the respective roles of these two kinds of explanation. Only the application of the model to a few different systems allows testing of the degree of generality of the laws introduced. On the other hand, the respective weights of these laws can be measured at the level of each agglomeration to establish a basis for inter-urban comparison. The application of Allen's intra-urban model to four French agglomerations (Rouen, Nantes, Bordeaux and Strasbourg) shows, for example, that general mechanisms explain about 60 per cent of the intra-urban change in the location of residential and economic activity. The residuals from the model have always been easily identified and correspond to micro-level decisions such as the siting of a university, the creation of service centres, or the building of collective housing (Pumain *et al.*, 1989).

- Master equation models can be used as spatio-temporal exploratory tools. A comparison of the estimated parameter values and observed variables characterising the spatial units and pairs of spatial units makes it possible to test hypotheses about the causes of change. An application to the system of French cities from 1954 to 1982 shows, for example, the importance of inertia in the system. Changes occur at the level of each city but have only slow effects and the top of the urban hierarchy remains very stable. The application of the model makes it possible to estimate the strength of this tendency of the system to reproduce its own structure and, according to the time period involved, it accounts for between 60 and 70 per cent of the evolution. So it seems that local specificities intervene only for about 20 to 30 per cent.

  In France these specificities have given a systematic advantage to southern cities at the very time northern cities tend to decline relative to what might be expected from their place in the urban hierarchy. This method makes it possible to determine a kind of confidence interval for what is plausible in the evolution of a city. Even if local factors can have real effects, they cannot possibly lead the evolution beyond a certain threshold. Dynamic models are more useful to provide ideas about these thresholds and the limits of a city in its growth or decline, than they are to predict the short-term evolution of the population. They are also more capable of showing future possible qualitative change in the whole system, such as a further concentration in the higher levels of the urban hierarchy than they are for predicting the evolution of an isolated city.
- A model using multiagent systems has to be built up progressively. The most basic rules are introduced first and tested, then the model is completed progressively. In this way it is possible to identify which conditions are necessary, and which are sufficient, to obtain a given pattern. For example, this kind of model makes it possible to find the minimal mechanisms necessary to simulate the emergence of a hierarchy of cities from a rural settlement pattern. If the preceding models show the mechanisms of continuous evolution, this one shows the processes which underlie the emergence of new states which are qualitatively different from those which preceded it. In this way, this kind of model is advantageous for describing discontinuous change. Cellular automata models have to some extent the same properties. Globally, the kind of learning produced by the application of such methods lies somewhere between those of Allen's and those of multiagent systems.

17.6 Coupling dynamic modelling and GIS

This comparison of modelling approaches shows them to have quite different and complementary properties. Given a geographical problem in a dynamic framework the choice of one method rather than another is not always easy. Their complementarities should lead to their integrated use. The Weidlich and Haag model (1988) is poor in geographical theory but is easy to calibrate, so that it is useful as an exploratory tool. For prediction, Allen's model is richer from a theoretical point of view but the instability inherent in its mathematics does not fit many geographical systems where inertia is an important evolutionary constraint and calibration problems reduce its predictive capability. The properties of cellular automata and multiagent systems are similar in spirit to Allen's model. They make it possible to introduce and associate many hypotheses about the functioning of the geographical system to be modelled but have difficulties in calibrating the dynamics. Compared with approaches based on differential equations, space and spatial interactions are more fundamentally considered.

Because of these respective advantages, older and newer methods should be used together and integrated in a chain of treatments at different stages in the same project. A Weidlich and Haag type model could be used to find the structure of the spatial dynamics and to understand the process of change. The form of the hypotheses about the evolution could then be refined for introduction into a general model. Some mechanisms, such as evolution of population, activities and income, could be modelled through differential equations whilst more qualitative change, such as the emergence of a city or change of function, could be modelled by cellular automata or multiagent systems. A coupling of the different approaches would give the richest approach.

This kind of integration would be especially useful if it was developed in a GIS framework and the uses of a GIS in this context are many:

1. *Handling tool.* Building a dynamic model involves many difficult steps and the suggested integration of different kinds of approaches make it even more difficult. Visualisation, in particular, is important in several different steps. First, it can be used in a static way to examine the initial situation, the observed and simulated situations at different dates, and predictions according to different scenarios. Secondly, it can also be used dynamically to show the observed and simulated system evolutions. A GIS is the perfect tool for handling the different databases, observed and simulated, and the various dynamic modelling procedures. So, coupling dynamic modelling with GIS gives a great flexibility which makes it possible to gain time in the elaboration phase of the model and to improve it sensibly. Some hypotheses can easily be tested in a first exploratory step, using all the properties of GIS. The degree of generality of any global dynamic laws can be investigated, eventual bifurcation points can be identified, and the appearance and disappearance of spatial discontinuities can be detected. Besides, GIS make it easy to show the behaviour of different subsets of spatial units and so provide insights into the respective weights of quite general dynamics and of more local specificities. All these explorations, necessary in the elaboration phase of a dynamic model, are considerably improved when they are done with a GIS.
2. *A calibration aid.* As already pointed out, there is a lack of automatic calibration procedures for spatio-temporal systems. The simultaneous dynamic visualisation of the observed phenomena, the simulated one, and the distribution of the residuals of the model is a help during its elaboration. It helps to assess the rhythms of convergence or divergence between simulated and observed dynamics, and to identify the break points in space and time where the model fails to reproduce the observed dynamics (Mathian and Sanders, 1993). Calculating and storing numerical indicators of such phenomena are difficult and they are only partially used in applications. They are almost impossible to handle in an exhaustive way but dynamic visualisation helps to find out very quickly the critical zones where further investigation should be undertaken to improve the model.
3. *Handling of various scales.* Many models need to take into account various scales. The dynamics of an inter-communal distribution is inscribed in the dynamics of the region which itself depends largely on those in other regions with which it has interactions. So all scales are interconnected and should be considered in a global dynamic model, but each level is driven by different kinds of processes. For example,

if one needs to model the evolution of the population over the last century at the level of the communes for a region or a country, one faces two different, interfering logics. First, there is a process of concentration of the population in the urban areas, with a desertion of the most rural and peripheral areas. Secondly, there is also a more recent process of diffusion from the biggest units to the nearest surroundings in periurbanisation. Another example would be the coupling of the inter-urban and the intra-urban systems, strongly related but each with its own logic of evolution. Handling these different scales is difficult if the model is not integrated in a GIS.

Since a GIS may be used at all the steps of dynamic modelling (elaboration, calibration, validation and communication of the results), a complete integration of model-building capabilities in the GIS is more useful than a loose coupling. Some research has already been reported which adopted this approach in the development of what has been called time-GIS (Mikula, 1993) and shows promising perspectives. On the other hand, the integration of dynamic modelling in GIS will considerably raise the interest of such tools for researchers or planners. A well-calibrated model makes it possible to test different scenarios, to get an insight into future evolution of a system, to assess the effects on the whole system of different kinds of external interventions at a local level, and, more generally, better understand the mechanisms of change of a geographical system. Hence research on the full integration of many different and complementary dynamic approaches in a GIS is essential both from the perspective of the dynamic modelling and from that of the GIS.

REFERENCES

ALLEN, P. (1978). Dynamique des centres urbains, *Sciences et Techniques*, **50**, 15–19.

ALLEN, P. (1991). Evolutionary models of human systems: Urban and rural landscapes as self-organizing systems, in Lepetit, B. and Pumain, D. (Eds) *Temporalites Urbaines*, Paris: Economica-Anthropos.

ALLEN, A., BOON, F., DENEUBOURG, J.L., DE PALMA, A. and SANGLIER, M. (1982). *Models of Urban Settlement and Structure as Self-Organizing Systems*, Washington DC: Department of Transportation.

ALLEN, P. and SANGLIER, M. (1979). A dynamic model of growth in a central place system, *Geographical Analysis*, **11**(3), 256–72.

BERRY, B.J.L. (1964). Cities as systems within systems of cities, *Papers of the Regional Science Association*, **13**, 147–63.

BURA, S., GUERIN-PACE, F., MATHIAN, F., PUMAIN, D. and SANDERS, L. (1996). Multiagent systems and the dynamics of a settlement system, *Geographical Analysis*, **28**, 161–78.

COUCLELIS, H. (1985). Cellular worlds: a framework for modelling micro–macro dynamics, *Environment and Planning*, **17**, 585–96.

DENDRINOS, D.S. and SONIS, M. (1990). *Chaos and Socio-Spatial Dynamics*, Berlin: Springer-Verlag.

FERBER, J. (1995). *Les systèmes multi-agents, vers une intelligence collective*, Paris: Interéditions.

FERBER, J. (1994). Reactive multi-agent systems: principles and applications, in Jennings, N. (Ed) *Fundamentals of Distributed Artificial Intelligence*, North Holland.

FRANKHAUSER, P. (1990). Fractal structures in urban systems, *Methods of Operations Research*, **60**.

GUERIN-PACE, F. (1993). *Deux siècles de croissance urbaine*, Paris: Anthropos.

HAAG, G. (1989). *Dynamic Decision Theory: Applications to Urban and Regional Topics*, Dordrecht: Kluwer.

HAAG, G., MUNZ, M., PUMAIN, D., SAINT-JULIEN, T. and SANDERS, L. (1992). Inter-urban migration and the dynamics of a system of cities, *Environment and Planning*, A, 24, 181–198.

HÄGERSTRAND, T. (1953). *Innovation diffusion as a spatial process*. Trans. A. Pred Chicago: University of Chicago Press, 1967.

HAKEN, H. (1983). *Advanced Synergetics, Volume 20*, Berlin: Springer Verlag.

LEPETIT, B. and PUMAIN, D. (Eds) (1993). *Temporalités Urbaines*. Paris: Economica-Anthropos.

LOMBARDO, S., PUMAIN, D., RABINO, G., SAINT-JULIEN, T. and SANDERS, L. (1987). Comparing urban dynamic models: the unexpected differences in two similar models, *Sistemi Urbani*, **2/3**, 213–28.

MATHIAN, H. and SANDERS, L. (1993). Modélisation dynamique et système d'information geographique, *Mappemonde*, **4**, 38–39.

MIKULA, B. (1993). 'Development of a time GIS'. Paper presented at the 8th European Colloquium in Theoretical and Quantitative Geography, Budapest.

PRIGOGINE, I. and STENGERS, I. (1979). *La Nouvelle Alliance*, Paris: Gallimard.

PUMAIN, D. (1982). *La Dynamique Des Villes*, Paris: Economica.

PUMAIN, D. and SAINT-JULIEN, T. (1978). *Les Dimensions du Changement Urbain. Memoires et Documents de Geographie*, Editions du CNRS.

PUMAIN, D. and SAINT-JULIEN, T. (1989). Migration et changement urbain en France 1975–1982, *Revue d'Economie Regionale et Urbaine*, **3**, 509–30.

PUMAIN, D., SANDERS, L. and SAINT-JULIEN, T. (1987). Application of a dynamic model, *Geographical Analysis*, **1-2**, 152–68.

PUMAIN, D., SANDERS, L., and SAINT-JULIEN, T. (1989). *Villes et Auto-Organisation*, Paris: Economica.

ROBSON, B.T. (1973). *Urban Growth: An Approach*, London: Methuen.

ROSSER, J.B. (1990). Approaches to the analysis of the morphogenesis of regional systems, *Socio-Spatial Dynamics*, **2**, 75–102.

ROY, G.C. and SNICKARS, F. (1996). CityLife: a study of cellular automata in urban dynamics, in Fischer, M., Scholten, H. and Unwin, D. (Eds) *Spatial Analytical Perspectives on GIS*, pp. 213–228, London: Taylor & Francis.

SANDERS, L. (1990). Modèles de la dynamique urbaine: une approche critique, in Lepetit, and Pumain, (Eds). *Temporalités Urbaines*, Paris: Economica-Anthropos.

SANDERS, L. (1992). *Systèmes de Villes et Synergètique*, Paris: Economica-Anthropos.

SANDERS, L., PUMAIN, D. and SAINT-JULIEN, T. (1991). Inter-urban migration and dynamics of the French urban system: socio-economic components, in Pumain, D. (Ed.). *Spatial Analysis and Population Dynamics*. Paris: J. Libbey- INED, Congres et colloques.

TOBLER, W.R. (1979). Cellular geography, in Gale, G. and Olsson, G. (Eds). *Philosophy of Geography*, pp. 279–386, Dordrecht: Reidel.

WEIDLICH, W. and HAAG, G. (1983). *Concepts and Models of a Quantitative Sociology*. Berlin: Springer Verlag.

WEIDLICH, W. and HAAG, G. (Eds). (1988). *Inter-regional Migration, Dynamic Theory and Comparative Analysis*. Berlin: Springer Verlag.

WHITE, R. and ENGELEN, G. (1994). Urban systems dynamics and cellular automata: fractal structures between order and chaos, *Chaos, Solitons and Fractals*, **4**, 563–583.

WHITE, R. and ENGELEN, G. (1995). Cellular automata as the basis of integrated dynamic regional modelling, *Environment and Planning*.

WILSON, A. (1981). *Catastrophe Theory and Bifurcation: Applications to Urban and Regional Systems*, London: Croom Helm.

CHAPTER EIGHTEEN

Conclusions

HENK J. SCHOLTEN

18.1 Introduction

GIS has an exceptionally broad base of interested users covering a wide spectrum of disciplines, and has been successful because it meets basic common needs for geographic information technology. However, as mentioned in the introduction and by several authors in this book, this geo-processing environment can only make progress when we are able to add spatial analysis functionality.

It is inescapable that modelling and other spatial analytical uses of the data will gradually become more in demand. However, this is not a one-way process, and there is a need both for new analysis functions to be added to the GIS toolbox and for new data structures to support their development and application. There is a danger that the growing imbalance between the availability of geographic data and the limited range of analytical technology will slow the growth of GIS and result in a failure to make full use of the information being collected. The problem exists at a time when the increasing convergence between cartographic and office computer systems (made possibly by GIS) is also creating many new opportunities for spatial analysis. In many instances, the purpose behind such analysis is vague and relates mainly to the fact that the data exist rather than to a more traditional form of highly focused scientific inquiry. The emerging mountain of real and potentially creatable geographical referenceable data challenges the conventional manner by which statistical analysis and modelling have been performed in the past, particularly in a geographical context. The challenge can be viewed as involving the need for an automated and more exploratory *modus operandi* in an environment which is data-rich but theory-poor. In other words, the historic emphasis on deductive approaches is becoming less practicable because of the increasing dominance of data-led rather than theory-driven questions. There is a need for a more creative approach and for analytical tools that can suggest new theories and generally support data exploration.

18.2 Position of spatial analysis procedures

It is important not to neglect the lessons from the first spatial analysis-driven quantitative revolution in geography. Spatial data analysis is extremely limited in what it can offer. It is important also to understand that spatial analysis is not an end in itself but a tool that is only useful in the context of some kind of application. As such, spatial analysis is problem-driven and is really part of what is often called 'spatial decision support techniques'. It follows, therefore, that spatial analysis is merely another class of map data processing technology which, if generic functions can be defined, has the potential further to enrich the GIS ensemble.

In the book, several authors described the importance of an exploratory approach to spatial information, and, given the theory-poor background of geography, this is understandable. For example, in the field of environmental impacts on health, a quantitative prediction of the effect of pollution is difficult to make, because of the absence of detailed knowledge concerning ambient air pollution concentrations, population and other fundamental issues such as human dose and exposure–response relationships. The contributions in several chapters have shown how relevant GIS tools can be developed.

18.3 Spatial information architecture

There have been dramatic changes in the information technology environment. In less than twelve years the price/performance ratios of computing and communications systems have improved by a factor of 1,000 (Donovan, 1994). Ratios of such magnitude encourage decentralised computing. Local systems can grow in smaller increments than they could in the era of mainframes, and thus respond to our requirements more effectively. When we look to the price/performance of local-area computer communications we see within the last eight years an improvement in the order of 1,000. What does this mean for the next five years? Are we able to handle these revolutionary improvements in the order of 1,000?

What is happening in the field of standards? We all remember the difficulties in exchanging text files between systems. Life has changed, and open standards have been set by vendor-independent bodies. These technologies are open as well as standard, and these standards are rapidly penetrating the product lines of all computing and communications vendors, as demonstrated by the popularity of ISDN, DCE (Distributed Computer Environment). Open systems standards are becoming a dominant force in the market. What does this mean for spatial information architecture? Let us look at two examples: today there is an 'Open GIS Consortium' (OGC) dedicated to open system approaches to geoprocessing. By means of consensus building and technology development activities, OGC has already had an impact on geodata and geoprocessing standards, and has promoted a vision of openness that integrates geoprocessing with the distributed architecture of the worldwide infrastructure for information management.

OGC is creating the Open Geodata Interoperability specification, a computing framework and detailed software specification that makes true geoprocessing interoperability possible. OGC maintains close ties with the standards community through ANSI and ISO technical committees and working groups (for more information: http://www.ogis.org). There are two fundamental types of interoperability mechanisms (Pulusani, 1995). The first is based on open data — formats, translation standards, data exchange and data models. The second is based on open process — service requests, query mechanisms and spatial data processing. If these are in place, then open GIS applications will be able to transparently share data.

There are all kinds of other initiatives moving in the same direction. The focus of promoting interagency co-operation and data sharing is being led in the USA by the Federal Geographic Data Committee (FGDC). The FGDC has established the National

Spatial Data Infrastructure (NSDI) and is proceeding to establish a meta-data standard and data-content standards. In Europe as well, several activities are moving in the direction of open GIS. In 1994 the European Umbrella Organisation for Geographical Information, EUROGI, was formed with a mission, amongst other things, to facilitate the development of standards in Europe. In 1996, the European Union is launching a programme called INFO2000. Spatial Information is a part of it, and the aims are comparable with NSDI. In The Netherlands and Portugal there are already national clearinghouses which make spatial data exchange between government agencies and the public possible.

18.4 Conclusions

As has been stated by Fabbri and Chung (Chapter 11), it can be argued that the programming of a satisfactory set of analytical functions within a GIS is probably of lesser difficulty than the construction of a model rationale adequate to predict natural complexity. Furthermore, as argued by Goodchild (1993) and Douven (1996) if the aim of spatial information is the use in the decision-making process:

- models must be spatially disaggregated, to allow policy to be evaluated at a detailed spatial scale using accurate data
- results must be displayed in a form that is understandable by the policy maker
- results need to be brought into consideration with economic and social data for comparison and correlation
- specific spatial decision support tools need to be developed (spatial evaluation, scenarios and weighing of alternatives)
- easy access to the information with the use of graphical user interfaces is a necessity.

Streit and Weismann discussed in Chapter 12 the four approaches to the integration of GIS and spatial models: isolated applications, loose coupling, tightly-coupled applications and integrated applications. As has already been stated, IT developments are moving fast, and at present it seems that there is only one alternative: integrated applications.

In, for example, the EUPHIDS project (Beinat and Van de Berg, 1996), it has been demonstrated how different process models are integrated in a decision-support environment for admission of new pesticides for the European market. Integration based on the components discussed:

- data sharing, not based on data transferring by tempory files, but on direct access
- modelling and mapping techniques, based on direct interfaces for standard computer languages
- user interface — one standardised graphical user interface.

For further development, technology will not be a limiting factor: it will be our ability to understand spatial processes, and to translate them into computer algorithms in a computer environment which will be developed for use by policy makers. Geographical information systems will be open and will continue to evolve to harmonise with our ever-changing needs. We need to look at the journey, not the destination, that spatial information technology has to make.

REFERENCES

Beinat, E. and **van de Berg, R.** (Eds) (1996). *EUPHIDS, a decision support system for the admission of pesticides*. Report no. 712405002, National Institute of Public Health and the Environment (RIVM), Bilthoven, The Netherlands.

Douven, W. (1996). *Enhancing the accessibility of geographic knowledge for the environmental manager; an application to pesticide decision making*. PhD thesis, Amsterdam: Free University (in press).

Goodchild, M. (1993). The state of GIS for environmental problem solving, in Goodchild, M.F., Parks, B.O. and Steyaert, L.T. (Eds) *Environmental modelling with GIS*. New York: Oxford University Press.

Pulusani, P. (1995). The Integration and Interoperability Challenge: the need for Open GIS. Huntsville, USA: Intergraph.

Index